Advances in Hydraulics and Hydroinformatics

Advances in Hydraulics and Hydroinformatics

Special Issue Editors

Jian Guo Zhou
Jianmin Zhang
Yong Peng
Alistair G. L. Borthwick

MDPI • Basel • Beijing • Wuhan • Barcelona • Belgrade • Manchester • Tokyo • Cluj • Tianjin

Special Issue Editors
Jian Guo Zhou
Manchester Metropolitan University
UK

Jianmin Zhang
Sichuan University
China

Yong Peng
Sichuan University
China

Alistair G. L. Borthwick
The University of Edinburgh
UK

Editorial Office
MDPI
St. Alban-Anlage 66
4052 Basel, Switzerland

This is a reprint of articles from the Special Issue published online in the open access journal *Water* (ISSN 2073-4441) (available at: https://www.mdpi.com/journal/water/special_issues/Hydraulics_Hydroinformatics).

For citation purposes, cite each article independently as indicated on the article page online and as indicated below:

LastName, A.A.; LastName, B.B.; LastName, C.C. Article Title. *Journal Name* **Year**, *Article Number*, Page Range.

Volume 2
ISBN 978-3-03936-126-7 (Hbk)
ISBN 978-3-03936-127-4 (PDF)

Volume 1–2
ISBN 978-3-03936-128-1 (Hbk)
ISBN 978-3-03936-129-8 (PDF)

Contents

About the Special Issue Editors

Jian Guo Zhou graduated from Wuhan University with a BSc in River Mechanics and Engineering and subsequently completed his MSc in Fluvial Mechanics at Tsinghua University. He received his PhD in Fluid Mechanics from the University of Leeds. He specialises in formulating mathematical models and developing numerical methods for flow problems in fluids and water engineering, such as coastal, estuary, hydraulic, river and environmental engineering.

Jianmin Zhang graduated from Shihezi University with a BSc, and subsequently completed his MSc at Xinjiang Agricultural University. In 2000, he received his PhD in Hydraulics and River Dynamics from Sichuan University. His core research focuses on engineering hydraulics, flood discharge, energy dissipation, hydraulic experiments, two-phase flows and numerical simulations of turbulence.

Yong Peng graduated from Northwest A&F University with a BSc in Hydraulic and Hydropower Engineering, and subsequently completed his MSc in Hydraulics and River Dynamics at Sichuan University. He received his PhD in Computational Hydraulics from the University of Liverpool. His main research focuses on flooding, contaminant transport, sediment transport, the lattice Boltzmann method, shallow water flow, two-phase flow, cavitation, and hydraulic experiments.

Alistair G. L. Borthwick has more than 40 years' experience in civil engineering. He holds a BEng and a PhD from the University of Liverpool, a DSc degree from Oxford University, and an honorary degree from the Budapest University of Technology and Economics. His research interests include environmental fluid mechanics, coastal and ocean engineering, and marine renewable energy. In 2019, he was awarded the Gold Medal of the UK Institution of Civil Engineers.

Preface to "Advances in Hydraulics and Hydroinformatics"

Great progress has been made in the research on hydraulics, hydrodynamics and hydroinformatics over the past few decades. This includes theoretical, experimental and numerical studies, leading to a new understanding and knowledge of water-related problems, covering a wide range of topics and applications such as hydrology, water quality, river and channel flows. For example, coherent vortex structures in a backward-facing step flow and secondary flow in an open channel bend are measured using PIV; the flow through a Y-shaped confluence channel partially covered with rigid vegetation on its inner bank is measured by ADV. Meanwhile, the velocity decay of an offset jet in a narrow and deep pool, and the skimming flow over a pooled stepped spillway, are studied numerically. In addition, mechanisms for sediment transport due to riverbank failure and a method of predicting maximum scour depth are investigated, and cavitation bubble collapse and its impact on a wall are studied. In order to accelerate knowledge transfer in resolving engineering problems, this Special Issue reports both on the state-of-the-art of the aforementioned research topics, and on advances in sediment transport dynamics, two-phase flows, flow-induced vibration, and hydropower station hydraulics. The goals of this Special Issue are to improve our understanding of the foregoing flow problems and to stimulate future research in these areas, aiming for an improved quality of life.

Jian Guo Zhou, Jianmin Zhang, Yong Peng, Alistair G. L. Borthwick
Special Issue Editors

Article

Influence of Flushing Velocity and Flushing Frequency on the Service Life of Labyrinth-Channel Emitters

Zhangyan Li [1], Liming Yu [1,*], Na Li [1], Liuhong Chang [2] and Ningbo Cui [3]

[1] Faculty of Modern Agriculture Engineering, Kunming University of Science and Technology, Kunming 650500, China; zhangyanli4851@sina.com (Z.L.); kjclina@163.com (N.L.)

[2] School of Hydraulic Engineering, Changsha University of Science and Technology, Changsha 410114, China; claire886@163.com

[3] State Key Laboratory of Hydraulics and Mountain River Engineering and College of Water Resource and Hydropower, Sichuan University, Chengdu 610065, China; cuiningbo@scu.edu.cn

* Correspondence: liming16900@sina.com

Received: 19 October 2018; Accepted: 8 November 2018; Published: 12 November 2018

Abstract: Dripline flushing is an effective way to relieve emitter clogging and extend the longevity of drip irrigation systems. This laboratory study was conducted at Kunming University of Science and Technology to evaluate the effect of three targeted flushing velocities (0.3, 0.6, and 0.9 m/s) and four flushing frequencies (no flushing, flushing daily, and flushing every three or five days) on the emitter's service life and the particle size distribution of the sediment discharged from emitters and trapped in an emitter channel. The gradation of particle size was analyzed by a laser particle size analyzer. The experiment results suggested that flushing velocity and flushing frequency had a significant effect on the service life of emitters, and the emitter's service life was extended by 30.40% on average under nine different flushing treatments. Flushing can effectively reduce the accumulation of sediments in the dripline and decrease the probability of coarse particles flowing into emitters and fine particles aggregating and cementing in the labyrinth channel, thus relieving the emitter clogging. Therefore, dripline flushing can effectively slow down clogging in muddy water drip irrigation system. The recommended flushing velocity should be set at 0.6 m/s, and the flushing intervals should be shortened.

Keywords: clogging; drip irrigation; flushing; particle size distribution; sediments

---

1. Introduction

Dripline flushing is a necessary maintenance practice for micro-irrigation systems. It removes particles that are not strained by the micro-irrigation system filters and that accumulate in the driplines [1,2]. Physical clogging caused by solid particles is considered the most common emitter clogging category of emitters [3–5]. For drip irrigation systems in the Yellow River irrigation areas of Ningxia and Inner Mongolia, the average sediment concentration in the water abstracted from the Yellow River reaches 35 kg/m^3. A large quantity of sand enters into drip irrigation systems and results in emitter clogging even after deposition and prefiltration treatment measures are taken [6]. Therefore, flushing is required to ensure a long economic service life of drip irrigation systems [7]. On the other hand, numerical simulations on shallow water flows and river flushing have been carried out by Peng et al. [8–14].

In order to achieve an optimal flushing effect, drip irrigation systems should be designed properly. Flushing must be done often enough and at an appropriate velocity to dislodge and remove the accumulated sediments [15]. A minimum flushing velocity of 0.3 m/s was recommended by the

American Society of Agricultural and Biological Engineers Engineering Practice, EP-405 [16], but some researchers have advised that a flushing velocity of 0.5–0.6 m/s is necessary when larger particles need to be discharged; for example, it is useful when coarser filters are used in drip irrigation systems [17,18]. In a 30-day field study in which the target flushing velocities were ranged from 0.23 to 0.61 m/s, Puig-Bargués et al. [18] did not find remarkable effects of flushing velocity on the emitter discharge. However, the higher the flushing velocities were set, the more solids were removed from the laterals. Increasing the velocity of flushing may need more costly system designs (e.g., larger supply pipe main, higher pumping requirements, reduced zone sizes) and labor requirements for flushing in irrigation systems should be increased [14].

Different studies explored different flushing frequencies: daily [19], weekly [20], every two weeks [17,21], and monthly [18]. Li et al. [21] claimed that emitter clogging was minimized in cases of biweekly flushing. However, Puig-Bargués et al. [22] reported no obvious differences between flushing frequencies at a velocity of 0.6 m/s, and emitter clogging was primarily affected by the interactions between emitter type, emitter location, and frequency of flushing. Thus, a general agreement on the optimum flushing frequency is lacking at this time, and related studies on flushing frequency and flushing velocity are scarce.

Therefore, the objective of this study was to analyze the influence of three flushing velocities and four flushing frequencies on emitter clogging in drip irrigation systems when using muddy water with full tests. Additionally, particle size distributions of the discharged sediments and residual sediments in the emitter were investigated to determine appropriate flushing schemes to extend the service life of drip irrigation systems.

2. Materials and Methods

2.1. Emitter Characteristics

The drip tape with non-pressure-compensating emitter manufactured by Dayu Water Conservation Ltd. in Jiuquan City, Gansu Province, China, which is widely used in agricultural irrigation, was applied in the experiments. Its external diameter was 16 mm, and the thickness of the wall was 0.40 mm. Figure 1 presents the structure of the labyrinth-channel emitter. The working pressure was 0.1 MPa. The rated discharge (q) was 1.37 L/h. The flow path had a width (W) of 0.94 mm, length (S) of 37.8 mm, and depth (D) of 0.60 mm. The sectional area (A) was 0.56 mm^2. The angle between the bevel edge and baseline was 38.2°. Space width (L) and height (H) of the tooth were 1.50 and 0.88 mm, respectively. The manufacturing variation of the emitter was 0.028. The flow exponent (x) was 0.49, and the discharge coefficient (k) was 1.12.

The test driplines were cut randomly from the same roll of irrigation drip tapes. The emitter's discharge was measured using pressures of 0.01, 0.02, 0.03, 0.04, 0.05, 0.06, 0.08, and 0.10 MPa to calculate the discharge coefficient and flow exponent.

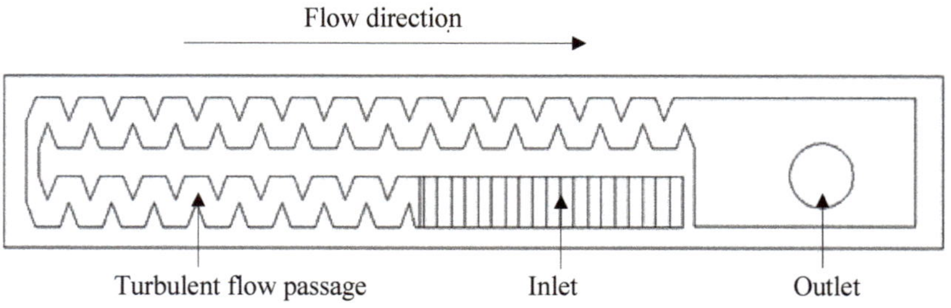

(a)

Figure 1. *Cont.*

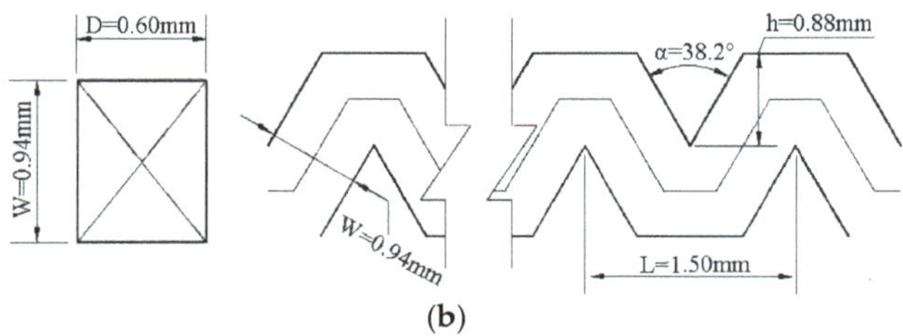

Figure 1. Emitter used in drip tape. (**a**) Labyrinth-channel emitter; (**b**) parameters of the flow path in the emitters. W, tooth space width; D, depth; α, angle of the bevel edge and baseline; L, tooth space width; h, tooth height.

2.2. Experimental Setup

Figure 2 shows that clean water was stored in a 100-L tank and muddy water was stored in a 150-L tank. Each was equipped with a 1.8-kW submersible pump with a 42-m rated head and 1.8-m^3/h rated discharge to provide the working pressure. The pressure adjustment valve and manometer I were installed on the three parallel tubes, which were connected to the submerged pump in the clean water tank, to adjust the different targeted flushing velocities. Manometer II was used to monitor the irrigation pressure. The pressure in manometer I and manometer II was 0.06 and 0.25 MPa, respectively, and their precision was 0.25%. A test platform (1.5 m in width, 4.8 m in length, and 1.0 m in height) was applied to support the laterals and emitters. Ten driplines were arranged on the test platform and all of them were equipped with control valves at the front and back ends. The spacing between each lateral was 0.16 m. Each lateral pipe had 15 emitters, with a spacing of 0.3 m.

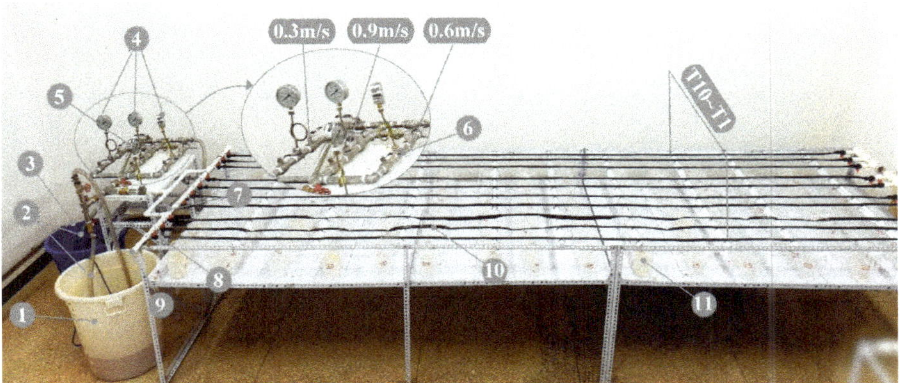

Figure 2. Experimental setup. 1, muddy water tank; 2, submersible pump; 3, clean water tank; 4, manometer I; 5, manometer II; 6, pressure adjustment valve; 7, polyvinyl chloride (PVC) pipe; 8, control valve; 9, PVC choke plug; 10, lateral; 11, measuring cup.

2.3. Measurement of Particle Size Distribution and Water Source

The distribution of particles size was analyzed by the Malvern laser particle size analyzer 2000 (Malvern Instruments Ltd., Malvern, UK). The measuring range of the particle analyzer was 0.02–2000 μm, and the particles were arranged in order of increasing size. When the accumulation volumes of sediment reached 10%, 50%, and 90%, the largest values of particle size were noted as D10, D50, and D90, respectively.

Tap water was used for lateral flushing in this study. Based on the maximum sediment concentration of irrigation water (0.8 g/L), the sand content in the muddy water was set at 2.5 g/L to accelerate the clogging process and reduce the experimental period. The test sediment was sandy loam soil obtained from Kunming, Yunnan Province of China, which was filtered using screens with 160 meshes after natural air drying. (In this study, 120 mesh screens were not selected for filtration to avoid insufficient test numbers caused by untimely clogging and the deposition of coarse particles.) Five sediment samples were randomly taken from the filtered sand after mixing well and the test sediment particle size distribution was presented based on their average values (see Table 1). The cumulative distribution and differential distribution of the tested sediments are shown in Figure 3.

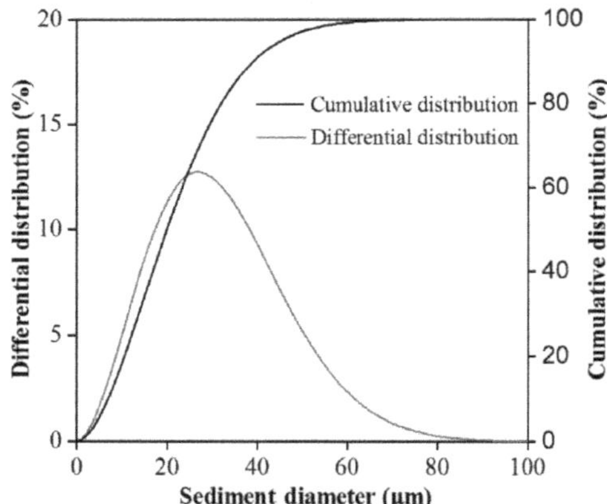

Figure 3. Differential distribution and cumulative distribution of sediments.

Table 1. Size distribution of sediment particles.

	D10	D20	D30	D40	D50	D60	D70	D80	D90	D100
Particle size	6.92	10.54	13.72	16.79	19.93	23.35	27.29	32.18	39.51	97.45
Standard deviation	0.34	0.45	0.55	0.63	0.72	0.82	1.00	1.27	1.65	1.33

2.4. Test Procedures

An experimental setup was conducted in a laboratory at Kunming University of Science and Technology, Kunming, Yunnan Province of China. In order to prevent the emitter flow rate and anti-clogging performance from being influenced by temperature variation, the experiments in this study were conducted between 13:00 and 16:00 CST, from 19 April to 7 September in 2017. Repeat experiments were carried out in three periods due to the limited width of the platform. The first repeated test was conducted between 19 April and 2 June, the second between 5 June and 19 July, and the third between 25 July and 7 September. The average temperature of the environment for the three test periods was 21.39 °C, 22.27 °C, and 22.08 °C, respectively.

2.4.1. Muddy Water Irrigation

The test conditions were as follows: normal irrigating pressure, 0.1 MPa; duration of irrigation event of each test, 20 min; initial volume of muddy water, 100 L; and total volume for each irrigation event, approximately 68.5 L. The muddy water was well stirred manually to avoid sedimentation. The emitter's discharge was measured every day using a marked 1000-mL measuring cup which was placed exactly beneath all the emitters. After each irrigation event, the weight of each measuring cup containing emitted muddy water was measured by a digital balance. The resolution of the digital balance was 0.01-g.

The emitter discharge was recorded as volume per unit time (L/h). For the emitter discharge less than 75% of the nominal rated flow, it was deemed to be clogged. The tested muddy water was replaced with new muddy water every day.

2.4.2. Sampling and Testing

The irrigation test was conducted with 45 events and 10 driplines, and the average discharge of all emitters in each lateral was taken into consideration. Therefore, 10 samples of each irrigation were measured three times. In all the tests, a total of 1350 (i.e., $45 \times 10 \times 3 = 1350$) muddy water samples were evaluated. The distribution of particle size was measured with all emitters in each dripline. Ten samples were assigned to each irrigation event, and the process was repeated three times. A total of 1350 samples were measured.

2.4.3. Clean Water Flushing

The present study evaluated the effect of two factors: (1) the flushing frequency, either daily (F_1), once per three days ($F_{1/3}$), or once per five days ($F_{1/5}$); and (2) the flushing velocity, at either 0.3 m/s ($V_{0.3}$), 0.6 m/s ($V_{0.6}$), or 0.9 m/s ($V_{0.9}$). Table 2 shows 10 treatments consisting of a control non-flushed treatment (T10) and a combination of targeted flushing velocities and flushing frequencies (T1–T9). There was only one dripline being flushed in every experiment, and each flushing event lasted for 5 min.

Table 2. Flushing velocity and flushing frequency treatment.

Treatment No.	Flushing Velocity (V)	Flushing Frequency (F)	Average Irrigation Events Before Emitter Clogging (d)
T1	$V_{0.3}$	F_1	32.67
T2	$V_{0.3}$	$F_{1/3}$	30.67
T3	$V_{0.3}$	$F_{1/5}$	27.33
T4	$V_{0.6}$	F_1	39.67
T5	$V_{0.6}$	$F_{1/3}$	33.67
T6	$V_{0.6}$	$F_{1/5}$	29.00
T7	$V_{0.9}$	F_1	41.33
T8	$V_{0.9}$	$F_{1/3}$	36.67
T9	$V_{0.9}$	$F_{1/5}$	30.67
T10	No flushing		23.33

2.5. Statistics

The data were analyzed using SPSS software for Windows, version 22.0 (IBM Corp., Chicago, IL, USA). Based on main effects analysis of variance (ANOVA), the significance of the differences between treatments for different response variables was evaluated. Furthermore, the results were categorized as "descriptive statistics" (such as mean) and Levene's test of equality of error variances. In addition, multiple comparisons were carried out based on Fisher's least significant difference tests because significant differences ($p < 0.05$) were indicated by ANOVA.

3. Results

3.1. Variations in Emitter Discharge

Figure 4 illustrates the value changes of the emitter discharge versus irrigating events. The level of straight line was 75% of the initial discharge, which was used as the criterion for emitter inefficiency or severe emitter clogging. Ten discharges of the emitters treated under diversified conditions showed a descendant trend with increasing irrigation events, suggesting that clogging with different degrees was seen in emitters until their service life ended. In the figure, the bold black line shows changes of the emitter discharge in T10, the control experiment. The emitter discharge in treatment T10 reached 75% of the rated discharge first, and its average discharge at the end of the test was the minimum without any flushing treatments made after irrigation. According to the irrigating events (Table 2), T10 provided only 23.33 irrigation events before emitters were severely plugged. However, for the remaining nine flushing treatments, the average normal irrigating event was 33.52, which indicated that the emitter's average service life in the flushed driplines after irrigation increased by 30.40% in comparison with those in the non-flushed T10. In the nine flushing treatments, treatment T7 had the longest irrigation event 41.33 and the universally largest average discharge; treatment T3 had the shortest irrigation event 27.33 and the universally lowest average discharge after 28 irrigation events. The service life of treatments T7 and T3 increased by 43.55% and 14.64%, respectively, compared with the non-flushing group T10. Additionally, treatments T7 and T3 were flushed at the frequency of F_1 and $F_{1/5}$ and at the velocity of $V_{0.9}$ and $V_{0.3}$, respectively, demonstrating that the emitters had a longer service life under high flushing velocities and high flushing frequencies.

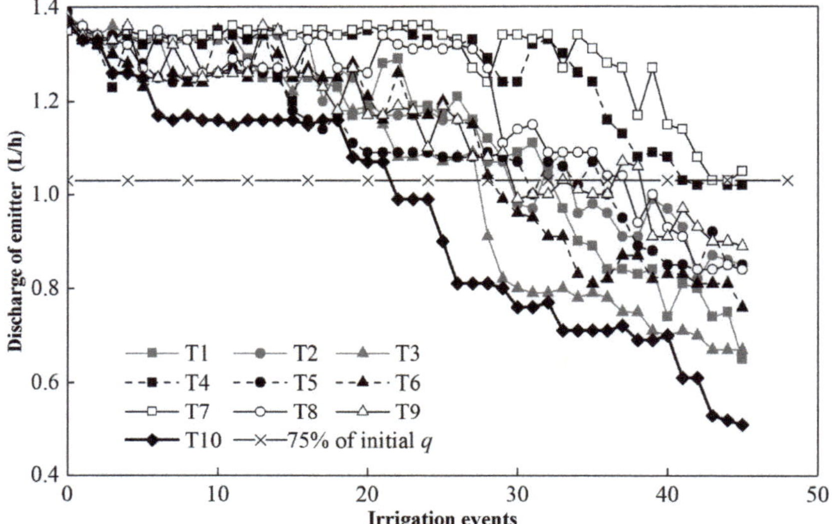

Figure 4. Average change in the emitter discharge versus irrigating events.

3.2. Variance Analysis of the Two Flushing Factors on the Service Life of Emitters

Flushing velocity and flushing frequency have significant effects on the service life of emitters, and the interactions were negligible (see Table 3). Figure 5a shows that the service life of emitters increased with the increase in the flushing velocity. The average service life of emitters was 22.80%, 31.60%, and 35.59% higher under $V_{0.3}$, $V_{0.6}$, and $V_{0.9}$ conditions, respectively, compared with the non-flushing conditions. Moreover, the service life of emitters under $V_{0.6}$ and $V_{0.9}$ conditions was similar, which meant that the flushing velocities in excess of 0.6 m/s would only help improve the service life to an insubstantial degree. Therefore, 0.6 m/s proved to be the optimal flushing velocity.

Table 3. Variance analysis of the effects of flushing velocity and flushing frequency on the service life of emitters.

Factors	F Value
Flushing velocity	21.94 **
Flushing frequency	46.82 **
Flushing velocity × Flushing frequency	2.03

$* p < 0.05; ** p < 0.01.$

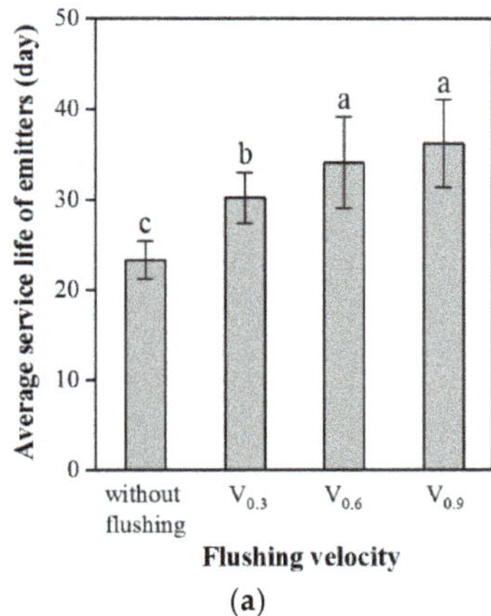
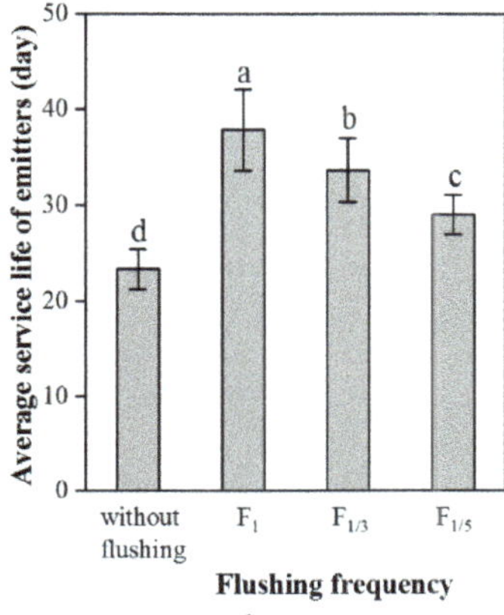

Figure 5. Average service life and standard errors of emitters at different flushing (**a**) velocities and (**b**) frequencies.

On the contrary, the service life of emitters was significantly reduced as the flushing frequency decreased (Figure 3b). The average service life of emitters was 38.43%, 30.71%, and 19.55% higher under F_1, $F_{1/3}$, and $F_{1/5}$ conditions, respectively, compared with the non-flushing treatment, indicating that a decrease in the flushing frequency effectively extended the service life of emitters.

3.3. Particle Size of Discharged Sediments

Table 4 shows that the effect of flushing velocity on D90 was significant, indicating that coarse particles passing through the labyrinth channel were affected by the flushing velocity, whereas fine particles were not. The flushing frequency had no effect on the particle size of discharged sediments. Figure 6 shows that D10, D50, and D90 mean values under non-flushing measures were 3.00, 7.16, and 12.65 µm, respectively, which were significantly higher than the different processing levels of flushing velocity and flushing frequency. This can be attributed to the fact that residual sediments in the lateral were discharged by flushing, resulting in a reduced concentration of sediments at the inlet of the labyrinth channel. Figures 7 and 8 show large amounts of residual sediments in the non-flushing lateral. The sediments mixed together, and more coarse particles flew into the labyrinth channel with the turbulence of the flow. Therefore, flushing effectively reduced the concentration of residual sediments in the lateral and decreased the probability of large particles entering into the labyrinth channel. Meanwhile, Figure 6b shows that the degree of D90 increased slightly with the decrease in the flushing frequency for the same reason.

Table 4. Variance analysis of the effects of flushing velocity and flushing frequency on the particle size of discharged sediments.

Factors	*F* Value		
	D10	**D50**	**D90**
Flushing velocity	2.15	3.38	4.18 *
Flushing frequency	0.66	1.20	3.41
Flushing velocity × Flushing frequency	0.14	0.39	0.91

* $p < 0.05$; ** $p < 0.01$.

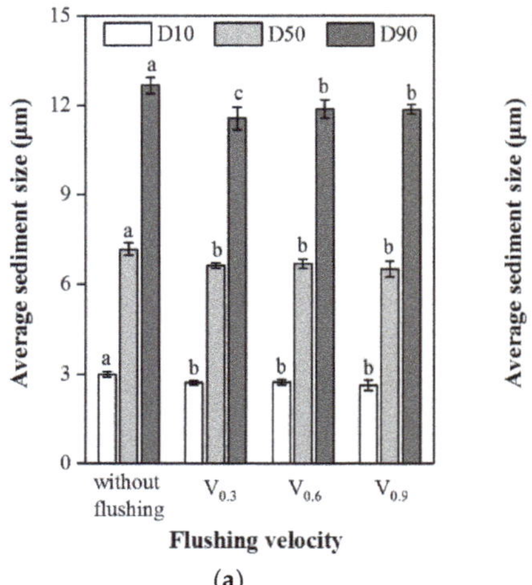
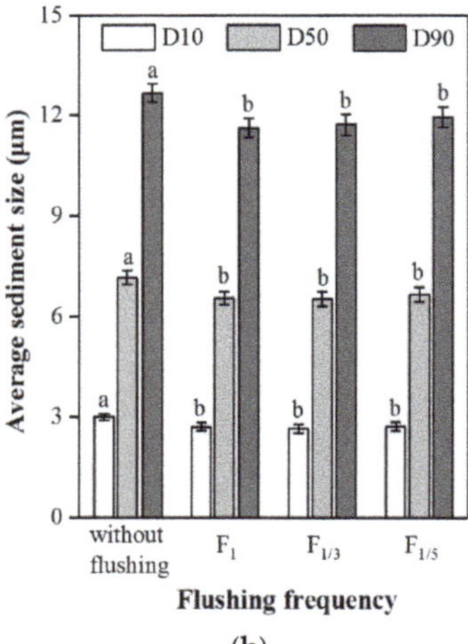

Figure 6. Average and standard errors of D10, D50, and D90 at different flushing (**a**) velocities and (**b**) frequencies.

3.4. Residual Sediments in Drip Tape and Emitters

Emitters and pipe sampled from the driplines were cut open to provide visual proof of clogging when experiments were finished (Figures 7 and 8). A few sediments were observed on the inner walls of the lateral with flushing treatment, whereas substantial sediment accumulations were observed in the lateral without flushing (Figure 5). Hence, flushing can effectively prevent particles from accumulating in the drip tape and emitters. Figure 8 shows that the sediment particles were seldom deposited in the labyrinth channel and the outlet of emitters due to flow turbulence with flushing, whereas clogging particles were found along the labyrinth channel of emitters without flushing, especially at the outlet. Additionally, other impurities such as plant root debris, which were not or seldom observed in the labyrinth channel of emitters with flushing treatment, were widely distributed in the labyrinth channel of emitters without flushing. In summary, flushing could effectively remove residual sediments and other impurities in the labyrinth channel, thus enhancing the anti-clogging performance of emitters.

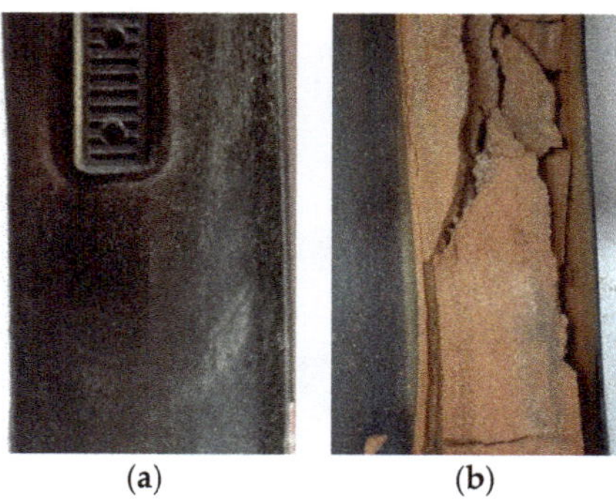

Figure 7. Residual sediments in the drip tape after tests. (**a**) Drip tape with flushing treatment; (**b**) drip tape without flushing treatment.

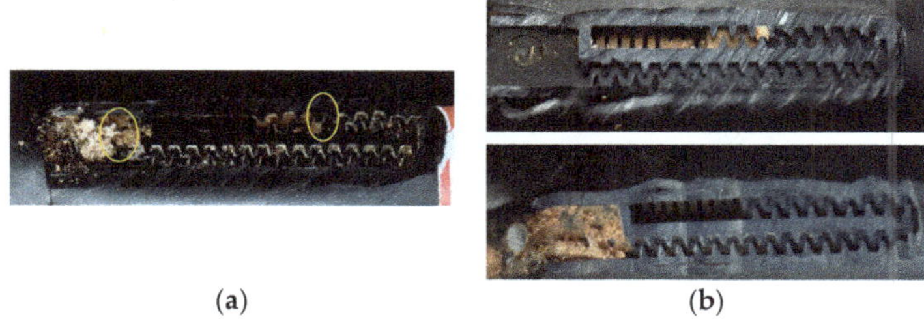

(a)   (b)

Figure 8. Residual sediments in the labyrinth channel of emitters (yellow parts indicate that the clogging substances are other impurities, not sediments). (**a**) Emitters with flushing treatment; (**b**) emitters without flushing treatment.

Table 5 shows the variance analysis of clogging substances in emitters. The flushing frequency and flushing velocity had significant effects on D10, D50, and D90. Figure 9 shows that D50 and D90 were the highest under non-flushing conditions, being 30% and 52% higher than the averages under the corresponding flushing conditions, respectively, indicating that flushing prevented large particles from entering into the labyrinth channel. The figure also shows that D50 and D90 decreased as the flushing velocity increased. For instance, D10, D50, and D90 under $V_{0.9}$ conditions were the minimum, being, respectively, 25.76%, 38.55%, and 47.14% lower than those under non-flushing conditions. In summary, a higher flushing velocity contributed to the discharge of coarse particles in emitters, thereby lowering risks of emitter clogging.

Table 5. Variance analysis of the effects of flushing velocity and frequency on the particle size of residual sediments in emitters.

Factors	*F* Value		
	D10	D50	D90
Flushing velocity	18.07 **	7.48 **	10.85 **
Flushing frequency	10.06 **	10.36 **	19.58 **
Flushing velocity × Flushing frequency	2.81	2.47	2.67

* $p < 0.05$; ** $p < 0.01$.

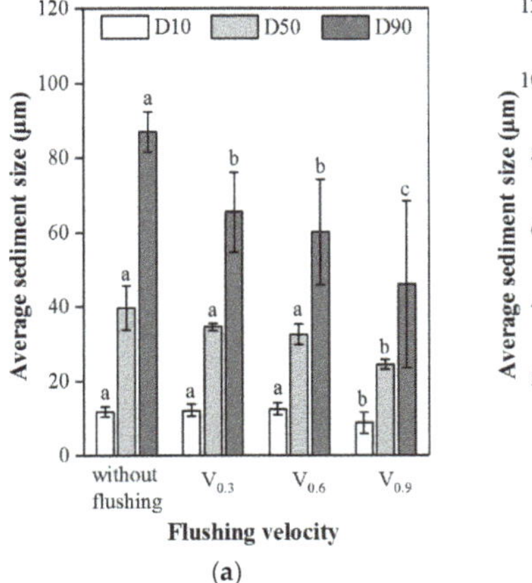
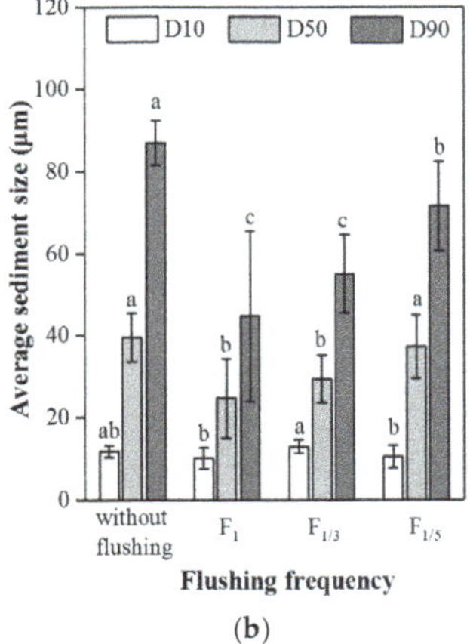

(a)   (b)

Figure 9. Average and standard error of D10, D50, and D90 at different flushing (**a**) velocities and (**b**) frequencies.

Figure 9 further shows that D50 and D90 increased as the flushing frequency decreased and was maximized under non-flushing conditions. D90 under F_1, $F_{1/3}$, and $F_{1/5}$ conditions was 48.50%, 36.63%, and 17.66%, respectively, significantly lower than that under non-flushing conditions. D50 under F_1 and $F_{1/3}$ was 37.60% and 25.85%, respectively, significantly lower than that under non-flushing conditions. However, D10 differed slightly between the F_1, $F_{1/3}$, and $F_{1/5}$ conditions. A low flushing frequency caused the accumulation of sediments in the lateral and increased the probability of large particles being trapped in the labyrinth channel, resulting in increased particle sizes of residual sediments in the channel. It also favored the agglomeration of sediments, resulting in a higher degree of D50 and D90. Combining these two effects, the flushing frequency had a significant effect on the size distributions of residual sediments in the labyrinth channel.

4. Discussion

4.1. Influence of Flushing Treatment on Emitter Clogging

For drip irrigation systems, appropriate flushing can effectively prevent emitter clogging by hindering the agglomeration of sediment particles or their adhesion to organic residuals to generate large particles [23,24]. This study demonstrated that lateral flushing using clean water after the irrigation of muddy water could enhance the service life of emitters by 14.64–43.55% compared with non-flushing cases, and both the flushing velocity and flushing frequency significantly affected the service life of emitters. However, Puig-Bargués et al. [18] reported that the flushing frequency and flushing velocity had little effect on the resulting emitter discharges in subsurface drip irrigation systems measured at the end of the study. The results were different from the findings of the present study mainly due to different irrigation systems and flushing operations.

Recent studies found that the service life of emitters increased with the increase in the flushing velocity. For instance, the service life of emitters flushed at the velocities of 0.6 and 0.9 m/s was extended by more than 30% compared with that without flushing. This paper held that the flushing velocity of 0.6 m/s would appear to be adequate. In a laboratory study simulating a dripline with a transparent PVC pipe, Puig-Bargués and Lamm [25] suggested that the minimum flushing velocity of 0.3 m/s appeared sufficient for most micro-irrigation systems working under typical conditions. This was not consistent with the findings in this paper, which could be attributed to the fact that the researchers did not fully consider the complexities of the flow regime that might occur in the emitter. Zhang et al. [26] compared the influence of constant pressure and pulse pressure on the anti-clogging performance of the labyrinth emitter. According to their findings, the pulse pressure could alleviate clogging and reduce discharge of the emitter. Some studies [27,28] demonstrated that the emitter's anti-clogging performance could be enhanced by increasing the flushing pressure. This was consistent with the conclusion of this study because the pressure of flushing was proportional to the velocity of flushing.

In this study, the service life of emitters degraded significantly in the order of $F_1 > F_{1/3} > F_{1/5} >$ without flushing. Li et al. [21] conducted an in situ surface drip irrigation experiment under three conditions of dripline flushing frequency, in which the recycled water from a sewage disposal plant was used. They reported that lateral flushing could significantly relieve the emitter clogging of the irrigation systems using reclaimed water. Elberry et al. [23] reported that clogging at a high flushing frequency was significantly lower than that at a low flushing frequency. The aforementioned study results are consistent with the findings of this study. Nevertheless, Feng et al. [29] found that emitter clogging was not prevented with five different frequencies of flushing in the drip irrigation experiment using salty groundwater, which is primarily affected by flushing operations and water quality.

4.2. Influence of Flushing Treatment on Particle Size Distributions

Flushing can effectively enhance the anti-clogging performance of emitters by removing residual sediments from the dripline and maintaining the cleanliness and the passability of the labyrinth channel [29,30]. The results of this study indicated that the passability of large sediment particles in the labyrinth channel was the dominant factor for emitter clogging. In this study, the corresponding pressure in emitters was 0.004, 0.008, and 0.016 MPa, which was much less than 0.1 MPa, the normal irrigation pressure under the flushing velocity of 0.3, 0.6, and 0.9 m/s, respectively. Nevertheless, this study demonstrated that these relatively low pressures still had a significant effect on the particle size of residual sediments in the labyrinth channel. Kou et al. [31] reported that a disordered system of granular materials, such as powders, sand, and foams, formed stable structures when unperturbed, but that they "relax" in the presence of external influences such as shear or tapping, becoming fluid in nature. Similarly, the slightest perturbation of a particle system due to flushing destabilized some of the particles trapped in the labyrinth channel of emitters, increasing the flowability of particles in the channel. Hence, the passability of the labyrinth channel was enhanced, thus extending the service life of emitters. In this study, D90 of discharged sediments was significantly affected by the flushing velocity. Specifically, D90 under $V_{0.9}$ was significantly larger than that under $V_{0.3}$ and $V_{0.6}$ and without flushing, indicating that the perturbation of the particles increased with the increase in the flushing velocity and resulted in the increased trafficability of coarse particles.

A certain concentration of particles is indispensable for the follow-up collision and flocculation of particles. As particle concentration in the influent water increased, the local concentration distribution of particles in the labyrinth channel was improved [32–34]. In this study, the particle size of discharged sediments increased with the increase in the flushing frequency and reached its maximum under non-flushing conditions. As the flushing frequency decreased, fine particles were continuously discharged from the emitters, resulting in an increase in the relative contents of coarse particles and concentration of sediments, thereby improving the probability of collision, sedimentation, and accumulation of large sediment particles. This was why D50 and D90 of residual sediments in emitters increased with the increase in the frequency of flushing.

4.3. Influence of Particle Size on Emitter Clogging

The clogging mechanism in the emitter is dependent on the particle size of sediments. Coarse particles readily underwent collision and deposition in the vortices of the water flow due to bigger sizes and the larger drag force of the particles. Therefore, the particles could not escape from the vortices. On the other hand, the major forces causing the fine particles to clog came from their surface tension and adhesion to surrounding substances. In this study, D10 and D50 of discharged sediments were not affected by flushing velocity or flushing frequency, indicating that fine particles had better flow characteristics due to their lower drag force in the flow. In other words, this study demonstrated that coarse particles tended to be trapped in the labyrinth channel, which was consistent with the findings of previous studies [35–38].

Wu et al. [39] pointed out that it was difficult to cause clogging when the particle size was less than 20 mm. Particles <20 mm took up 50% of the muddy water (see Table 1), whereas D90 of discharged sediments was significantly lower than 20 μm (Table 6). Moreover, D100 of discharged sediments was only 27.18 μm. This indicated that particles larger than 20 μm tended to stay in the labyrinth channel and that particles smaller than 20 μm could easily flow out with the water. Table 6 shows that D50 of residual sediments in emitters was 24.38–39.67 μm, which was 22–99% higher than that in the initial water. In addition, D90 of residual sediments in emitters was 44.74–86.89 μm, which was 13–120% higher than that in the initial irrigation water resource.

Generally, the agglomeration and flocculation of fine particles occurred during the flow, and the upper-limit particle size was 10 μm [40]. Table 6 shows that the percentage of particle sizes below 10 μm was 10% in the residual sediments in the labyrinth channel but 90% in the discharged sediments. Hence, the emitter plugging in this study was caused mainly by large residual sediment particles in the emitter's labyrinth channel, instead of the agglomeration and flocculation of small sediment particles. This could be attributed to high flushing velocity and high flushing frequency.

Table 6. Average particle size of residual sediments in emitters and discharged sediments (μm).

		D10	D50	D90	D100
Discharged sediments	With flushing	2.68 ± 0.07	6.56 ± 0.14	11.73 ± 0.25	25.30 ± 0.50
	Without flushing	2.98 ± 0.29	7.17 ± 0.42	12.70 ± 0.74	27.18 ± 3.05
Residual sediments in emitters	With flushing	11.14 ± 2.30	30.49 ± 8.14	57.12 ± 14.68	69.63 ± 9.67
	Without flushing	11.74 ± 1.41	39.67 ± 5.92	86.88 ± 5.32	89.68 ± 6.71

5. Conclusions

The nine flushing measures led to an enhancement of 30.40% in the service life of emitters on average, and both flushing velocity and flushing frequency had significant effects on the service life of emitters. Although the life of emitters flushed at 0.9 m/s was longer than that of emitters flushed at 0.6 m/s, the numerical differences were small (<4%). Therefore, the flushing velocities setting at around 0.6 m/s appeared to be sufficient. Increasing the flushing frequency would be an economical way to achieve the optimal flushing effect without providing a higher flushing velocity, which might increase the system cost. Moreover, the inner walls of the lateral were clean in the presence of flushing and only local clogging was observed in the labyrinth channel. Severe sedimentation was observed in the lateral in the absence of flushing, and the clogging by sediments was observed along the labyrinth channel. Furthermore, flushing decreased the accumulation of sediments in the lateral, which prevented large sediment particles from entering into the labyrinth channel and lowered the probability for small sediment particles to agglomerate and flocculate into large particles, thus reducing the risks of emitter clogging.

Author Contributions: Software, Z.L.; Methodology, N.L.; Writing—original draft preparation, Z.L.; Writing—review and editing, L.Y.; Funding acquisition, L.Y., L.C., and N.C.

Funding: This research was funded by the National Natural Science Foundation of China (NSFC) (Nos. 51769009, 51379024, 51809022) and the Funding Program for University Engineering Research Center of the Yunnan Province in China.

Acknowledgments: The authors would like to express their appreciation for the financial support received from the National Natural Science Foundation of China (NSFC) (Nos. 51769009, 51379024, 51809022) and the Funding Program for University Engineering Research Center of the Yunnan Province in China.

Conflicts of Interest: The authors declare no conflict of interest.

References

1. Adin, A.; Sacks, M. Dripper-clogging factors in wastewater irrigation. *J. Irrig. Drain. Eng.* **1991**, *117*, 813–826. [CrossRef]
2. Ravina, I.; Paz, E.; Sofer, Z.; Marcu, A.; Shisha, A.; Sagi, G. Control of emitter clogging in drip irrigation with reclaimed wastewater. *Irrig. Sci.* **1992**, *13*, 129–139. [CrossRef]
3. Nakayama, F.S.; Bucks, D.A. Water quality in drip/trickle irrigation: A review. *Irrig. Sci.* **1991**, *12*, 187–192. [CrossRef]
4. Pitts, D.J.; Haman, D.Z.; Smajstrla, A.G. Causes and prevention of emitter plugging in micro-irrigation systems. *Bull. Fla. Cooperative Ext. Serv.* **2011**, *40*, 201–218.
5. Yan, D.Z.; Bai, Z.H.; Mike, R.; Gu, L.K.; Ren, S.M.; Yang, P.L. Biofilm structure and its influence on clogging in drip irrigation emitters distributing reclaimed wastewater. *J. Environ. Sci.* **2009**, *21*, 834–841. [CrossRef]
6. Li, Y.K.; Liu, Y.Z.; Li, G.B.; Xu, T.W.; Liu, H.S.; Ren, S.M.; Yan, D.Z.; Yang, P.L. Surface topographic characteristics of suspended particulates in reclaimed wastewater and effects on clogging in labyrinth drip irrigation emitters. *Irrig. Sci.* **2012**, *30*, 43–56. [CrossRef]
7. Lamm, F.R.; Camp, C.R. Chapter 13: Subsurface drip irrigation. In *Micro-Irrigation for Crop Production: Design, Operation, and Management*, 1st ed.; Lamm, F.R., Ayars, J.E., Nakayama, F.S., Eds.; Elsevier: Amsterdam, The Netherlands, 2007; pp. 473–551.
8. Peng, Y.; Zhou, J.G.; Burrows, R. Modelling the free surface flow in rectangular shallow basins by lattice Boltzmann method. *J. Hydraul. Eng.* **2011**, *137*, 1680–1685. [CrossRef]

9. Peng, Y.; Zhou, J.G.; Burrows, R. Modelling solute transport in shallow water with the lattice Boltzmann method. *Comput. Fluids* **2011**, *50*, 181–188. [CrossRef]

10. Peng, Y.; Zhang, J.M.; Zhou, J.G. Lattice Boltzmann model using two-relaxation-time for shallow water equations. *J. Hydraul. Eng.* **2016**, 142–149. [CrossRef]

11. Peng, Y.; Zhang, J.M.; Meng, J.P. Second order force scheme for lattice Boltzmann model of shallow water flows. *J. Hydraul. Res.* **2017**, *55*, 592–597. [CrossRef]

12. Peng, Y.; Zhou, J.G.; Zhang, J.M.; Burrows, R. Modeling moving boundary in shallow water by LBM. *Int. J. Mod. Phys. C* **2013**, *24*, 1–17. [CrossRef]

13. Peng, Y.; Zhou, J.G.; Zhang, J.M.; Liu, H.F. Lattice Boltzmann modelling of shallow water flows over discontinuous beds. *Int. J. Numer. Meth. Fluids* **2014**, *75*, 608–619. [CrossRef]

14. Peng, Y.; Zhou, J.G.; Zhang, J.M. Mixed numerical method for bed evolution. *Proc. Inst. Civ. Eng. Water Manag.* **2015**, *168*, 3–15. [CrossRef]

15. Nakayama, F.S.; Boman, B.J.; Pitts, D.J. Chapter 11: Maintenance. In *Micro-Irrigation for Crop Production: Design, Operation, and Management*, 1st ed.; Lamm, F.R., Ayars, J.E., Nakayama, F.S., Eds.; Elsevier: Amsterdam, The Netherlands, 2007; pp. 389–430.

16. ASAE Standards. *EP405.1: Design and Installation of Micro-Irrigation Systems*; ASAE: St. Joseph, MI, USA, 2003.

17. Brenes, M.J.; Hills, D.J. Micro-irrigation of wastewater effluent using drip tape. *Appl. Eng. Agric.* **2001**, *17*, 303–308.

18. Puig-Bargués, J.; Lamm, F.R.; Trooien, T.P.; Clark, G.A. Effect of dripline flushing on subsurface drip irrigation systems. *Trans. ASABE* **2010**, *53*, 147–155. [CrossRef]

19. Ravina, I.; Paz, E.; Sofer, Z.; Marm, A.; Schischa, A.; Sagi, G.; Yechialy, Z.; Leve, Y. Control of clogging in drip irrigation with stored treated municipal sewage effluent. *Agric. Water Manag.* **1997**, *33*, 127–137. [CrossRef]

20. Hills, D.J.; Tajrishy, M.A.; Tchobanoglous, G. The influence of filtration on ultraviolet disinfection of secondary effluent for micro-irrigation. *Trans. ASAE* **2000**, *43*, 1499–1505. [CrossRef]

21. Li, Y.; Song, P.; Pei, Y.; Feng, J. Effects of lateral flushing on emitter clogging and biofilm components in drip irrigation systems with reclaimed water. *Irrig. Sci.* **2015**, *33*, 235–245. [CrossRef]

22. Puig-Bargués, J.; Arbat, G.; Elbana, M.; Duran-Ros, M.; Barragán, J.; Ramírezde Cartagena, F.; Lamm, F.R. Effect of flushing frequency on emitter clogging in micro-irrigation with effluents. *Agric. Water Manag.* **2010**, *97*, 883–891. [CrossRef]

23. Elberry, A.M.; Abuarab, M.E.; Elebaby, F.G. Effect of flushing frequency with sulfuric and phosphoric acids on emitter clogging. In Proceedings of the ASABE Annual International Meeting, Louisville, KY, USA, 7–10 August 2011.

24. Li, Y.F.; Li, J.S.; Zhao, W.X.; Wang, Z. Review on irrigation technology applying sewage effluent-advances and prospects. *Trans. Chin. Soc. Agric. Mach.* **2015**, *46*, 102–110.

25. Puig-Bargués, J.; Lamm, F.R. Effect of flushing velocity and flushing duration on sediment transport in micro-irrigation driplines. *Trans. ASABE* **2013**, *56*, 1821–1828.

26. Zhang, L.; Wu, P.T.; Zhu, D.L. Effect of pulsating pressure on labyrinth emitter clogging. *Irrig. Sci.* **2017**, *35*, 267–274. [CrossRef]

27. Yu, L.; Li, N.; Liu, X.; Yang, Q.; Long, J. Influence of flushing pressure, flushing frequency and flushing time on the service life of a labyrinth-channel emitter. *Biosyst. Eng.* **2018**, *172*, 154–164. [CrossRef]

28. Yu, L.; Li, N.; Yang, Q.; Liu, X. Influence of flushing pressure before irrigation on the anti-clogging performance of labyrinth channel emitters. *Irrig. Drain.* **2017**, *67*, 191–198. [CrossRef]

29. Feng, D.; Kang, Y.; Wan, S.; Liu, S. Lateral flushing regime for managing emitter clogging under drip irrigation with saline groundwater. *Irrig. Sci.* **2017**, *35*, 217–225. [CrossRef]

30. Song, P.; Li, Y.; Li, J.; Pei, Y. Chlorination with lateral flushing controlling drip irrigation emitter clogging using reclaimed water. *Trans. Chin. Soc. Agric. Eng.* **2017**, *33*, 80–86.

31. Kou, B.; Cao, Y.; Li, J.; Xia, C.; Li, Z.; Dong, H.; Zhang, A.; Zhang, J.; Kob, W.; Wang, Y. Granular materials flow like complex fluids. *Nature* **2017**, *551*, 360–363. [CrossRef] [PubMed]

32. Wei, Z.; Tang, Y.; Wen, J.; Lu, B. Two-phase flow analysis and experimental investigation of micro-PIV and anti-clogging for micro-channels of emitter. *Trans. Chin. Soc. Agric. Eng.* **2008**, *24*, 1–9.

33. Wei, Q.; Lu, G.; Liu, J.; Shi, Y.; Dong, W.; Huang, S. Evaluation of emitter clogging in drip irrigation by two-phase flow simulation and laboratory experiments. *Comput. Electron. Agric.* **2008**, *63*, 294–303.

34. Li, J.; Chen, L.; Li, Y. Field evaluation of emitter clogging in subsurface drip irrigation system. *J. Hydraul. Eng.* **2008**, *39*, 1272–1278.

35. Niu, W.Q.; Liu, L.; Chen, X. Influence of fine particle size and concentration on the clogging of labyrinth emitters. *Irrig. Sci.* **2013**, *31*, 545–555. [CrossRef]

36. Zhang, J.; Zhao, W.; Wei, Z.; Tang, Y.; Lu, B. Numerical and experimental study on hydraulic performance of emitters with arc labyrinth channels. *Comput. Electron. Agric.* **2007**, *56*, 120–129. [CrossRef]

37. Wang, W.; Wang, F.; Niu, W.; Hu, X. Numerical analysis of influence of emitter channel structure on suspended granule distribution. *Trans. Chin. Soc. Agric. Eng.* **2009**, *25*, 1–6.

38. Liu, L.; Niu, W.; Bob, Z. Influence of sediment particle size on clogging performance of labyrinth path emitters. *Trans. Chin. Soc. Agric. Eng.* **2012**, *28*, 87–93.

39. Wu, Z.; Zhang, Z.; Zhang, K.; Luo, C.; Niu, W.; Yu, L. Influence of particle size and concentration of sediment on clogging of labyrinth channels emitters. *Trans. Chin. Soc. Agric. Eng.* **2014**, *30*, 99–108.

40. Qian, N.; Wan, Z. *Kinetics of Sediments*; Science Press: Beijing, China, 1983.

Article

Effect of Particle Size and Shape on Separation in a Hydrocyclone

Zhaojia Tang [1], Liming Yu [1,*], Fenghua Wang [1], Na Li [1], Liuhong Chang [2] and Ningbo Cui [3]

[1] Faculty of Modern Agriculture Engineering, Kunming University of Science and Technology, Kunming 650500, China; tangzhaojia01@sina.com (Z.T.); wangfenghua018@163.com (F.W.); kjclina@163.com (N.L.)

[2] School of Hydraulic Engineering, Changsha University of Science and Technology, Changsha 410114, China; claire886@163.com

[3] State Key Laboratory of Hydraulics and Mountain River Engineering and College of Water Resource and Hydropower, Sichuan University, Chengdu 610065, China; cuiningbo@scu.edu.cn

* Correspondence: liming16900@sina.com

Received: 24 October 2018; Accepted: 18 December 2018; Published: 21 December 2018

Abstract: Given the complex separation mechanisms of the particulate mixture in a hydrocyclone and the uncertain effects of particle size and shape on separation, this study explored the influence of the maximum projected area of particles on the separation effect as well as single and mixed separations based on CFD–DEM (Computational Fluid Dynamics and Discrete Element Method) coupling and experimental test methods. The results showed that spherical particles flowed out more easily from the downstream as their sizes increased. Furthermore, with the enlargement of maximum projected area, the running space of the particles with the same volume got closer to the upward flow and particles tended to be separated from the upstream. The axial velocity of the combined separation of 60 µm particles and 120 µm particles increased by 25.74% compared with that of single separation of 60 µm particles near the transition section from a cylinder to a cone. The concentration of 60 µm particles near the running space of 120 µm particles increased by 20.73% and those separated from the downstream increased by 4.1%. This study showed the influence of particle size and maximum projected area on the separation effect and the separation mechanism of mixed sand particles in a hydrocyclone, thereby providing a theoretical basis for later studies on the effect of particle size and shape on sedimentation under the cyclone action in a hydrocyclone.

Keywords: CFD–DEM coupling; hydrocyclone; particle shape; particle size; water and sediment separation

---

1. Introduction

A hydrocyclone is widely used for various separation purposes in different fields, such as mining, chemical industry, and agriculture. In the engineering field, it is often adopted to separate and classify the sand in water. The specific working process is as follows. As the water and sand mixture enters from the inlet, the water flows out from the upstream and downstream. The sand particles with strong water-following ability are carried out from the upstream and those with poor water-following ability are carried out from the downstream under the internal swirling flow.

In recent years, the separation mechanism of hydrocyclone has been mainly studied from the following aspects:

(1) The separation effect of various solid particle properties and different external conditions: The equation of particle motion in a hydrocyclone is established and the formula for calculating the optimum separation medium density is deduced according to the force analysis on different particles in the separation [1]. The calculation method and equation are put forward based on the relationship between particle size and fluid drag force [2]. The separation effect of submicron particles in a microclone at different temperatures and water pressures [3], the effect of internal flow field on the separation of different-sized particles [4], and the effect of inlet velocity and particle shape on the separation effect [5] are also studied.

(2) The effect of a hydrocyclone structure on separation: The optimal particle separation effect is determined by the orthogonal analysis on the optimal insertion depth, wall thickness, and optimal inlet flow [6]. The influence of the inlet structure in the performance of a hydrocyclone is explored [7]. Three kinds of hydrocyclones are designed to improve the particle separation efficiency, and the optimum separation structure is determined by studying the pressure and velocity distribution inside a hydrocyclone [8]. The effect of cone angle structure on the classification of fine particles [9] and of the inlet angle, cylinder height, and inner cone on separation [10,11] are discussed. In addition, studies are conducted on the separation results with hydrocyclones of diverse structures and different working conditions. Factors such as feed flow rate, size of the downstream, isolation tank, suspension concentration, viscosity, and pressure drop are included [12].

The separation effect of solid particles has been investigated by observing changes in the external conditions, optimization of hydrocyclone structure, and so on. Only a few basic studies have been conducted on the shape and size of sand particles. Thus, more efforts are called for on the study of the interaction between particles and the complex motion of different-shaped particles in a hydrocyclone. Due to the differences in particle sizes and shapes in nature, the separation becomes more complex, with unpredictable separation effects. In this study, the CFD–DEM coupling method and experimental testing method were employed to study the effect of maximum projected area on the separation performance with the same volume of sand. The mechanism underlying the separation of water and sand in a hydrocyclone was proposed by comparing the separation of single 60 μm spherical sand particles and 60 and 120 μm mixed particles.

2. Experimental Modes and Methods

2.1. Hydrocyclone Model

The parameters of the hydrocyclone used in this study were as follows. The diameters of the inlet, upstream, and downstream are 2, 3.4, and 2.2 mm, respectively. The cylinder section was 40 mm in height and 10 mm in diameter. The cone angle was 4 degrees. The shape and mesh conditions are shown in Figure 1. Based on the separation characteristics of hydrocyclones, it is generally believed that the sand particles moving against the wall where the cylinder and cone connect can be separated from the downstream. Therefore, the D-D section was set on the model to divide the area between the radial center and the wall into five regions. Figure 1a shows the model segmentation and meshing, and Figure 1b provides a magnified view of the D-D section 1:5. The three-dimensional structure was drawn using the UG10.0 (Unigraphics NX10.0) software, the meshing was done using the GAMBIT 2.4.6 software, and the numerical simulation was based on the ANSYS 16.0 and DEM 2.6.1 coupling method.

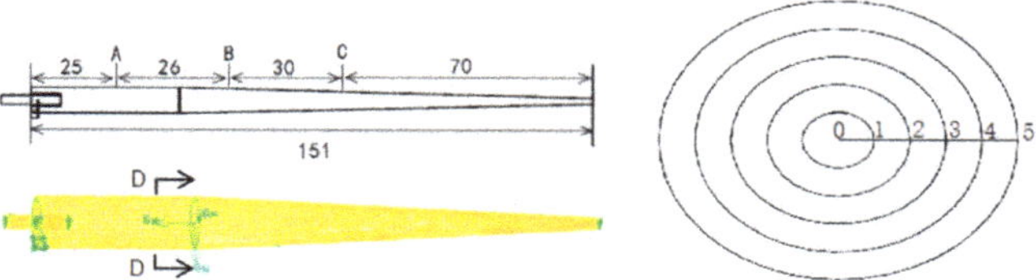

Figure 1. The axial length of Z axis and cross section of the cylinder of the hydrocyclone.

2.2. Numerical Simulation

2.2.1. Particle Characteristics

In this study, a single separation simulation of five spherical sand particles with diameters of 60, 70, 90, 106, and 120 µm was performed. A single separation simulation was conducted on three plate-like particles with the same volume as spherical sand particles of 60 µm (in Figure 2, the maximum projected area is S), and their maximum projected area was 3.11 S (type A), 2.24 S (type B), and 1.36S (type C), respectively. A mixed separation simulation was conducted on spherical sand particles with diameters of 60 and 120 µm. Table 1 shows the deformation amount, thickness, maximum projected area, and equivalent diameters of particles are shown in Table 1. The type C, B, and A particles share the maximum projected area with particles of 70 µm, 90 µm, and 106 µm, respectively.

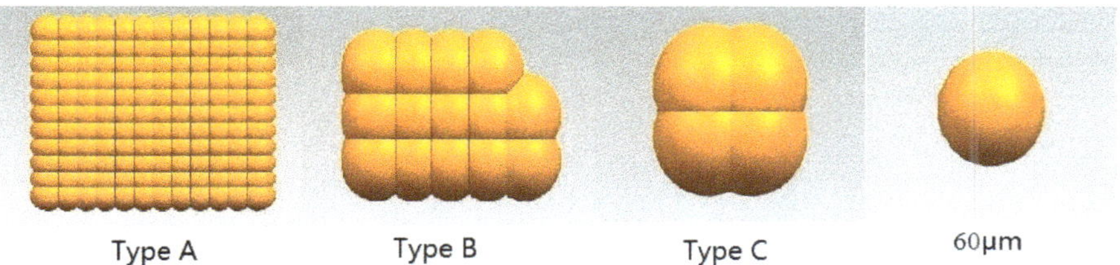

Figure 2. Particle shapes in numerical simulation.

Table 1. Parameters of particles in the numerical simulation.

Title 1	Type A	Type B	Type C	60 µm	70 µm	90 µm	106 µm	120 µm
Mass (µg)	0.28	0.28	0.28	0.28	0.44	0.95	1.54	2.23
Thickness (mm)	0.015	0.030	0.045	0.060	0.070	0.090	0.106	0.120
Maximum projected area (S)	3.11	2.24	1.36	1.00	1.36	2.24	3.11	4.00

In simulation experiments, the recorded residence time of sand particles in the cyclone equaled the period from their entering the inlet to their leaving the downstream or upstream. The data of particle velocity, position, and time were exported from the software DEM and edited by the MATLAB 2017b (Matrix Laboratory) postprocessor program. The values between the third and fourth seconds were collected as the analysis data.

2.2.2. Mathematical Model and Simulation Method

Flow in the hydrocyclone is considered to be an incompressible viscous fluid. The effect of gravity and the roughness of the hydrocyclone walls were taken into account, though surface tension was neglected. In the present study, the inlet velocity of the hydrocyclone was 2.00 m s^{-1}, The continuous medium flow is calculated from the continuity and the Navier–Stokes equations based on the local mean variables over a computational cell, which are given by:

$$\frac{\partial \varepsilon}{\partial t} + \nabla \cdot (\varepsilon u) = 0 \tag{1}$$

$$\frac{\partial (\rho_f \varepsilon u)}{\partial t} + \nabla \cdot (\rho_f \varepsilon u u) = -\nabla P - F_{p-f} + \nabla \cdot (\varepsilon \tau) + \rho_f \varepsilon g \tag{2}$$

F_{p-f} is Interaction forces between fluid and solid phases, equal to $\sum_{i=1}^{k_c} f_{p-f,i}/\Delta V_c$, N/m^3; the flow solved in Equations (1) and (2) represents the mixture flow of medium and sand, and was obtained by use of the models in a commercial ANSYS software package, i.e., Fluent. The details of the medium flow calculation and its validation can be found elsewhere [13,14]. In this work, we only give an overall description of the ANSYS model for the hydrocyclone.

The governing equations that are described above were discretized by the control volume numerical technique, and then the SIMPLE pressure–velocity coupling technique, with a second-order upwind scheme for the convection terms that were employed to solve the discretized equations over the computational domain.

The maximum particle volume fraction for the sand used in this study was less than 1%, which means the mixture of water and sand belonged to dilute phase flow. The Lagrangian coupling method was employed and Table 2 summarizes the parameter settings for the sand [15], Table 3 summarizes the boundary condition in the model. Ansys16.0 and DEM2.6.1 were employed during this study. The discrete approach was used to simulate sand movement, collisions among sand particles and between the sand particle and the hydrocyclone wall, and the effects of sand movement on the surrounding continuous phase, energy and momentum exchange. Collisions among sand particles and between sand particles and the wall did not lead to significant plastic deformation, which in consequence were attributed to hard particle contact, which is a wet grain contact model. The 'Hertz-Mindlin (no slip) built-in' model was utilized in this work [16].

Table 2. Parameters used in the model.

Phase	Parameter	Symbol	Units	Value
Solid	Density distribution	ρ	$kg \cdot m^{-3}$	2500
	Rolling friction coefficient	μ_r	-	0.01
	Sliding friction coefficient	μ_s	-	0.30
	Poisson's ratio	V	-	0.40
	Young's modulus	E	$N \cdot m^{-2}$	2×10^{-7}
	Coefficient of Restitution	c_r	-	0.55
fluid	Density	ρ	$kg \cdot m^{-3}$	998.20
	Viscosity	μ	$kg \cdot m^{-1} \cdot s^{-1}$	0.001

Table 3. Boundary condition.

	Symbol	Units	Value
Particle velocity at inlet	v	$m \cdot s^{-1}$	2.00
Viscosity of Water Phase	v	$m \cdot s^{-1}$	2.00
Turbulent intensity	I	-	5%
Hydraulic radius	D	mm	2.00
Pressure at upstream	-	Pa	0.00
Pressure at downstream	-	Pa	0.00
Back-flow turbulence intensity	I_h	-	5%
Number of particle	-	$N \cdot s^{-1}$	1000
Particle diameter	d_i	μm	-

In the simulation, the contributions of viscous drag force, pressure gradient force and gravity were considered. Other additional forces such as virtual mass force and Saffman force were not considered, as they are an order of magnitude smaller compared with the foregoing [17].

The CFD-DEM coupling process is described as follows: the continuous phase is resolved by ANSYS to acquire fluid drag force with sand, which was transformed from flow field information through the drag force model. The stress state of sands was measured by DEM to obtain new information such as the position and velocity of sands and to the flow field. ANSYS was used to model the flow and the most representative stress condition of sediments. The two approaches were coupled with a compound model that included the particle–fluid interaction force [16]. The relevant equations can be seen in Table 4.

Table 4. Equations and specifications.

S/N	Name	Formula	Description
1	The normal force (F_n)	$F_n = \frac{4}{3}E^*(R^*)^{\frac{1}{2}}\alpha^{\frac{3}{2}}$	R^* is the equivalent radius α is the normal overlap
2	The equivalent elastic modulus (E^*)	$\frac{1}{E^*} = \frac{1-V_1^2}{E_1} + \frac{1-V_2^2}{E_2}$	E_1, v_1 and E_1, v_1 are elastic modulus and Poisson's ratio of sand 1 and sand 2
3	The damping force ($F_n{}^d$)	$F_n^d = -2\sqrt{\frac{5}{6}}\beta\sqrt{S_n m^*}V_n^{rel}$	V_n^{rel} is the normal relative velocity; S_n is the normal stiffness β is coefficient
4	The equivalent mass (m^*)	$m^* = \frac{m_1 m_2}{m_1 + m_2}$	m_1 and m_2 are the mass of sand 1 and sand 2
5	The tangential force among the sands (F_t)	$F_t = -S_t\delta$	δ is the tangential overlap
6	The tangential stiffness (S_t)	$S_t = 8G^*\sqrt{R^*\alpha}$	
7	The equivalent shear modulus (G^*)	$G^* = \frac{2-V_1^2}{G_1} + \frac{2-V_2^2}{G_2}$	G_1 and G_2 are shear modulus of sand 1 and sand 2
		-	V_1 and V_2 are velocity of sand 1 and sand 2
8	The tangential damping force among sand particles (F_t)	$F_t = -2\sqrt{\frac{5}{6}}\beta\sqrt{S_t m^*}V_t^{rel}$	V_t^{rel} is the tangential relative velocity
9	The rolling friction (T_i)	$T_i = -\mu_r F_n R_i \omega_i$	μ_r is coefficient of rolling friction R_i is the distance between the center of mass to the point of contact; ω_i is unit angular velocity vector of object at the contact point

Particle–fluid interaction, force ($f_{p-f,i}$):

$$f_{D,i} = \left(0.63 + \frac{4.8}{Re_{p,i}^{0.5}}\right)^2 \frac{\rho_f |u_i - v_i|((|u_i - v_i|)}{2} \frac{\pi d_i^2}{4} \varepsilon_i^{-\beta}$$

$$Re_{p,i} = \frac{d_i \rho_f \varepsilon_i |u_i - v_i|}{\mu_f}, \quad \beta = 3.7 - 0.65\exp\left[-\frac{\left(1.5 - \log Re_{p,i}\right)^2}{2}\right] \tag{3}$$

$$\varepsilon = 1 - \frac{\sum_{i=1}^{k_C} V_i}{\Delta V_C}$$

$$f_{pg,i} = V_{p,i}\nabla P \tag{4}$$

The free settling velocity (u_t)

$$u_t = \frac{gd^2}{18\upsilon}\left(\frac{\rho_p - \rho}{\rho}\right) \tag{5}$$

Here, the drag coefficient (C) is

$$C = \frac{(\rho_P - \rho)Vg}{A\rho\left(\frac{u^2}{2}\right)} \tag{6}$$

And the particle Reynolds number (Re) is

$$Re = \frac{D_{1p} u \rho}{\mu} \tag{7}$$

The drag coefficient varies depending on the inverse of the particle Reynolds number, thus the approximated drag coefficient was calculated as a function of the particle Reynolds number. The approximated drag coefficient Ca is expressed as

$$C_a = C_0 + C_1 \left(\frac{1}{Re} \right) \tag{8}$$

and C_0 and C_1 were determined from the correlation of the particle Reynolds numbers and measured drag coefficients using the ratio of the particle diameter D_2 to thickness T as the particle shape factor

$$C_0 = 0.4555 \mathrm{Ln} \left(\frac{D_2}{T} \right) + 0.4687 \tag{9}$$

$$C_1 = 19.285 \tag{10}$$

Finally, the approximated drag coefficient was expressed as [18]:

$$C_a = 19.285 \left(\frac{1}{Re} \right) + 0.4555 \mathrm{Ln} \left(\frac{D_2}{T} \right) + 0.4687 \tag{11}$$

Pressure gradient force [19]: Due to the uneven distribution of hydraulic pressure on the surfaces of particles, the force is given by:

$$F_p = -\frac{\nabla p}{\rho_p} \tag{12}$$

The total drag force of fluid on sand particles (F_D)

$$F_D = \xi A_p \frac{\rho u^2}{2} \tag{13}$$

When a particle accelerates in the viscous fluid, it will cause the surrounding fluid to accelerate, too. Due to the inertia effect, the fluid has a reaction force to the particle, which makes the inertia of the particle virtually increase. The related force is given by [2]:

$$F_A = \frac{1}{2} \frac{\rho}{\rho_p} \left(\frac{du}{dt} - \frac{du_p}{dt} \right) \tag{14}$$

When a particle moves in a flow field with velocity gradient, it will experience a lateral force from the low flow rate side to the high flow rate side, given by:

$$F_S = \frac{9.69}{\pi d_p \rho_p} (\mu \rho |\gamma|)^{\frac{1}{2}} C_{LS} (u - u_p) \tag{15}$$

when a particle rotates at an angular velocity of ω, a lateral force perpendicular to the relative velocity and the rotation axis of the particle is generated from the upstream side to the downstream side, given by:

$$F_M = \frac{3}{4} \frac{\rho}{\rho_p} \frac{(u - u_p)^2}{d_p} C_{LM} \frac{\omega_r \times u_r}{|\omega_r||u_r|}, \quad \omega_r = \omega - \Omega \tag{16}$$

When a particle moves in a viscous fluid, the boundary layer on the surface of the particle will carry a certain amount of fluid. Due to the inertia of the fluid, the fluid cannot immediately follow the acceleration or deceleration of the particle, so this boundary layer is unstable, and the particle will be affected by a time-dependent force, given by:

$$F_B = \frac{9}{d_p \rho_p} \sqrt{\frac{\rho \mu}{\pi}} \int_{t_0}^{t} \frac{(du/d\tau) - (du_p/d\tau)}{\sqrt{t - \tau}} d\tau \tag{17}$$

In the Lagrangian coordinate system, the motions of particles are predicted based on the forces acting on them. In our studied system, the governing equation of particles can be given by

$$\frac{du_p}{dt} = \left(1 - \frac{\rho}{\rho_p}\right)g + F_D + F_P + F_A + F_S + F_M + F_B \tag{18}$$

The formula for the separation effect of the hydrocyclone:

$$P_{\text{downstream}} = \frac{N_{\text{downstream}}}{N_{\text{downstream}} + N_{\text{upstream}}}\%$$

$$P_{\text{upstream}} = \frac{N_{\text{upstream}}}{N_{\text{downstream}} + N_{\text{upstream}}}\% \tag{19}$$

2.3. Separation Experiments

Figure 3 provides the flow chart of separation experiment in the hydrocyclone. The water pressure was controlled at 20 kPa by adjusting the feed valve (consistent with the pressure set in numerical simulation).

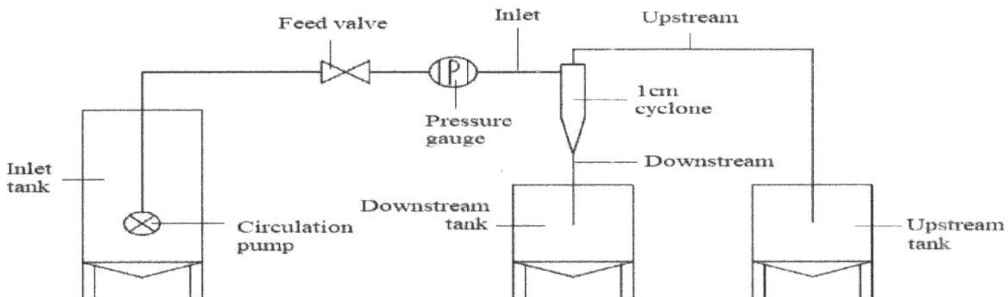

Figure 3. Schematic diagram of the hydrocyclone test.

Sand particles with D50 of 120 μm were screened using a 115 to 125 mesh screen to substitute the 120 μm particles in numerical simulation, and particles with D50 of 60 μm were screened using a 230 to 250 mesh screen to substitute the 60 μm particles in numerical simulation.

The volume of 120 μm particles was eight times that of 60 μm particles. The volume of 120 μm particles put into the hydrocyclone was eight times that of 60 μm particles to ensure an equal number of both particles.

Table 5 shows the standard volume concentration adopted. The concentrations for 60 and 120 μm particles were 5 kg·m^{-3} and 40 kg·m^{-3}, respectively.

Table 5. Experiments for particle separation.

Particles	Particle Size (μm)	Concentration (kg·m^{-3})	Separation Time (s)
Singe	60	5	16
	120	40	16
Mixed	60 + 120	5 + 40	16

Experiment 1. *Sand particles (61–106 µm; 9 kg) were screened out using a 150 to 240 mesh screen to substitute the 60 to 106 µm irregular particles in numerical simulation, and a separation simulation was performed on them.*

Experiment 2. *Three experiments were carried out, including two independent single separations of 60 and 120 µm particles and one mixed separation of both particles.*

3. Analysis of Experimental Results

3.1. Flow Field, Flow Rate and Particle Distribution

The separation process of water and sand in a hydrocyclone is as follows: the turbulent flow of water drives the sediment particles to move, as a result of which the latter eventually flow out from the upstream and downstream separately. It can be seen that turbulent flow of water plays an important role in the separation of sediment particles. After it flows in from the inlet, the water flows into a swirling turbulent flow in the cavity, and finally carries the sediment particles out of the upstream or downstream separately.

Figure 4 shows the velocity of water flow in the hydrocyclone along the X-, Y- and Z-axes, the resultant velocity of water flow (scalar quantity obtained by calculating the values of three coordinate axis), and the simulated result of water flow pressure distribution.

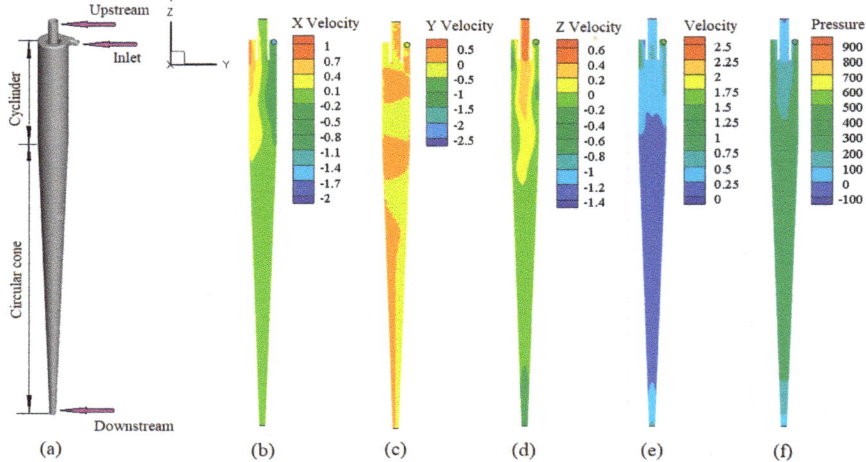

Figure 4. Structure of the cyclone, water flow velocity (m·s^{-1}) and pressure (Pa) distribution: (**a**) structure of hydrocyclone; (**b**) flow velocity in X-axis; (**c**) flow velocity in Y-axis; (**d**) flow velocity in Z-axis; (**e**) water flow velocity distribution; (**f**) water flow pressure distribution.

Figure 4a shows the structure of the hydrocyclone, Figure 4b, 4c and 4d are the water flow velocity (m/s) distributions along the X-, Y- and Z-axes, respectively. Figure 4e and 4f presents the water flow velocity distribution and the water flow pressure distribution, separately. The velocity distribution along the X-axis is symmetrically distributed on both sides centered with the Z axis. The velocity distribution along the Y-axis exhibits periodic positive and negative phases. With regard to the speed distribution along the Z-axis, the velocity near the wall of the cylindrical section is low, and particles there flow from the downstream. The water flows through a zero-speed buffer zone when approaching the axis center, and eventually flows to the upstream. The general trend is that the water flow velocity in the cavity changes significantly at the intersection between cylinders, and the flow velocity increases near the downstream (as shown in Figure 4e). The pressure distribution in the cavity did not show obvious changes, but changes significantly when near the outlet. By observing the water flow velocity and pressure distribution, it can be known that the water flow carries the sediment particles from the inlet, and being blocked by the wall, particles rotate in circles in the cyclone chamber, and finally flow out from the upstream or the downstream. Since the water carrying capacity is influenced by the shape and size of the particles themselves, the hydrocyclone is usually used for sediment particle separation.

Figure 5 shows the position distribution of the particles in the hydrocyclone during the simulation. Figure 5a shows the separation simulation of 60 and 120 μm particles, and Figure 5b shows the simulation of the effect of the cross-section change of sand particles in the separation. Figure 5c shows the distribution of sand particles along the Z-axis and Figure 5d shows the particle distribution at the D-D section in Figure 5b.

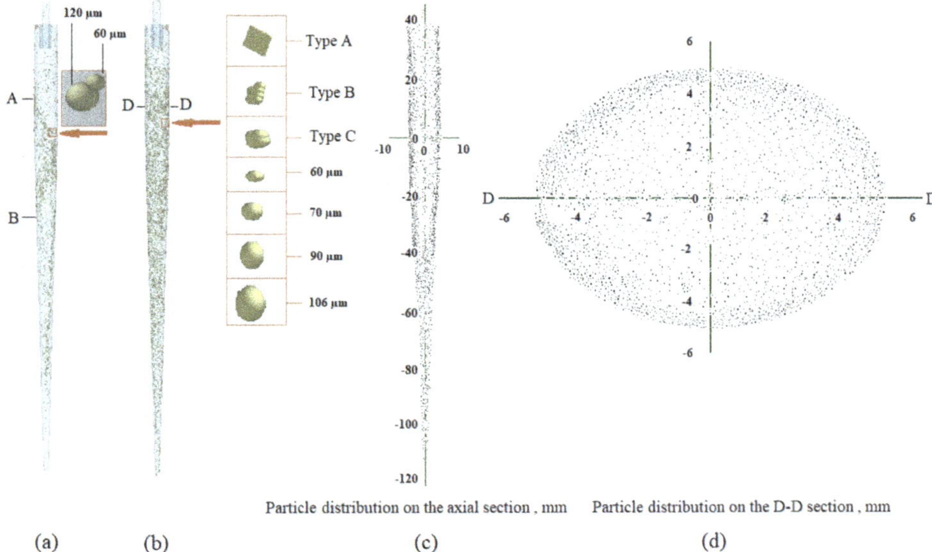

Figure 5. Sand particle distribution in simulation: (**a**) the separation simulation of 60 and 120 μm particles; (**b**) the simulation of the effect of the cross-section change of sand particles in the separation; (**c**) the distribution of sand particles along the Z-axis; (**d**) the particle distribution at the D-D section in (**b**).

From Figure 5c,d, the overall distribution of sand particle in the cylindrical section and part of the conical section can be obtained, and the closer of particles are to the wall, the higher the density is. The radial (X-axis) concentration distribution, as well as the velocity and the separation results in Figure 5b are shown in Tables 6 and 7, respectively. The axial velocity of the Z-axis, the radial (X-axis) distribution percentage, and the separation results in Figure 5a are shown in Figure 6, Tables 8 and 9, respectively.

3.2. Relationship between Maximum Projected Area and Separation Results

3.2.1. Radial Concentration Distribution of Particles

Table 6 shows the concentration distribution of different sand particles in the radial area of Figure 1. The general trend of sand particle distribution was that the higher the concentration distribution and the nearer the sands to the wall, the more concentrated the large sand particles, especially the 106 μm sand particles, which accounted for 86.93% of the total area.

Table 6. Concentration distribution of different sand particles in the cylindrical section.

Radial Distance	Type A	Type B	Type C	60 μm	70 μm	90 μm	106 μm
	Percentage of Content (%)						
0–1	8.10	6.82	3.96	2.65	2.11	1.81	1.01
1–2	9.01	7.12	5.39	5.14	3.23	2.34	1.96
2–3	12.87	11.12	9.01	8.63	7.05	5.91	3.21
3–4	18.59	17.36	16.20	13.85	11.61	7.32	6.89
4–5	51.43	57.58	65.44	69.73	76.00	82.62	86.93

The volume of 60 μm particles was the same as that of particle types C, B, and A. Their concentrations were 69.73%, 65.44%, 57.58%, and 51.43%, respectively, in the area with 4 to 5 mm radial distance near the wall. Therefore, the concentration of sand particles with the same volume tended to decrease gradually near the wall of the hydrocyclone with the increase in the maximum projected area.

The concentrations of 60, 70, 90, and 106 μm particles in the area with 4 to 5 mm radial distance were 69.73%, 76.00%, 82.62%, and 86.93%, respectively. That is to say, the concentration of spherical particles tended to increase in the direction of the hydrocyclone wall with the increase in the volume.

Taking the 2-3 area as the center, the concentrations of types C, B, and A in the area with 0 to 2 mm radial distance were 4.01%, 9.79%, and 14.14% higher than those of 70, 90, and 106 μm particles, respectively. Furthermore, their concentrations in the area with 3 to 5 mm radial distance were 5.97%, 15.00%, and 23.80% lower, respectively. Therefore, for the particles with the same volume and maximum projected area, the higher the concentration in the center of the hydrocyclone, the more likely the central part (the upwelling) flowing out from the upstream; for the particles with the same maximum projected area, the bigger the volume, the higher the concentration near the hydrocyclone wall, and the closer to the side wall, the more likely they might flow out from the downstream.

3.2.2. Separation Results in Numerical Simulation

Table 7 shows the separation results of the upstream and downstream with changes in the volume and maximum area.

Table 7. Separation results of particles from the upstream.

Radial Distance	Maximum Projected Area (S)	$t_{upstream}$ (s)	$t_{downstream}$ (s)	$V_{upstream}$ (m s^{-1})	$V_{downstream}$ (m s^{-1})	$L_{upstream}$ (m)	$L_{downstream}$ (m)	p (%)
Type A	3.11	0.63	2.21	0.41	0.59	0.26	1.30	24.0
Type B	2.24	0.66	2.15	0.41	0.59	0.27	1.27	17.8
Type C	1.36	0.69	2.11	0.41	0.59	0.28	1.25	14.9
60 μm	1.00	0.72	2.08	0.40	0.58	0.29	1.21	9.4
70 μm	1.36	0.75	2.02	0.39	0.59	0.29	1.19	7.4
90 μm	2.24	0.79	1.98	0.39	0.60	0.31	1.19	3.2
106 μm	3.11	0.81	1.92	0.39	0.60	0.32	1.15	0.9

$t_{upstream}$, the average residence time of particles crossing the upstream; $t_{downstream}$ is the average residence time of particles crossing the downstream; $V_{upstream}$ is the average absolute velocity of particles crossing the upstream; $V_{downstream}$ is the average absolute velocity of particles crossing the downstream; $L_{upstream}$ is the average path length of particles crossing the upstream; $L_{downstream}$ is the average path length of particles crossing the downstream; and p is the percentage of the particles crossing the upstream in the separated particles, which was obtained using Equation (19).

As shown in Table 7, 9.4%, 7.4%, 3.2%, and 0.9% of spherical particles with diameters of 60, 70, 90, and 106 μm flowed out from the upstream, respectively. Therefore, it was harder for larger spherical particles to be separated from the upstream. The 60 μm particles had the same volume as types C, B, and A particles. As the maximum projected area increased gradually, 9.4%, 14.9%, 17.8%, and 24% of them flowed out from the upstream, accordingly. Thus, it was easier for the particles with the same volume to be separated from the upstream as the maximum projected area increased.

The 60 μm particles and types C, B, and A particles flowed out from the upstream in less time and traveled a shorter distance with the increase in the maximum projected area. For instance, the particles of type A took 0.09 s less compared with 60 µm particles and the average distance decrreased by 0.03 m.

However, the 60, 70, 90, and 106 µm particles needed more time to flow out from the upstream and travel a longer distance with the increase in their size. Among them, the 106 µm particles took the most time, 0.09 s more compared with 60 µm particles on average, and the average distance increased by 0.03 m.

For particles obtained from the upstream, the average velocity of 60 µm spherical particles and types A, B, and C particles was greater than that of 70, 90, and 106 µm particles. However, the average velocity of particles obtained from the downstream was in contrast to those from the upstream. The absolute velocity of different particles was less than that of water flow. Therefore, the smaller the particles or the larger the projected area, the stronger the ability to flow.

3.3. Separation Mechanism of Single and Mixed Particles

3.3.1. Distribution of Axial Velocity and Radial Concentration

Figure 6 reflects the radial velocity of 60 and 120 µm particles. Three curves are seen from the top downward. The first one is about the average velocity of 120 µm particles in a positive direction, the direction of downstream. (The average velocity of 120 µm particles in single and mixed separations is shown by the same curve as it is the same in different separations, and 60 µm particles have no influence on 120 µm particles.) The second curve shows the average velocity of 60 µm particles in mixed separation in a positive direction (the direction of downstream). The third one is about the average velocity of 60 µm particles in single separation in a positive direction of Z axis. Figure 6b shows the partially enlarged image of Figure 6a (the former is five times bigger than the latter).

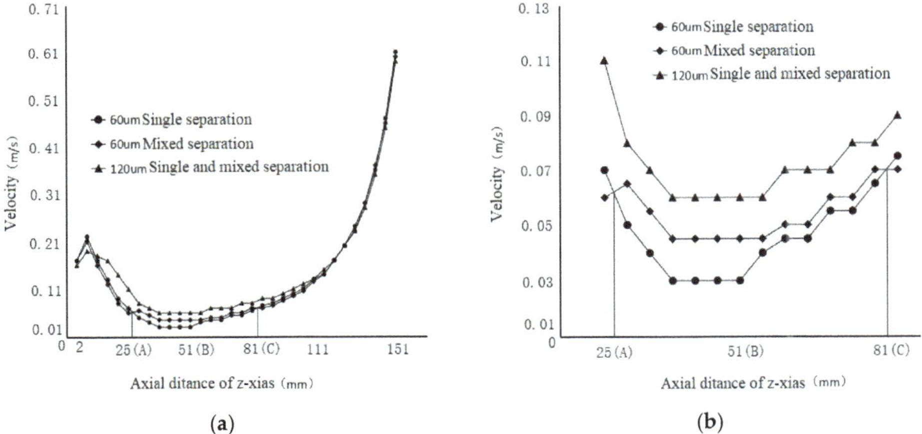

Figure 6. Average velocity distribution in a positive direction of Z axis: (**a**) radial velocity of 60 and 120 µm particles; (**b**) partially enlarged image of (**a**).

The velocity and direction of water flow significantly changed at 25–81 mm in Figure 6, which is the transition from a cylinder to a cone. Particles slowed down in a positive direction on the Z axis due to a lower water velocity, and the average velocity of different-sized particles also decreased. Thus, particles in this section mostly flow slowly in the whole separation process.

The velocity of 60 and 120 µm particles on Z axis had four changing stages: rising, falling, flattening, and rising again. Compared with a single separation, the average velocity of 60 µm particles increased by 25.74% at 25–81 mm in mixed separation, which was quite obvious at 25–51 mm. However, the velocity tended to be consistent at 0–25 mm and 81–151 mm.

The velocity was the lowest at 25–81 mm in the whole process. As shown in Figure 1, it was the border of cylinder and cone sections, 25–51 mm in the cylinder section and the rest in the cone section.

Table 8 shows the distribution of particle concentration at A to B in Figure 4. It was found that 93.23% of 120 μm particles in single separation were concentrated 4–5 mm away from the center of the circle in the radial direction. The percentage of particles nearest to the center of the circle was 1.33%. However, 69.73% of 60 μm particles in single separation were concentrated at 4–5 mm, and the percentage of particles nearest to the center of the circles was 2.65%. When the particles of these two sizes were mixed, the radial concentration distribution of 120 μm particles was in accordance with that of single separation generally. However, the radial concentration distribution of 60 μm particles in single separation was quite different from that of mixed separation. The difference was reflected in the section of 4–5 mm, and the percentage reached 20.73%.

In other words, influenced by 120 μm particles, 20.73% of 60 μm particles were concentrated in the aforementioned section in mixed separation.

Table 8. Concentration distribution of 60 and 120 μm particles in the cylinder section.

		0–1 mm	1–2 mm	2–3 mm	3–4 mm	4–5 mm
	Single	2.65	5.14	8.63	13.85	69.73
60 μm particles	Mixed	1.41	1.16	1.19	5.78	90.46
	Difference	−1.24	−3.98	−7.44	−8.07	20.73
	Single	1.33	1.50	1.64	2.30	93.23
120 μm particles	Mixed	1.21	1.32	1.79	2.95	92.73
	Difference	−0.12	−0.18	0.15	0.65	−0.5

3.3.2. Particle Separation Results

The calculation of particle separation took 4 s (see Table 9). The number of particles that flowed into the hydrocyclone was more than the sum of particles separated from upstream and desilting port in 0–3 s; the separation was unstable. However, it reached a dynamic balance because the number of particles flowing into the hydrocyclone was close to that of the particles separated from it in 3–4 s. When 60 μm particles were mixed with others, the average time spent in separation from the downstream was 0.2 s less than that in single separation while the separation increased by 4.1%. Furthermore, 120 μm particles were not separated in either single separation or mixed separation from the upstream.

Table 9. Separation of 60 and 120 μm particles.

		$t_{avearge}$	$N_{downstream}$	$N_{upstream}$	p (%)
Single	X	2.00	906	94	90.6
	D	1.70	1000	0	100
Mixed	X	1.80	947	53	94.7
	D	1.70	1000	0	100

X and D, the number of X (60 μm particles) and D (120 μm particles) was 1000 per second; $t_{avearge}$ is the average time duration in which particles were separated from the downstream staying in the hydrocyclone in 3–4 s; $N_{downstream}$ is the number of particles passing the downstream; $N_{upstream}$ is the number of particles passing the upstream in 3–4 s; and p is the percentage of particles in the downstream, which was calculated using Formula (19).

4. Experimental Tests

4.1. Test on the Effect of Cross-Section Change on Separation

Figure 7 depicts the observation results of sand samples in the experiment. Figure 7a shows the representative sand particles screened using a vibrating screen. Both particles 1 and 2 could pass the 110 μm hydrocyclone screen, but they were not typical spherical particles, and their volume was less than that of spherical particles of the same size. Figure 7b shows the typical comparison of thickness. Obviously, particles 1 and 3 were thicker than particle 2 in terms of light transmittance. Figure 7c shows the particles with different projected area and thickness. Among them, the volumes of particles 1, 2, 3, and 4 might be the same. Figure 7d shows statistics on the maximum projected area of typical lamelliform particles separated from the upstream. The maximum projected area of particles 1, 2, 3, and 4 in Figure 4 was 7968.39, 8346.47, 8596.55, 3826.37, and 3949.87 μm^2, respectively. Accordingly, the size of spherical particles was 100.75, 103.11, 104.65, 69.81, and 70.93 μm, whose volume was less than that of spherical particles with the same projected area. Smaller particles were more than larger particles. Moreover, most were lamelliform particles with a similar size.

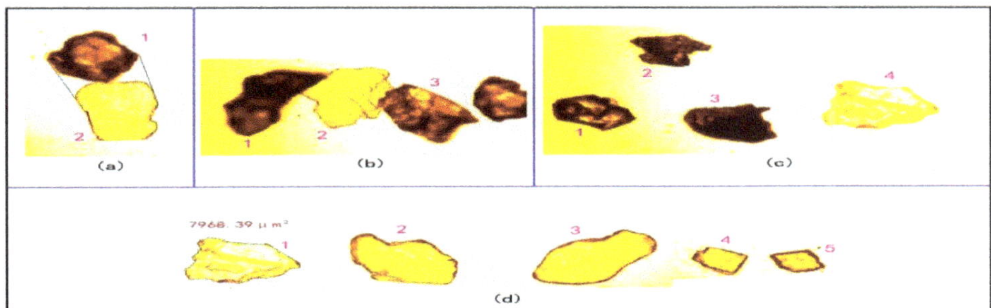

Figure 7. Distribution of particle characteristics ($\times 1$ k).

Different particles (10 groups) were observed with a transflective polarized microscope whose number was 59XC-PC before the experiment. It was found that most of 500 particles had a bad light transmittance and only a small number of them (10–30 particles) had a good light transmittance. However, about 160–210 among 500 particles obtained from the upstream were good in light transmittance (because they were relatively thin) after the experiment.

The results indicated that the larger the size of particles, the less the particles were separated from the upstream. As shown in Figure 8, the percentage of 60 and 106 μm particles was 27.75 and 1.91%, respectively.

Therefore, smaller and thinner particles were separated from the upstream more easily. In other words, the process of separating sand particles in the hydrocyclone led to the classification of particles of different sizes, which was equivalent to the filtering of particles of different shapes (lamelliform particles). The experimental result was consistent with the regularities reflected in numerical simulation.

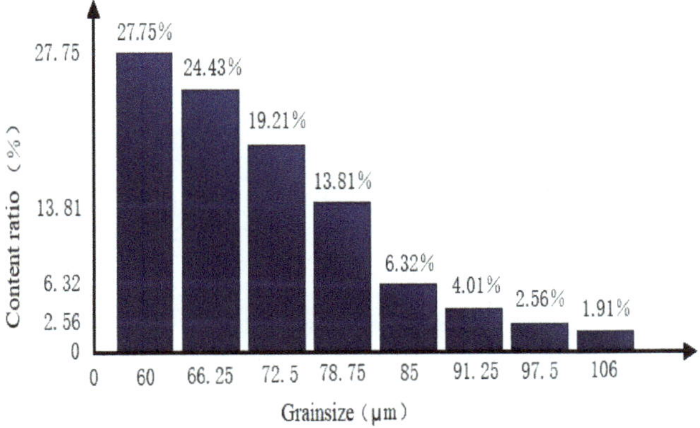

Figure 8. Percentage of particles with different sizes on the upstream.

4.2. Separation Test of Sand Particles 60 and 120 µm

The concentration of 60 and 120 µm sand particles was set as 5 kg·m^{-3} and 40 kg·m^{-3}, respectively, to ensure an equal number of sand particles put into the hydrocyclone. The capacity of the container was 1 m^3. The separation experiment lasted for 16 min.

Table 10 shows the results: 120 µm sand particles were separated from the downstream in both single and mixed separations. However, 312 g of 60 µm sand particles were separated from the upstream in single separation, and 255 g were separated in mixed separation. It was concluded that larger particles improved the separation of smaller ones from the upstream in mixed separation, which was consistent with the regularities reflected in numerical simulation.

Table 10. Separation experiments for 60 and 120 µm sand particles.

	Single		Mixed	
	60 µm	120 µm	60 µm	120 µm
Upstream	0.312	0	0.255	0
Downstream		28.85		29.25

5. Discussion

5.1. Influence of Maximum Projected Area and Volume Changes on Separation Results

Several parameters affect the performance of hydrocyclones, including the inlet velocity, solid content, liquid-phase viscosity, particle size and shape, and so forth. Saber et al. [20] believed that the separation effect would improve with the increase in the size of spherical particles, which was obviously better than that of plate-like particles. Also, the separation effect declined with the increase in the size of plate-like particles with diameters more than 60 µm. Abdollahzadeh et al. [21] thought that the separation effect decreased along with the increase in flatness when the particle diameter was more than 10 µm. Endoh et al. [22] found that particles with a large size and high proportion were separated from the upstream. Wills et al. [23] held the view that plate-like particles, such as mica, were often discharged upstream even though they were relatively coarse. Kashiwaya et al. [17] believed that the shape of particles would influence the drag coefficient and separation effect. The drag coefficient would increase when the particles became flatter. Then, the increased drag force would increase the influence of water movement on particles. Wang et al. [24] found that the change in particle velocity fluctuated evidently as the size of particles became smaller and the drag force increased. Some particles passed through the locus of zero vertical velocity, joined the upward flow, and then moved out from the upstream. In this study, particles of types C, B, and A, corresponding to 70, 90, and 106 µm particles, respectively, could enter the upward flow and then flow out of the upstream easily as particles became smaller and drag force increased. Therefore, the corresponding percentage dropped sharply in the area with 3 to 5 mm radial distance and increased in the area with 1 to 3 mm radial distance.

Drag force is proportional to the maximum projected area of particles along the direction of fluid flow (Formula (19)), which means that the drag force is strengthened with the increase in the maximum projected area of particles. Hence, the maximum projected area of particles has a certain influence on the separation results of the hydrocyclone. Kashiwaya et al. considered that when Reynolds number reached a certain level (above 50), the maximum projected area of particles was a major influencing factor. They also claimed that the radial variation in the tangential velocity in the hydrocyclone gave rise to a moment on plate-like particles and the particles settled, maintaining the orientation as their broad planes became perpendicular to the radial direction. In this study, the pressure gradient force in the upwelling direction toward the center of the hydrocyclone in the same position increased with the increase in the cross-sectional area. Also, the particles could be easily pushed into the upwelling and then out of the upstream. Figure 9 shows that the volume of 60 µm spherical particles was the same as that of particles of types C, B, and A, but the cross-sectional area was larger and hence the track was simpler in the hydrocyclone.

Figure 9 shows the representative motion tracks of sand particles with different cross-sectional areas when they moved from the upstream. The figure shows that 60 μm spherical particles had the same volume as that of particles of types C, B, and A. With the increase in the maximum projected area, the tracks became simpler and the time needed for separation became shorter. Furthermore, 60, 70, 90, and 106 μm particles were all spherical. The tracks separated from the upstream become more complex with the increase in the volume of particles. That is, it was harder for particles to separate from the upstream as the time taken for separation increased. Tracks (2) and (8), (3) and (7), and (4) and (6) belong to the particles with the same maximum projected area. The smaller the volume of particles, the simpler the tracks when separated from the upstream, indicating that the time taken for separation was shorter, and it was harder for particles to separate from the upstream. Hence, the maximum projected area might have an obvious and a stable influence on the motion tracks of particles, hence affecting the separation effect.

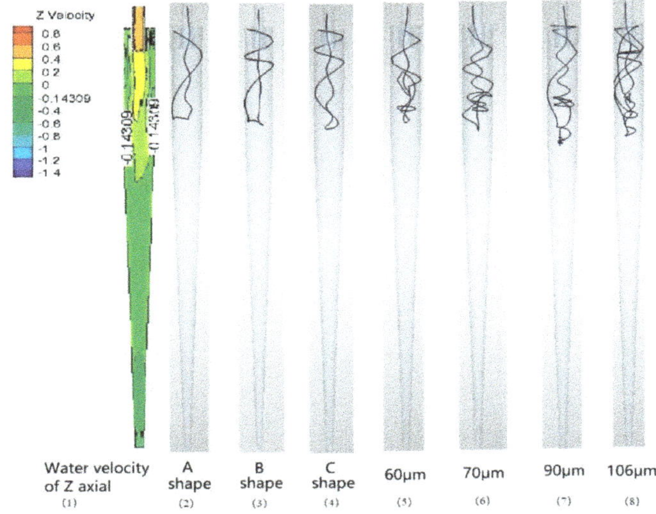

Figure 9. Representative motion tracks of sand particles.

5.2. Analysis of the Following Phenomenon of Fine Particles

Zhang et al. [19] believed that the movement of particles was controlled by the combined effect of the radial fluid drag force, pressure gradient force, and centrifugal force. Of these, the fluid drag force of the inward direction increased exponentially with decreasing size of small particles. A part of the small particles was pushed into the center of the hydrocyclone and then flowed out of upstream. Xu et al. [25] held the view that large particles moved toward the wall under the effect of inertial force and gravity. In this test, a part of 60 μm particles flowed out of the upstream, while the 120 μm particles did not. However, the following phenomenon enhanced the density of the 60 μm particles near the wall, where most 120 μm sand particles were present, thereby reducing the possibility of 60 μm particles flowing into the center of the hydrocyclone and out of the upstream. Therefore, the force on particles in the hydrocyclone was complicated. Zhu et al. [26] noted that the comprehensive effect of the fluid drag force, pressure gradient force, virtual mass force, and gravity on particles should be taken into consideration. Peng et al. used the lattice Boltzmann method to successfully simulate the shallow water flows and river bed evolution [27–34] as well as two-phase flows [35,36].

Tsuji et al. [37] believed that the large leading particles had a huge drag force on small trailing particles, which had little or even no drag force on leading particles in turn. The experiment showed that 120 μm particles had a great impact on 60 μm particles in terms of distribution, velocity, or separation effect. However, no 120 μm particles separated from the upstream. Furthermore, 120 μm sand particles could influence 60 μm particles to a great degree, but the latter's influence on the former could be neglected, which was consistent with the findings of other studies. Meanwhile, Zhu et al. [38] claimed that Re could affect drag, and the increase in Re could obviously weaken the influence of leading particles on small ones. Zhang et al. [18] held the view that the following drag of small particles toward large ones would decrease with the increasing distance, and it would be affected by particle's ratio and the local wake flow compared with that between the particles and the incoming fluid stream. Schubert et al. [39] considered that this phenomenon was due to particle interaction, while the finer particles were captured by the larger particles and the effect was more prominent with the increase in the fraction of large particles. In this study, Re did not change at all under the force of inlet pressure, which was the only one used in this experiment. However, the following phenomenon was quite obvious from the perspectives of process and results.

Under the ideal circumstances, the deposition velocity of spherical particles is proportional to particle diameter (Formula (13)). The velocity is high with the increase in the particle diameter. The numerical calculation and experimental results in this study showed that larger particles were faster than smaller particles in terms of deposition velocity in the hydrocyclone. CFD–DEM coupling was adopted to capture the following phenomena of the movement of small particles toward large ones (see Figure 10). Figure 10a,b presents the motion tracks of 120 and 60 μm particles in the downstream in mixed separation, whereas Figure 10c is the enlarged drawing of the complex segment in Figure 10b. In the figure, among particles around the transition section from a cylinder to a cone, 60 μm particles firstly entered the hydrocyclone, and their motion track was chaotic. After meeting with the 120 μm particles, the 60 μm particles changed their direction and were separated from the downstream together with the 120 μm ones. This phenomenon also explains why the average velocity of mixed separation of 60 and 120 μm particles near the boundary of cylinder and cone was higher than that of single separation (Figure 6).

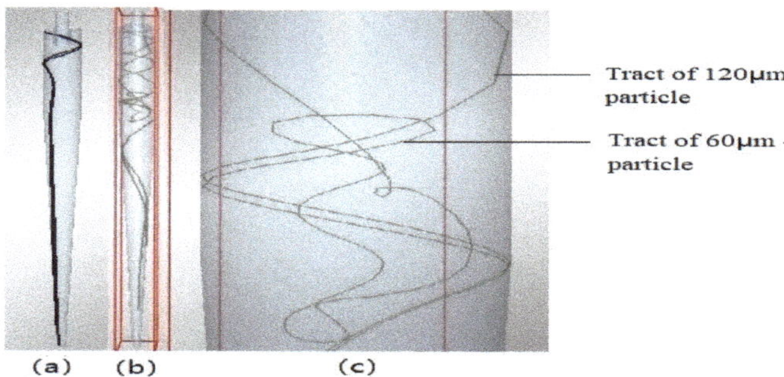

Figure 10. Representative motion tracks of particles.

6. Conclusions

In this study, numerical simulation and experimental methods were applied to reveal the separation mechanism of the hydrocyclone for different-sized particles and the influence of maximum projected area on the separation effect. The conclusions were as follows:

The percentage of 60, 70, 90, 106, and 120 μm spherical particles separated from the upstream was 9.4%, 7.4%, 3.2%, 0.9%, and 0%, respectively, indicating that spherical particles were more easily separated from the downstream as their size increased.

For 60 μm spherical particles and types C, B, and A particles with the same volume, the separation percentages from the upstream were 9.4%, 14.9%, 17.8%, and 24.0%, respectively, implying that the running space was closer to the upward flow in the hydrocyclone as the maximum projected area increased. Hence, particles with the same volume were easier to be separated from the upstream.

The axial velocity of 60 μm particles mixed with 120 μm particles in separation increased by 25.74% compared with that in the single separation of 60 μm particles near the transition section from a cylinder to a cone. The concentration near the wall, where mainly 120 µm particles were present, increased by 20.73%, and the separation effect in the downstream improved by 4.1%. Therefore, large particles could optimize the separation effect of small particles from the downstream.

Author Contributions: Software, Z.T.; methodology, N.L. and F.W.; writing—original draft preparation, Z.T.; writing—review and editing, L.Y.; funding acquisition, L.Y., L.C. and N.C.

Funding: This research was funded by the National Natural Science Foundation of China (NSFC) (No. 51769009, 51379024,51809022) and Funding Program for University Engineering Research Center of the Yunnan Province in China.

Acknowledgments: The authors appreciate the financial support received from the National Natural Science Foundation of China (NSFC) (No. 51769009, 51379024, 51809022) and Funding Program for University Engineering Research Center of the Yunnan Province in China.

Conflicts of Interest: Declare conflicts of interest or state.

Nomenclature

A	measure of area
A_p	maximum projected area of sand particles in the direction of fluid flow, m^2
C_{LS}	Saffman lit coefficient
C_{LM}	magnus lift coefficient
D_1	drag
D_2	particle diameter
d_p	particle diameter, m
d	damping
d_i	diameter of sand particles, m
f	fluid phase
F_A	added mass force per unit mass, $m·s^{-2}$
F_B	basset force per unit mass, $m·s^{-2}$
F_D	total drag force;
F_M	Magus force per unit mass, $m·s^{-2}$
F_P	pressure gradient force per unit mass, $m·s^{-2}$
$F_{p\text{-}f}$	interaction forces between fluid and solids phases
F_S	Saffman lift force per unit mass, $m·s^{-2}$
g	gravity acceleration vector, 9.81 $m·s^{-2}$
$i(j)$	corresponding to i(j)th particle
k_c	number of particles in a computational cell, dimensionless
$N_{downstream}$	number of sand particles separated from the downstream
$N_{upstream}$	number of sand particles separated from the upstream
$P_{downstream}$	fractional flow of the downstream
$P_{upstream}$	fractional flow of the upstream
P	pressure, Pa
pg	pressure gradient
p	particle phase
Re	Reynolds number, dimensionless
T	particle shape factor (thickness)
t	time, s
u	fluid velocity vector, $m·s^{-1}$
u_t	settling velocity, $m·s^{-1}$
u_r	relative velocity, $m·s^{-1}$
V	volume, m^3
v	fluid kinematic viscosity, $kg·m^{-1}\,s^{-1}$
ΔV_c	volume of a computational cell, m^3
ΔP	pressure drop, Pa
∇p	pressure gradient, $kg·m^{-2}s^{-2}$

Greek letters

ε	porosity, dimensionless
μ	fluid viscosity, $kg \cdot m^{-1} \cdot s^{-1}$
τ	viscous stress tensor, $N \cdot m^{-3}$
ρ	density, $kg \cdot m^{-3}$
β	Empirical coefficient defined in Equation (3), dimensionless
ρ_p	particle density, $kg \cdot m^{-3}$
γ	fluid strain rate, s^{-1}
ω_r	relative angular velocity, $rad \cdot s^{-1}$
ω	particle rotation angular velocity, $rad \cdot s^{-1}$
ξ	drag force coefficient
Ω	fluid rotation angular velocity, $rad \cdot s^{-1}$
f	force

References

1. Fu, S.; Yong, F.; Yuan, H.; Tan, W.; Dong, Y. Effect of the medium's density on the hydrocyclonic separation of waste plastics with different densities. *Waste Manag.* **2017**, *67*, 27–31. [CrossRef] [PubMed]
2. Zhang, Y.; Cai, P.; Jiang, F.; Dong, K.; Jiang, Y.; Wang, B. Understanding the separation of particles in a hydrocyclone by force analysis. *Powder Technol.* **2017**, *322*, 471–489. [CrossRef]
3. Neesse, T.; Dueck, J.; Schwemmer, H.; Farghaly, M. Using a high pressure hydrocyclone for solids classification in the submicron range. *Miner. Eng.* **2015**, *71*, 85–88. [CrossRef]
4. Yu, L.; Zou, X.; Hong, T.; Yan, W.; Chen, L.; Xiong, Z. 3D numerical simulation of water and sediment flow in hydrocyclone based on coupled cfd-dem. *Trans. Chin. Soc. Agric. Mach.* **2016**, *01*, 126–132.
5. Zhu, G.; Liow, J. Experimental study of particle separation and the fishhook effect in a mini-hydrocyclone. *Chem. Eng. Sci.* **2014**, *111*, 94–105. [CrossRef]
6. Liu, H.; Wang, Y.; Han, T. Influence of vortex finder configurations on separation of fine particles. *CIESC J.* **2017**, *05*, 1921–1931.
7. Tang, B.; Xu, Y.; Song, X.; Xu, J. Effect of inlet configuration on hydrocyclone performance. *Trans. Nonferrous Met. Soc. China* **2017**, *27*, 1645–1655. [CrossRef]
8. Hwang, K.-J.; Chou, S.-P. Designing vortex finder structure for improving the particle separation efficiency of a hydrocyclone. *Sep. Purif. Technol.* **2017**, *172*, 76–84. [CrossRef]
9. Vakamalla, T.R.; Koruprolu, V.B.R.; Arugonda, R.; Mangadoddy, N. Development of novel hydrocyclone designs for improved fines classification using multiphase cfd model. *Sep. Purif. Technol.* **2017**, *175*, 481–497. [CrossRef]
10. Zhang, C.; Wei, D.Z.; Cui, B.Y.; Li, T.S.; Luo, N. Effects of curvature radius on separation behaviors of the hydrocyclone with a tangent-circle inlet. *Powder Technol.* **2017**, *305*, 156–165. [CrossRef]
11. Zhou, Q.; Wang, C.; Wang, H.; Wang, J. Eulerian–lagrangian study of dense liquid–solid flow in an industrial-scale cylindrical hydrocyclone. *Int. J. Miner. Process.* **2016**, *151*, 40–50. [CrossRef]
12. Zhao, Q.; Li, W.; He, W. Experimental study of pressure drop characteristics of a novel solid/liquid hydrocyclone. *J. Eng. Thermophys.* **2010**, *01*, 61–63.
13. Wang, B.; Chu, K.W.; Yu, A.B. Numerical study of particle–fluid flow in a hydrocyclone. *Ind. Eng. Chem. Res.* **2007**, *46*, 4695–4705. [CrossRef]
14. Wang, B.; Chu, K.W.; Yu, A.B.; Vince, A. Modelling the multiphase flow in a dense medium cyclone. *Ind. Eng. Chem. Res.* **2009**, *48*, 3628–3639. [CrossRef]
15. Asakura, K.; Asari, T.; Nakajima, I. Simulation of solid–liquid flows in a vertical pipe by a collision model. *Powder Technol.* **1997**, *94*, 201–206. [CrossRef]
16. Chu, K.W.; Wang, B.; Yu, A.B.; Vince, A. Cfd-dem modelling of multiphase flow in dense medium cyclones. *Powder Technol.* **2009**, *193*, 235–247. [CrossRef]
17. Niu, W.; Liu, L.; Chen, X. Influence of fine particle size and concentration on the clogging of labyrinth emitters. *Irrig. Sci.* **2013**, *31*, 545–555. [CrossRef]
18. Kashiwaya, K.; Noumachi, T.; Hiroyoshi, N.; Ito, M.; Tsunekawa, M. Effect of particle shape on hydrocyclone classification. *Powder Technol.* **2012**, *226*, 147–156. [CrossRef]

19. Zhang, J.; Fan, L.S. A semianalytical expression for the drag force of an interactive particle due to wake effect. *Ind. Eng. Chem. Res.* **2002**, *41*, 5094–5097. [CrossRef]

20. Niazi, S.; Habibian, M.; Rahimi, M. A comparative study on the separation of different-shape particles using a mini-hydrocyclone. *Chem. Eng. Technol.* **2017**, *40*, 699–708. [CrossRef]

21. Abdollahzadeh, L.; Habibian, M.; Etezazian, R.; Naseri, S. Study of particle's shape factor, inlet velocity and feed concentration on mini-hydrocyclone classification and fishhook effect. *Powder Technol.* **2015**, *283*, 294–301. [CrossRef]

22. Endoh, S. Study of the shape separation of fine particles using fluid fields-dynamic properties of irregular shaped particles in wet cyclones. *J. Soc. Powder Technol. Jpn.* **1992**, *29*, 125–132. [CrossRef]

23. Wills, B.A.; Napier-Munn, T. *Wills' Mineral Processing Technology*, 7th ed.; Elsevier Science & Technology Books: Amsterdam, The Netherlands, 2006; pp. 145–206.

24. Wang, B.; Yu, A.B. Computational investigation of the mechanisms of particle separation and "fish-hook" phenomenon in hydrocyclones. *AIChE J.* **2010**, *56*, 1703–1715. [CrossRef]

25. Xu, P.; Wu, Z.; Mujumdar, A.S.; Yu, B. Innovative hydrocyclone inlet designs to reduce erosion-induced wear in mineral dewatering processes. *Dry. Technol.* **2009**, *27*, 201–211. [CrossRef]

26. Zhu, C.; Liang, S.C.; Fan, L.S. Particle wake effects on the drag force of an interactive particle. *Int. J. Multiph. Flow* **1994**, *20*, 117–129. [CrossRef]

27. Peng, Y.; Zhou, J.G.; Burrows, R. Modelling the free surface flow in rectangular shallow basins by lattice Boltzmann method. *J. Hydraul. Eng. ASCE* **2011**, *137*, 1680–1685. [CrossRef]

28. Peng, Y.; Zhou, J.G.; Burrows, R. Modelling solute transport in shallow water with the lattice Boltzmann method. *Comput. Fluids* **2011**, *50*, 181–188. [CrossRef]

29. Peng, Y.; Zhang, J.M.; Zhou, J.G. Lattice Boltzmann Model Using Two-Relaxation-Time for Shallow Water Equations. *J. Hydraul. Eng. ASCE* **2016**, *142*, 06015017. [CrossRef]

30. Peng, Y.; Zhang, J.M.; Meng, J.P. Second order force scheme for lattice Boltzmann model of shallow water flows. *J. Hydraul. Res.* **2017**, *55*, 592–597. [CrossRef]

31. Peng, Y.; Zhou, J.G.; Zhang, J.M.; Burrows, R. Modeling moving boundary in shallow water by LBM. *Int. J. Mod. Phys. C* **2013**, *24*, 1–17. [CrossRef]

32. Peng, Y.; Zhou, J.G.; Zhang, J.M.; Liu, H.F. Lattice Boltzmann Modelling of Shallow Water Flows over Discontinuous Beds. *Int. J. Numer. Method Fluids* **2014**, *75*, 608–619. [CrossRef]

33. Peng, Y.; Zhou, J.G.; Zhang, J.M. Mixed numerical method for bed evolution. *Proc. Inst. Civ. Eng. Water Manag.* **2015**, *168*, 3–15. [CrossRef]

34. Peng, Y.; Meng, J.P.; Zhang, J.M. Multispeed lattice Boltzmann model with stream-collision scheme for transcritical shallow water flows. *Math. Probl. Eng.* **2017**, *2017*, 8917360. [CrossRef]

35. Peng, Y.; Mao, Y.F.; Wang, B.; Xie, B. Study on C-S and P-R EOS in pseudo-potential lattice Boltzmann model for two-phase flows. *Int. J. Mod. Phys. C* **2017**, *28*, 1750120. [CrossRef]

36. Peng, Y.; Wang, B.; Mao, Y.F. Study on force schemes in pseudopotential lattice Boltzmann model for two-phase flows. *Math. Probl. Eng.* **2018**, *2018*, 6496379. [CrossRef]

37. Tsuji, Y.; Morikawa, Y.; Terashima, K. Fluid-dynamic interaction between two spheres. *Int. J. Multiph. Flow* **1982**, *8*, 71–82. [CrossRef]

38. Zhu, G.; Liow, J.L.; Neely, A. Computational study of the flow characteristics and separation efficiency in a mini-hydrocyclone. *Chem. Eng. Res. Des.* **2012**, *90*, 2135–2147. [CrossRef]

39. Schubert, H. On the origin of "Anomalous" shapes of the separation curve in hydrocyclone separation of fine particles. *Part. Sci. Technol.* **2004**, *22*, 219–234. [CrossRef]

Case Report

A Comprehensive Method of Calculating Maximum Bridge Scour Depth

Rupayan Saha [1], Seung Oh Lee [2] and Seung Ho Hong [1],*

[1] Department of Civil and Environmental Engineering, West Virginia University, 1306 Evansdale Drive, Morgantown, WV 26506, USA; rs0002@mix.wvu.edu

[2] Department of Civil Engineering, Hongik University, 94 Wausan-ro, Mapo-gu, Seoul 04066, Korea; seungoh.lee@hongik.ac.kr

* Correspondence: sehong@mail.wvu.edu; Tel.: +1-304-293-9926

Received: 4 October 2018; Accepted: 1 November 2018; Published: 3 November 2018

Abstract: Recently, the issues of scour around a bridge have become prominent because of the recurrent occurrence of extreme weather events. Thus, a bridge must be designed with the appropriate protection measures to prevent failure due to scour for the high flows to which it may be subjected during such extreme weather events. However, the current scour depth estimation by several recommended equations shows inaccurate results in high flow. One possible reason is that the current scour equations are based on experiments using free-surface flow even though extreme flood events can cause bridge overtopping flow in combination with submerged orifice flow. Another possible reason is that the current practice for the maximum scour depth ignores the interaction between different types of scour, local and contraction scour, when in fact these processes occur simultaneously. In this paper, laboratory experiments were carried out in a compound shape channel using a scaled down bridge model under different flow conditions (free, submerged orifice, and overtopping flow). Based on the findings from laboratory experiments coupled with widely used empirical scour estimation methods, a comprehensive way of predicting maximum scour depth is suggested which overcomes the problem regarding separate estimation of different scour depths and the interaction of different scour components. Furthermore, the effect of the existence of a pier bent (located close to the abutment) on the maximum scour depth was also investigated during the analysis. The results show that the location of maximum scour depth is independent of the presence of the pier bent but the amount of the maximum scour depth is relatively higher due to the discharge redistribution when the pier bent is absent rather than present.

Keywords: bridge scour; sediment transport; submerged flow; physical hydraulic modeling

---

1. Introduction

When a bridge is constructed in a river, the flow pattern around the bridge changes because a unique flow field develops locally around the bridge pier and abutment. Furthermore, reduced flow area through the bridge opening by the existence of embankment/abutments at both and/or one side of the river results in higher flow velocity due to the acceleration. This unique flow field with the higher velocity can seriously damage bridge foundations. Thus, if the depth of the foundation is not deep enough, the chance of bridge failure becomes higher.

A bridge can fail because of several causes, such as earthquake, wind, and flooding. Among them, bridge scour is the biggest reason of bridge failure [1,2]. For example, about 60% of bridge failures of total bridge collapse in the United States since 1950 have been related to the scour of bridge foundations [3]. The Colorado Department of Transportation (CDOT) estimated a minimum of 30 state highway bridges were destroyed and twenty were seriously damaged by floods in the year of 2013 [4]. In Nepal, due to the degradation of bed materials during the 2014 flooding, the foundation of the highway bridge over the Tinau River was seriously exposed [5]. As explained in the above examples, it is justified to say that bridge scour is one of the main bridge safety problems all over the world. Thus, accurate prediction of scour at the bridge foundation becomes the primary aim of engineers for the safety of a bridge.

Since the late 1950s, numerous studies on scour around bridges have been conducted and have generated formulas for equilibrium scour depth estimation at bridge foundations [6–9]. Although using equilibrium scour depth around a bridge foundation is a reasonable practice to design a bridge, sometimes it shows an overly conservative design compared to the field measurements. Contrary to the conservative design, a study in South Carolina found that the observed scour depth was greater than the equilibrium scour depth based on using 100-year flooding even if there had not been a 100-year flooding since the bridge was built [10,11]. Repeated occurrences of smaller flooding events might cause scour that was greater than the scour estimation in South Carolina. Moreover, some researchers have used computational fluid dynamics (CFD) simulations and found that the scour depth was under-predicted by the steady-state calculations while it was over-predicted by the unsteady-state [12,13].

One of the possible reasons of lacking an accurate scour prediction method is that many of the scour predictor equations were derived from simplified laboratory studies under free flow cases. The extreme amount of water associated with huge flood events can result in a complex flow field around the bridge under bridge overtopping flow in combination with submerged orifice flow. In addition to the complex flow field, the irregular shape of river geometry as well as non-uniform sediment distribution cannot be reproduced in a simplified-idealized laboratory setting such as in rectangular flume as in the previous research. Also, most of the scour predicting equations are 2nd or 3rd order which may not accurately predict scour depth as the scouring process is a complicated scholastic phenomenon.

Another possible reason for the inaccurate prediction of scour depth is that the current practice of total scour depth assumes that contraction and local scour are independent processes. Contraction scour is the consequence of flow acceleration due to contraction in the flow area, while local scour is caused by local vortex structures around the base of the obstruction. However, when the bridges are constructed in the river, these two-flow patterns tend to occur concurrently, which make local scour and contraction scour time dependent [14]. Thus, to predict maximum scour depth around a bridge foundation, a single equation should be developed rather than two separate equations for different types of scour.

Hence, the main objective of this study is to develop a single equation that can be used to predict maximum scour depth where different types of scour occur simultaneously. A scale down laboratory experiment was conducted to find the interaction between different scour components and the result was applied to and compared with the most widely used scour equations in the US (Colorado State University (CSU) and Melville-Sheppard (M/S) equations) to suggest an improved way of calculating the maximum scour depth. These equations predict maximum pier scour depths. Basic applications include simple pier substructure configurations and riverine flow situations in alluvial sand-bed channels. The CSU equation has been used for bridge scour evaluations and bridge design for countless bridges in the U.S. and worldwide. The M/S equation combines pier geometry, shape, and angle of attack to compute an effective pier width, a^*, and also distinguishes between clear-water and live-bed flow conditions.

2. Materials and Methods

2.1. Physical River Modeling

As shown in Figure 1, a 1:60 scale down physical model was constructed in the hydraulics laboratory including full river bathymetry of the Towaliga River bridge at Macon, Georgia by using Froude number similarities between the model and the prototype [15–17]. A relatively small bridge was selected in this experiment because a large number of smaller bridges can fail during extreme hydrologic events. The drainage area of the selected site at the bridge is 816 km^2. Discharge was estimated as 1700 m^3/s by the U.S. Geological Survey for Tropical Storm Alberto in 1994 which is larger than 500 year flooding. During this historic event, severe overtopping of the bridge and significant scour around the pier bents in the left floodplain occurred, but the embankment and bridge remained sufficiently intact until repairs could be made to the scoured area. The actual scour depth measured during the storm events can be found in Hong and Sturm [16,17]. As shown in Figure 1, each of the pier bents consists of two in-line rectangular columns having a width of 1 m, and they were modeled inclusive of pile caps and piles. A bridge deck width of 13 m, in accordance with standard two-lane roads was also constructed in the laboratory.

Figure 1. Towaliga River bridge in the field and model in the laboratory.

The approach section of the bridge was 7.3 m long followed by a 6.09 m long moveable working bed section. The approach section was filled with uniform size of small gravel (d_{50} = 3.3 mm) to make a fully rough turbulent approach flow. In the moveable working bed section, the full depth was filled with sand with median diameter size (d_{50}) of 0.53 mm. The bridge model was constructed within the moveable working bed section for the scour experiment.

For selecting sediment size in the laboratory study, recently developed scour modeling methodology was applied [2,18–20]. Selecting sediment size in the laboratory was mostly related to the ratio of pier size to sediment size (a/d_{50}). In the field, the effect of sediment size, d_{50}, has not been considered to be important to pier scour because of very large values of a/d_{50} [21]. However, in the laboratory, pier scour depth to pier size ratio (d_{ps}/a) tends to increase with a/d_{50} up to a maximum at $a/d_{50} \approx 25$ and seemingly becomes independent of the ratio when a/d_{50} is greater than 50 [19]. However, another researcher has suggested that d_{ps}/a may decrease at very large values of a/d_{50} based on experiments in a large flume [22]. Therefore, the sediment size, d_{50}, was chosen such that the ratio of pier size to sediment size, a/d_{50} was in the range of 25–50 where it has negligible influence on pier scour. Furthermore, the ratio of the approach velocity to the critical velocity which concludes the condition of clear water scour regime was also an important factor to choose sediment size in the experiment. Considering all the above conditions, the sediment size for this experiment was chosen as d_{50} = 0.53 mm with σ_g = 1.2, where σ_g is geometric standard deviation of the particle size distribution.

Within the working moveable bed section, the erodible embankment/abutment was constructed at both sides of the channels by using erodible fill with rock-riprap protection in order to reproduce the same conditions as in the field [16,23]. Furthermore, the existing roadway and bridge deck were modeled based on the scale-ratio and put on top of the embankments. With this approach, submerged orifice flow with and without overtopping during the extreme amount of flooding as observed in the field could be simulated in the laboratory. As shown in Figure 2, eight pier bents were also constructed through the bridge to support the bridge deck and for the local scour experiment. Each of the pier bents consisted of two in-line rectangular columns. Abutment structures were constructed and buried at both sides of the erodible embankments.

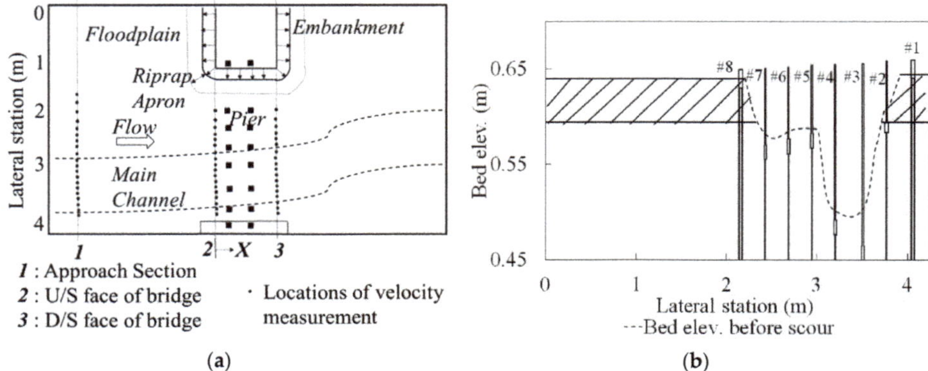

Figure 2. Geometry of compound channel for (**a**) plan view with velocity measurement locations; (**b**) cross section view at bridge.

2.2. Experimental Procedure

After completion of the model structure, the flume was slowly filled with water from a downstream supply hose to saturate the sand without disturbing the initial bottom contours. Then bottom elevations were measured in detail throughout the entire working section using an Acoustic Doppler Velocimeter (ADV). After that, a larger flow depth than the required value was set by the tailgate, then discharge was increased slowly to prevent initial scour while setting up the test discharge. Then the tailgate was lowered to achieve the desired depth of flow. In the meantime, a point gauge was used to measure the flow depth to measure the targeted water depth. Once the desired flow rate and water depth were achieved, scouring was continued for 5 to 6 days until equilibrium (change in scour depth less than 2% within 24 h) was reached. After reaching equilibrium condition, the entire bed elevation (bed elev.) was measured by a point gauge and the ADV in detail to obtain accurate contours after scour. After finishing the moveable bed experiment, the surface of the movable bed was fixed by spraying polyurethane. In the fixed bed conditions, the velocities were measured by ADV in the approach section and bridge upstream (U/S) and downstream (D/S) section. During the velocity measurements, typical correlation values in the experiments were greater than 90% and the Signal Noise Ratio (SNR) was greater than 15. The sampling frequency of the ADV was chosen to be 25 Hz with a sampling duration of 2 min at each measuring location. More detailed measurements techniques using ADV can be found in several other articles [24–29].

2.3. Assessment of Reference Scour Depth

Two well established theoretical pier scour equations were used to decide the reference scour depth at the pier for this research. One of the most commonly used pier scour equation in the United State is the CSU equation (also known as the HEC-18 equation). The CSU equation includes a correction factor for pier shape, angle of attack of flow, and the bed conditions. The CSU equation was initially developed from a laboratory data set measured by several researchers [6,30,31]. After the initial development, the CSU equation has been progressively modified over the years and is currently recommended by the Federal Highway Administration (FHWA) for estimating equilibrium scour depths at simple piers as follow [6]:

$$\frac{d_{csu}}{Y_2} = 2K_1K_2K_3\left(\frac{a}{Y_2}\right)^{0.65}Fr_2^{0.43} \tag{1}$$

where, K_1, K_2, K_3 are the correction factors for pier nose shapers, for angle of attack of flow, and for bed conditions, respectively; Fr_2 = Froude number at the pier face = $V_2/(gY_2)^{0.5}$; a = Pier width; d_{csu} = Equilibrium pier scour depth; Y_2 = Flow depth directly upstream of the pier; V_2 = Mean velocity of flow directly upstream of the pier; g = Acceleration of gravity. Several evaluations of the CSU equation through laboratory and field datasets have shown this equation may perform better than other bridge predictive equations [32–34]

Another commonly used pier scour prediction method is the Melville–Sheppard or M/S equation [6,34,35]. A new variable is introduced in the M/S equation, effective pier width, a^*, which shows the combined effect of pier geometry, shape, and angle of attack. The equation is as follows:

$$\frac{d_{ms}}{a^*} = 2.5f_1f_2f_3 \tag{2}$$

where, f_1, f_2, f_3 are the factors for flow-structure interactions, for flow-sediment interactions, and for sediment-structure interactions, respectively. Here,

$$f_1 = \tanh\left[\left(\frac{Y_2}{a^*}\right)^{0.4}\right] \tag{3}$$

$$f_2 = \left[1 - 1.2\left[\ln\left(\frac{V_2}{V_c}\right)\right]^2\right] \tag{4}$$

$$f_3 = \left[\frac{\left(\frac{a^*}{d_{50}}\right)^{1.13}}{10.6 + 0.4\left(\frac{a^*}{d_{50}}\right)^{1.33}}\right] \tag{5}$$

where, V_c is the critical velocity for movement of d_{50}, calculated by Keulegan's equation.

Using the provided theoretical pier scour equation, pier scour depth was predicted with the variables measured in the experiments. Because theoretical equations were derived from a simple laboratory set-up in a rectangular flume, the result can be used as a reference pier scour depth without any effect of flow contraction. Then, the reference pier scour depth was compared with the measured maximum scour depth around the pier where the flow contraction and local vortex structure interact and lead to the maximum scour depth. The results showed the effect of flow contraction on the pier scour and were used to suggest precise estimation of pier scour depth [8].

3. Results and Discussion

A total of eight experiments were conducted for this research and the experimental conditions are summarized in Table 1 where Q is the total discharge and q_2/q_1 is the discharge contraction ratio which can be used as a key-independent variable that accounts for flow redistribution and resulting flow acceleration through a bridge section. V_1, Y_1, and V_2, Y_2 are cross-sectional mean velocity, and depth of floodplain in the approach and at the upstream face of the bridge section, respectively. Also, maximum scour depth expressed as flow depth measured at the location of maximum scour (Y_m) and longitudinal (flow direction) distance measured from the upstream face of the bridge to the location of maximum scour depth (X) (see in Figure 2a) are presented in Table 1. A better idea of the scour distribution at the bridge cross section in these experiments can be seen in Figure 6 as a reference. Run 9 and 10 were selected from the previous study [25] and used to validate the results in this research so that the suggested method can be used for different hydraulic and geometry conditions. As shown in Table 1, runs 1, 4, 5, and 6 were performed in free (F) flow condition; whereas, the remaining runs were conducted in the submerged flow conditions with overtopping (OT) and without overtopping (SO).

Close to the toe of the abutment where the abutment side slope ends and the first pier is introduced, these locations are the most vulnerable to scour because abutment scour, pier scour, and contraction scour occur simultaneously. Thus, to find the effect of pier on the maximum scour depth, pier bent #7 was removed from the river model and experiments were conducted in runs 7 and 8 with the exact same flow conditions as in runs 2 and 3. Even if the velocities and scour depths were measured in the entire working moveable bed section, only the floodplain flow variables and scour depths are presented in this paper because the maximum scour depth occurred on the floodplain in all our cases.

For the velocity measurements at each cross-section, point velocities were measured along multiple vertical transects through the entire cross-section, and at each vertical transect. Point velocities were measured at minimum of four points and maximum ten points vertically depending on the depth of water. Then, the depth-averaged velocity was determined by the best fit of the logarithmic velocity profile in each vertical transect. Based on the depth-averaged velocity, the cross-sectional mean velocity of the floodplain was calculated for the approach and bridge section as presented in Table 1.

Table 1. Experimental conditions and location of maximum scour depth measurement.

Run	Flow Type	Q (m^3/s)	q_2/q_1	V_1 (m/s)	Y_1 (m)	V_2 (m/s)	Y_2 (m)	X (m)	Y_m (m)
1	F	0.029	1.45	0.082	0.125	0.134	0.110	0.122	0.174
2	SO	0.038	1.53	0.082	0.122	0.177	0.088	0.094	0.201
3	OT	0.053	1.37	0.076	0.149	0.183	0.088	0.076	0.219
4	F	0.046	1.46	0.085	0.146	0.140	0.131	0.049	0.207
5	F	0.053	1.71	0.076	0.149	0.146	0.137	0.049	0.210
6	F	0.038	1.58	0.082	0.122	0.146	0.110	0.049	0.186
7	SO	0.038	1.53	0.082	0.122	0.177	0.088	0.186	0.202
8	OT	0.053	1.37	0.076	0.149	0.183	0.088	0.094	0.229
9	F	0.105	1.71	0.165	0.073	0.302	0.070	-	0.180
10	OT	0.198	1.05	0.219	0.149	0.390	0.088	-	0.274

Notes: F = free flow, SO = Submerged orifice flow, OT = Overtopping.

3.1. Measurement of the Maximum Scour Depths

Initial bottom elevations were measured throughout the test (moveable bed) section before the experiment and then the final bottom elevations were measured at the same locations in equilibrium conditions. Then, to find the location and magnitude of the maximum scour depth, bed elevations after and before the scouring were compared. The maximum scour depth in all of the laboratory experimental runs was found within the bridge and close to the bridge pier (Table 1).

3.2. Prediction of Maximum Scour Depth

The main assumption in this research is that maximum scour depth consists of theoretical pier scour depth and additional scour due to flow contraction. Based on this assumption, the maximum scour depth can be calculated by the equation,

$$\text{Max. Scour depth} = \text{Theoretical pier scour} + \text{Additional scour by flow contraction} \tag{6}$$

The schematic diagram is given in Figure 3 to define the variables used for the analysis. The theoretical pier scour depth can be decided using the CSU (d_{csu}) or M/S equation (d_{ms}) with measured flow variables and the results are shown in Table 2. Then, the "Additional scour by flow contraction" can be determined by measured the water depth at the deepest point and theoretical pier scour depth as follows;

$$Y_{m\text{-}csu} = Y_m - d_{csu} \tag{7}$$

$$Y_{m\text{-}ms} = Y_m - d_{ms} \tag{8}$$

where, $Y_{m\text{-}csu}$ and $Y_{m\text{-}ms}$ are the calculated water depths at the deepest location subtracted from the total water depth at the point by the theoretical pier scour depth of CSU (d_{csu}) or M/S equation (d_{ms}) respectively, which stands for the resulting water depth due to the additional scour. The variables required for calculation of the reference pier scour depth using the CSU and M/S equations are listed in Tables 1 and 2. Here, θ is the flow angle of attack at the pier face which is required to estimate correction factor, K_2 and effective pier width, a^* for the CSU and M/S equations, respectively. The additional scour components are also shown in Table 2 in terms of non-dimensional variables.

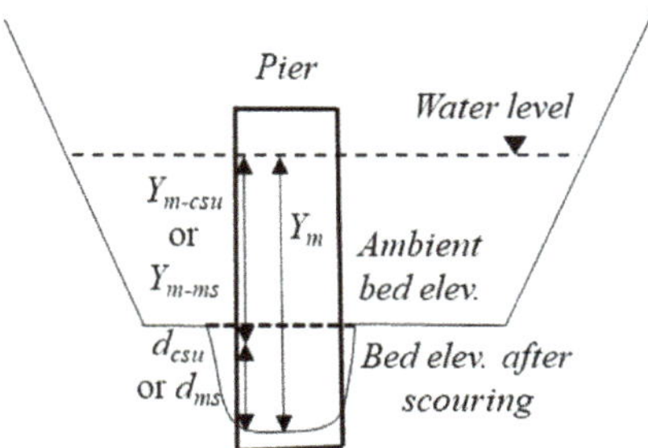

Figure 3. Schematic diagram for calculation of maximum scour depth.

Table 2. Summary of experimental results to calculate maximum scour depth.

Run	Fr_2	θ	a^* (m)	V_2/V_c	d_{csu} (m)	d_{ms} (m)	$\frac{Y_{m-ms}}{Y_1}$	$\frac{Y_{m-csu}}{Y_1}$
1	0.13	20	0.009	0.46	0.029	0.006	1.34	1.15
2	0.19	12	0.009	0.63	0.027	0.014	1.51	1.41
3	0.20	8	0.008	0.64	0.024	0.014	1.36	1.29
4	0.12	15	0.009	0.46	0.027	0.006	1.38	1.23
5	0.13	15	0.009	0.49	0.028	0.007	1.35	1.21
6	0.14	22	0.009	0.50	0.031	0.009	1.44	1.27
7	0.19	12	0.009	0.63	0.027	0.014	1.51	1.41
8	0.20	8	0.008	0.64	0.024	0.014	1.42	1.35
9	0.36	15	0.045	0.66	0.083	0.075	1.40	1.30
10	0.42	15	0.045	0.68	0.095	0.081	1.29	1.19

If the assumption is correct, the effect of additional scour due to flow contraction can be represented by the values of Y_{m-csu} and Y_{m-ms}. Thus, the non-dimensional value of additional scour components calculated using the CSU and M/S equation are compared with the measured discharge contraction ratio in Figure 4a,b respectively. As the value of flow contraction ratio increases, the normalized value of additional scour depth gradually increases. The results clearly reveal that the effect of flow contraction on the additional scour term becomes higher as the value of q_2/q_1 increases. Runs 9 and 10 were also plotted in Figure 4 to validate the findings in different hydraulic and geometry conditions and the result favors the validation as it shows a similar trend of runs 9 and 10 with pressure and free flow data respectively. For both cases using the CSU equation and M/S equation, the pressure flow line has a steeper slope than for the free flow cases and lies above. Because of the vertical flow contraction in addition to the existing lateral flow contraction in pressure flow, the resulting additional scour depth due to the flow contraction is higher than in the free flow cases. Currently, additional experiments are being conducted. With more experimental conditions as well as the data set in Figure 4, the best fit equation will be provided for calculation of scour due to flow contraction under free and pressure flow by a least square regression analysis

It is interesting to note that, the regression analysis results using only with the data set in Figure 4 show higher values of exponents of q_2/q_1 in the case with the M/S pier scour equation for both of the pressure and free flow cases. This finding illustrates that the M/S equation shows a smaller value of pier scour depth in the same flow conditions compared to the CSU equation. The ratios between the theoretical pier scour depths calculated using CSU (d_{csu}) and M/S (d_{ms}) are plotted with the value of flow intensity factor (V_2/V_c) in Figure 5. As shown in Figure 5, the M/S equation shows almost 50% to

75% underestimated values compared to the CSU results. The CSU pier scour equation was developed for live-bed scour and thus ignores the effect of different values of flow intensity and assumes the value is equal to 1; but in the M/S equation, V_2/V_c is considered as an individual factor to account for the effect of clear water conditions and varies from 0.3 to 1. So, it is obvious that the CSU pier scour equation shows a larger value when applied to the clear water condition. However, there is another important difference between the CSU and M/S equations, which is the consideration of bed material size. However, in our case, the effect of sediment size cannot be considered because all of the experiments are conducted with same sediment size.

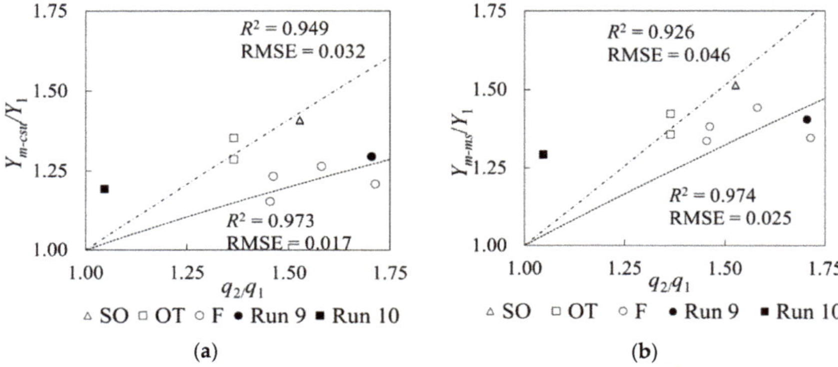

Figure 4. Effect of flow contraction on additional scour components using (**a**) Colorado State University (CSU) and (**b**) Melville-Sheppard (M/S) equations.

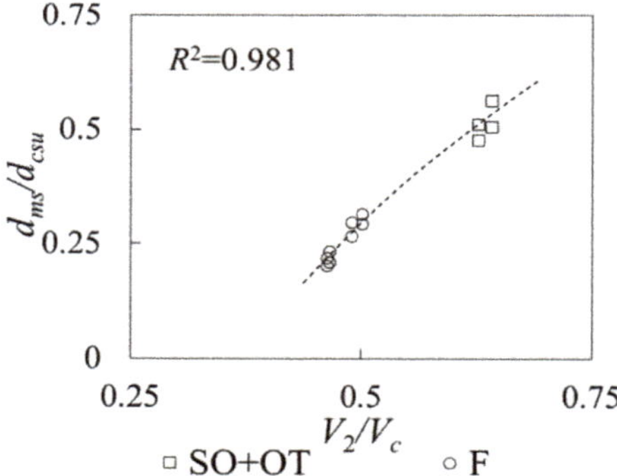

Figure 5. Comparison of CSU and M/S pier scour depth in terms of flow intensity.

Finally, a comprehensive procedure for predicting the maximum scour depth with respect to bridge design is introduced:

1. Collecting field geometry data including sediment size;
2. Calculate the flow variables using software or hydraulic laboratory modeling;
3. Compute the theoretical pier scour depth from established pier scour equations (CSU or M/S equations);
4. The additional scour depth due to the effect of flow contraction can be estimated with the equations developed by regression analysis with the data in Figure 4 as well as more laboratory/field data sets;

5.   Adding the results from steps 3 and 4 to predict the maximum scour depth for bridge design.

In addition to suggest a comprehensive procedure, the effect of pier bent (located near to the abutment) on maximum scour depth was investigated qualitatively. As shown in Table 1, runs 7 and 8 were conducted with the exact same flow condition as in runs 2 and 3 respectively, but removing pier bent #7 in the bridge section. Figure 6 shows the cross-section comparisons between runs 3 and 8 after scouring. The maximum scour depth occurred at pier bent #6 for both runs 3 and 8. So, the location was independent of the presence of the closest pier bent (#7). However, the amount of maximum scour depth was slightly higher in the case of run 8 where pier bent #7 was absent. A similar trend was observed for runs 2 and 8. Discharge redistribution over the time and interaction between each pier can be the result for this difference of maximum scour depth and needs further study.

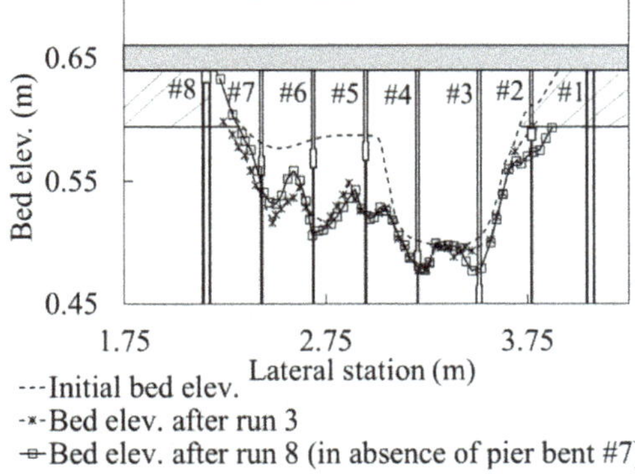

Figure 6. Comparison of cross-sections for runs 3 and 8.

4. Conclusions

Many investigations have been made attempting to estimate the maximum scour depth and to understand the mechanism of scour around bridge piers. Most of the previous investigations were based on lab experiments using a rectangular channel under free flow. However, due to the recent extreme rainfall events, submerged orifice flow and overtopping flow occur at the bridge frequently, where the flow field around the bridge substructure is more complex than in free flow because of vertical flow contraction in addition to existing lateral flow contraction. Furthermore, most of the natural channel shapes are not rectangular. Also, the current guidelines recommended by HEC-18 assumed that contraction and local scour processes are independent and so they can be determined separately and summed to estimate total scour depth. However, during large flooding events, local scour and contraction scour occur simultaneously and a separate calculation of local scour and contraction scour results in inaccurate scour depth. To overcome the weak points that the current methodology has, laboratory experiments were carried out in a scaled down physical model and a single equation was developed to predict maximum scour depth which can be used without separate calculation of different types of scour components in pressure flow as well as in free flow cases.

Based on the basic assumption of maximum scour depth as summation of theoretical pier scour depth and additional scour depth due to flow contraction, a comprehensive way of predicting maximum scour depth in clear water conditions was suggested. Furthermore, the results demonstrate that the contraction effect on maximum scour depth increases as the flow contraction ratio increases. Also, due to additional vertical contraction in the pressure flow case, the effect of flow contraction on the maximum scour depth shows up larger than in free flow. Another outcome from our investigation concludes that the location of the maximum scour depth is independent of the existence of the closest pier bent but the absence of the closest pier bent increases the scour depth.

Even though this study suggested an improved method for the scour depth prediction in clear water conditions, a well-designed physical model is recommended to investigate the scour characteristics under live bed conditions. In addition, different sediment sizes and non-uniform size sediment should be incorporated in the future research, as natural rivers generally consist of non-uniform sediment. Finally, scour modelling using computational fluid dynamics (CFD) should also be explored to refine the proposed equation for design purposes.

Author Contributions: S.H.H. and S.O.L. provided background data and motivation. S.H.H. and S.O.L. designed the experiments and determined the results; R.S. analyzed the data; R.S. and S.H.H. contributed to motivation, data analysis input, and interpretation of results; R.S. and S.H.H. interpreted the results and wrote the paper; and all authors participated in final review and editing of the paper.

Funding: National Research Foundation of Korea (NRF) grant funded by the Korea government (MSIT) (NRF-2017R1A2B2011990); West Virginia University internal grant.

Acknowledgments: The laboratory data used in this paper were measured in Georgia Tech and the permission was granted by Seung Ho Hong and Seung Oh Lee.

Conflicts of Interest: The authors declare no conflict of interest.

References

1. Kattell, J.; Eriksson, M. *Bridge Scour Evaluation: Screening, Analysis, and Countermeasures*; General Technical Reports 9877 1207-SDTDC; U.S. Department of Agriculture: San Dimas, CA, USA, 1998.
2. Melville, B.W.; Coleman, S.E. *Bridge Scour*; Water Resources Publications, LLC: Highlands Ranch, CO, USA, 2000.
3. Shirhole, A.M.; Holt, R.C. *Planning for A Comprehensive Bridge Safety Program*; Transportation Research Record No. 1290; Transportation Research Board, National Research Council: Washington, DC, USA, 1991.
4. Novey, M. Cdot Assessing 'Millions and Millions' in Road Bridge Damage. Available online: www. coloradoan.com (accessed on 15 September 2013).
5. Shrestha, C.K. Bridge Pier Flow Interaction and Its Effect on the Process of Scouring. Ph.D. Thesis, University of Technology Sydney (UTS), Ultimo, Australia, 2015.
6. Arneson, L.A.; Zevenbergen, L.W.; Lagasse, P.F.; Clopper, P.E. *Evaluating Scour at Bridges*, 15th ed.; FHWA-HIF-12-003, HEC-18; Department of Transportation, Federal Highway Administration: Washington, DC, USA, 2012.
7. Lee, S.O. Physical Modeling of Local Scour Around Complex Bridge Piers. Ph.D. Thesis, School of Civil and Environmental Engineering, Georgia Institute of Technology, School of Civil and Environmental Engineering, Georgia Institute of Technology, Atlanta, GA, USA, 2006.
8. Sheppard, D.; Melville, B.; Demir, H. Evaluation of Existing Equations for Local Scour at Bridge Piers. *J. Hydraul. Eng.* **2014**, *140*, 14–23. [CrossRef]
9. Sturm, T.W.; Ettema, R.; Melville, B.M. *Evaluation of Bridge-Scour Research: Abutment and Contraction Scour Processes and Prediction*; NCHRP 24-27; National Co-operative Highway Research Program: Washington, DC, USA, 2011.
10. Melville, B.; Chiew, Y. Time Scale for Local Scour at Bridge Piers. *J. Hydraul. Eng.* **1999**, *125*, 59–65. [CrossRef]
11. Shatanawi, K.M.; Aziz, N.M.; Khan, A.A. Frequency of discharge causing abutment scour in South Carolina. *J. Hydraul. Eng.* **2008**, *134*, 1507–1512. [CrossRef]

12. Alemi, M.; Maia, R. Numerical Simulation of the Flow and Local Scour Process around Single and Complex Bridge Piers. *IJCE* **2018**, *16*, 475. [CrossRef]
13. Sajjadi, S.A.H.; Sajjadi, S.H.; Sarkardeh, H. Accuracy of numerical simulation in asymmetric compound channels. *IJCE* **2018**, *16*, 155. [CrossRef]
14. Hong, S. Interaction of Bridge Contraction Scour and Pier Scour in a Laboratory River Model. Master's Thesis, School of Civil and Environmental Engineering, Georgia Institute of Technology, Atlanta, GA, USA, 2005.
15. Hong, S.; Lee, S.O. Insight of Bridge Scour during Extreme Hydrologic Events by Laboratory Model Studies. *KSCE J. Civ. Eng.* **2017**, *22*, 1–9. [CrossRef]
16. Hong, S.; Sturm, T.W. Physical modeling of abutment scour for overtopping, submerged orifice, and free surface flows. In Proceedings of the 5th Conference on Scour and Erosion, San Francisco, CA, USA, 7–10 November 2010.
17. Hong, S.; Sturm, T.W. Physical model study of bridge abutment and contraction scour under submerged orifice flow conditions. In Proceedings of the 33rd IAHR Congress: Water Engineering for a Sustainable Environment, Vancouver, BC, Canada, 9–14 August 2009.
18. Fael, C.M.S.; Simarro-Grande, G.; Martin-Vide, J.P.; Cardoso, A.H. Local scour at vertical wall abutments under clear-water flow conditions. *Water Res. Res.* **2006**, *10*, 1–12. [CrossRef]
19. Hong, S.; Abid, I. Physical Model Study of Bridge Contraction Scour. *KSCE J. Civ. Eng.* **2016**, *20*, 2578–2585. [CrossRef]
20. Lee, S.O.; Sturm, T.W.; Gotvald, A.; Landers, M. Comparison of laboratory and field measurements of bridge pier scour. In Proceedings of the Second International Conference on SCOUR and EROSION-ICSE, Meritus Mandarin, Singapore, 14–17 November 2004; pp. 231–239.
21. Lee, S.O.; Sturm, T.W. Effect of sediment size scaling on physical modeling of bridge pier scour. *J. Hydraul. Eng.* **2009**, *135*, 793–802. [CrossRef]
22. Sheppard, D.; Odeh, M.; Glasser, T. Large Scale Clear-Water Local Pier Scour Experiments. *J. Hydraul. Eng.* **2004**, *130*, 957–963. [CrossRef]
23. Ettema, R.; Kirkil, G.; Muste, M. Similitude of large-scale turbulence in experiments on local scour at cylinders. *J. Hydraul. Eng.* **2006**, *132*, 33–40. [CrossRef]
24. Hong, S.; Sturm, T.W.; Stoesser, T. Clear Water Abutment Scour in a Compound Channel for Extreme Hydrologic Events. *J. Hydraul. Eng.* **2015**, *141*, 1–12. [CrossRef]
25. Hong, S. Prediction of Clear Water Abutment Scour Depth in Compound Channel for Extreme Hydrologic Events. Ph.D. Thesis, School of Civil and Environmental Engineering, Georgia Institute of Technology, Atlanta, GA, USA, 2011.
26. Lane, S.N.; Biron, P.M.; Bradbrook, K.F.; Butler, J.B.; Chandler, J.H.; Crowell, M.D.; McLelland, S.J.; Richards, K.S.; Roy, A.G. Three-dimensional measurement of river channel flow processes using acoustic Doppler velocimetry. *Earth Surf. Process. Landf.* **1998**, *23*, 1247–1267. [CrossRef]
27. SonTek. *Acoustic Doppler Velocimeter (ADV) Principles of Operation*; SonTek Technical Notes; SonTek: San Diego, CA, USA, 2001.
28. Wu, P.; Hirshifield, F.; Sui, J. ADV measurements of flow field around bridge abutment under ice cover. CGU HS Committee on River Ice Processes and the Environment. In Proceedings of the 17th Workshop on River Ice, Edmonton, AB, Canada, 21 July 2013.
29. Ben Meftah, M.; Mossa, M. Scour holes downstream of bed sills in low-gradient channels. *J. Hydraul. Res.* **2006**, *44*, 497–509. [CrossRef]
30. Chabert, J.; Engeldinger, P. *Etude Des Affonillements Author Des Piles Des Ponts*; Laboratoire National d'Hydraulique de Chatou: Chatou, France, 1956.
31. Shen, H.W.; Schneider, V.R.; Karaki, S. Local scour around bridge piers. *J. Hydraul. Div.* **1969**, *95*, 1919–1940.
32. Gaudio, R.; Grimaldi, C.; Tafarojnoruz, A.; Calomino, F. Comparison of formulae for the prediction of scour depth at piers. In Proceedings of the 1st IAHR European Division Congress, Edinburgh, UK, 4–6 May 2010.
33. Gaudio, R.; Tafarojnoruz, A.; Bartolo, S.D. Sensitivity analysis of bridge pier scour depth predictive formulae. *J. Hydroinform.* **2013**, *15*, 939–951. [CrossRef]

34. Ferraro, D.; Tafarojnoruz, A.; Gaudio, R.; Cardoso, A.H. Effects of pile cap thickness on the maximum scour depth at a complex pier. *J. Hydraul. Eng.* **2013**, *139*, 482–491. [CrossRef]
35. Melville, B.W. Pier and abutment scour: Integrated approach. *J. Hydraul. Eng.* **1997**, *123*, 125–136. [CrossRef]

Article

Collapsing Mechanisms of the Typical Cohesive Riverbank along the Ningxia–Inner Mongolia Catchment

Guosheng Duan [1], Anping Shu [1,*], Matteo Rubinato [2], Shu Wang [1] and Fuyang Zhu [1]

[1] School of Environment, Key Laboratory of Water and Sediment Sciences of MOE, Beijing Normal University, Beijing 100875, China; duanguosheng2004@163.com (G.D.); wangshu861217@gmail.com (S.W.); fuyangzhu1017@gmail.com (F.Z.)

[2] Department of Civil and Structural Engineering, The University of Sheffield, Sir Frederick Mappin Building, Mappin Street, Sheffield S1 3JD, UK; m.rubinato@sheffield.ac.uk

* Correspondence: shuap@bnu.edu.cn; Tel.: +86-10-5880-2928

Received: 7 August 2018; Accepted: 11 September 2018; Published: 18 September 2018

Abstract: As one of the major sediment sources in rivers, bank collapse often occurs in the Ningxia–Inner Mongolia catchment and, to date, it caused substantial social, economic and environmental problems in both local areas and downstream locations. To provide a better understanding of this phenomenon, this study consisted of modifying the existing Bank Stability and Toe Erosion Model (BSTEM), commonly used to investigate similar phenomena, introducing new assumptions and demonstrating its applicability by comparing numerical results obtained against field data recorded at six gauging stations (Qingtongxia, Shizuishan, Bayan Gol, Sanhuhekou, Zhaojunfen, and Toudaoguai). Furthermore, the impact of multiple factors typical of flood and dry seasons on the collapse rate was investigated, and insights obtained should be taken into consideration when completing future projects of river adaptation and river restoration.

Keywords: Ningxia–Inner Mongolia; Yellow River; riverbank collapse; BSTEM model; flood & dry season; sediment transport

---

1. Introduction

Riverbank collapse is a phenomenon caused by several natural factors (e.g., intense rainfall, site topography, properties of the river bed and hydraulic conditions of the river flow) [1] and artificial factors (e.g., man-made bank undercutting, basal clean-out) [2]. The Yellow River is the China's second-longest river and previous studies [3] confirmed that 518.38 km^2 of the riverbank along the Ningxia–Inner Mongolia part of the Yellow River was eroded between 1958 and 2008. This elevated rate of riverbank erosion is due to continuous variations of flow levels, causing the deposition of loose riverbed material in alluvial plains, especially after heavier rainfall events typical of climate change [4,5]. To date, riverbank erosion has caused substantial social (e.g., traffic disruption due to flooding), economic (e.g., loss of farm land) and environmental problems (e.g., sediment dynamics and water quality) in both local areas and downstream locations [6,7].

Previous studies were completed to tackle this challenge and to identify what can induce riverbanks to collapse in catchments in both developing and developed countries. Grabowski [8] found that the soil texture, the soil structure, the unit weight and water composition [9,10] all play a principal role on the banks erosion. While non-cohesive banks are eroded through discrete particle entrainment that can be quantified using the magnitude of the shear stress and the particle size [11,12], cohesive banks are eroded through entrainment of aggregates [13] and this mechanism is very challenging to diagnose when considering the electrochemical forces acting between them [14,15]. The detachment and erosion phenomena of cohesive material by gravity and/or flowing water are typically controlled by a variety of physical, electrical and chemical forces, such as cohesion, electrochemical forces, pore-water pressure and matric suction [15]. These forces within and between aggregates cause their erosion to be complex. Furthermore, cohesive riverbanks are usually poorly drained due to their silt and clay composition, and thus may experience excess pore water pressures, one of the main agents of subaerial erosion. Clayey banks are also susceptible to desiccation cracking and slaking from wetting–drying cycles [16]. Numerous studies have examined the impact of water content on the riverbank erosion [17–22] while other studies have investigated the subaerial erosion due to the weakening and weathering of bank materials [23–25]. Additionally, Yu [26] completed a series of experimental tests to quantify the influence and the impact of the water infiltration on bank collapse and consequent sediment deposition. They identified that partial erosion can be due to fluvial hydraulic forces or gravitational forces. Moreover, total erosion of the bank can be generated if both forces are combined at their maximum magnitude, with a consequent disintegration of the bank and the transportation of fallen blocks within the flow, changing the bed elevation. Focusing on the Ningxia–Inner Mongolia banks, the erosion process has been found to be influenced by three specific factors: (i) the sand blown by the wind from the desert [27–30]; (ii) the sediment transported by the river from the upstream catchment [31,32]; and (iii) the material falling from the bank due to natural collapse [33]. It is fundamental to replicate all these governing processes within numerical models when assessing stability of the riverbanks. A large amount of literature is available and it covers all the numerical modeling aspects investigated to date [34–38]. One of the numerical tools most commonly used is the Bank Stability and Toe Erosion Model (BSTEM), developed by the National Sedimentation Laboratory in Oxford, Mississippi, USA [39], which has been continually modified and improved by the authors since its creation to improve its performance. To model the riverbanks' stability, BSTEM calculates a factor of safety (F_s) for multi-layers combining three equilibrium-method models: (i) the layers simulated are horizontal; (ii) vertical slices simulate tension crack; and (iii) the common failure mechanisms are cantilever failures. BSTEM also assumes that all the collapsed material is totally removed from the bank and does not accumulate on the toe [40]. BSTEM is frequently used to simulate: (i) the banks' stability and the consequent sediment loading inside the river [41]; (ii) stream rehabilitation projects [42]; and (iii) erosion and failure mechanisms [43,44]. Despite several findings that have resulted in a better understanding of this phenomenon, to the authors' knowledge there is a lack of studies which involve the evaluation of BSTEM for long-term streambank erosion and consequent long-term failure.

Therefore, to fill this gap, this work comprises a collection of field datasets in the Ningxia–Inner Mongolia area that were implemented on the BSTEM model to obtain a new solution to the mechanisms of long-term bank erosion. To achieve that, three hypothesis different from those applied on the original BSTEM model were executed: (i) when the soil collapses from the bank, part of it is washed away by the water, and the remaining part accumulates on the toe; (ii) the shape of the material that accumulates on the toe is considered to have a triangle form considering the cross-sectional view and the angle away from the river bank is the angle of repose of the sediment, which is taken as 30 degrees; and (iii) the downstream area of a bank is considered the first one to be affected by the collapse process.

The paper is organized as follows. In the next sections the study area, the procedure to collect field data and the numerical analysis applied to replicate the river bank collapse recorded (Section 2) are described. Section 3 presents the comparison between numerical and experimental datasets and the factors that could influence the rate of collapse are discussed. Finally, Section 4 reports the conclusions.

2. Materials and Methods

2.1. Case Study

The Ningxia–Inner Mongolia catchment (Figure 1) is located in the lower part of the upper Yellow River and has a length of 913.5 km, starting from Qingtongxia (Ningxia Hui autonomous region) and ending at Hekouzhen (Inner Mongolia autonomous region). This site includes six gauging stations (Qingtongxia, Shizuishan, Bayan Gol, Sanhuhekou, Zhaojunfen, and Toudaoguai) and it is characterized by four desert regions (HedongSandyland, Ulan Buh Desert, Tengger Desert and Hobq Desert) and two alluvial plains (Yinchuan Plain and Hetao Plain).

The climate of this region can be classified as temperate continental climate, which is characterized by long cold winters, and warm to hot summers [45]. The annual rainfall (maximum and minimum peaks commonly reached in summer and winter, respectively) ranges from 100 to 288 mm.

The case study selected is located in Dengkou which is in the Inner Mongolia Province. The gauging station Bayan Gol is adjacent to it. Being part of the Hetao plain, the Dengkou catchment is characterized by a low gradient ($\approx 0.17\%$) and loose riverbed materials. This site was carefully chosen based on three specific criteria. Firstly, preliminary monitoring datasets of this site have confirmed that the riverbank was actively retreating. Secondly, the site was sufficiently close to a gauging station (about 5 km), hence detailed rainfall readings including time-series, frequencies and water surface elevations associated with each flow condition flowing next to the bank site were available and could be collected. Last but not least, the morphology and the soil properties of this area are representative of the cohesive streambank along all the Ningxia–Inner Mongolia catchment which has been drastically eroded from 1958 to 2008 [3]. Therefore, it is fundamental to provide a better understanding of this phenomenon, especially in this zone, to provide mitigation measures and plan actions to reduce the effects due to erosion of the riverbanks.

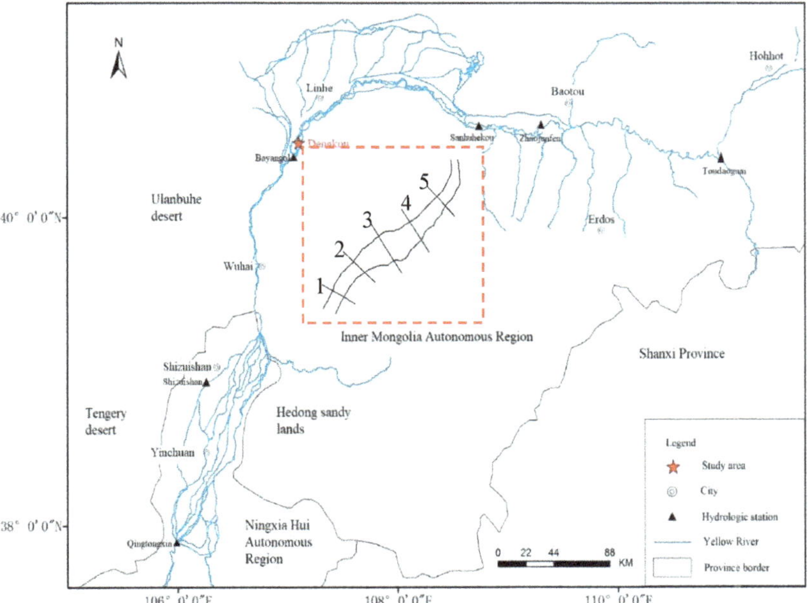

Figure 1. Location of the case study considered for this work. 1, 2, 3, 4 and 5 are the five surveying cross sections (500 m far from each other) selected for the collection of field-data.

2.2. Field Data

To analyze the complex processes of riverbank collapse affecting Dengkou, it was necessary to inspect the nature of the bank and the typical flow conditions in the parts of the Yellow river crossing this area. Five cross sections (1, 2, 3, 4 and 5, Figure 1), spaced 500 m from each other, were identified along the banks for the acquisition of field datasets. Parameters and methods selected by the authors for the monitoring and collection of datasets from the field study are displayed in Table 1.

Table 1. Indexes and methods for in-site measurement.

	Bank Shape	Soil Composition	Mechanical Properties
Parameter monitored	Bank height (m) Bank gradient (°) Length of bank toe (m)	Median diameter (μm) Unit weight (g/cm^3) Moisture content (%)	Critical shear stress (kPa) Internal friction angle (°) /
Equipment utilized for the monitoring	Gradometer ($\pm 1°$), Tape (± 1 cm)	Scale, Laser particle sizer (± 5 μm)	Direct shear apparatus

Additionally, to measure the magnitude of bank collapse along the area of study, five stakes were positioned vertically to the flow direction at each section selected for the monitoring (Figure 2). Five measurements were conducted respectively on 21 May, 13 July, 9 August, 23 September and 30 September in 2011 to quantify the collapse distance. The collapse distances between each section recorded are displayed in Table 2.

Table 2. Collapse distances (m) recorded in 2011.

Time Scale	Section 1	Section 2	Section 3	Section 4	Section 5
21 May–13 July	4.60	4.50	5.80	5.30	5.60
13 July–9 August	3.05	2.80	3.10	3.95	3.60
9 August–23 September	5.25	5.08	6.80	6.20	6.50
23 September–30 September	1.68	1.45	1.10	2.40	1.45
Total	14.58	13.81	16.80	17.85	17.15

Figure 2. Collapse distance measurement for surveying Section 1.

2.2.1. Hydrologic Data

Monthly average hydrologic datasets (2011, year of site inspection) obtained from the gauging station Bayan Gol used for this study are showed in Table 3.

Table 3. Monthly average hydrologic data (2011).

Month	Flow (m³/s)	Velocity (m/s)	Water Depth (m)	Sediment Concentration (kg/m³)	Sediment Transport Rate (kg/s)
1	494.23	0.68	3.09	0.21	102
2	568.11	0.74	3.02	0.22	123
3	495.58	1.06	1.94	1.23	615
4	684.77	1.08	1.34	2.16	1480
5	444.00	0.89	1.03	0.64	283
6	369.60	0.81	0.85	0.59	217
7	368.35	0.87	0.83	0.77	283
8	528.00	1.32	0.94	3.20	1690
9	817.30	1.42	1.40	1.95	1590
10	612.48	1.37	0.99	1.27	779
11	495.90	1.18	0.62	1.17	580
12	592.52	0.95	2.00	2.58	1530

2.2.2. Bank Characteristics of Each Section

Riverbank properties at each section (1, 2, 3, 4 and 5, Figure 1) were obtained through five field inspections in 2011 (1. 21 May, 2. 13 July, 3. 9 August, 4. 23 September and 5. 30 September), and details are showed in Table 4. During the characteristic long cold winters in this region, the Yellow River typically freezes, hence the summer season has been selected as sampling period to avoid the quantification of additional effects on river bed changes [46,47] which could have made the already complex comparison between experimental datasets and numerical results even more challenging.

Table 4. Characteristics of the sections selected for this study.

Section	Bank Height (m)	Bank Gradient (°)	Toe Length (m)	Median Diameter (μm)	Wet Unit Weight (g/cm³)	Critical Shear Stress (kPa)	Internal Friction Angle (°)
1	1.9	82	0.85	33	1.55	7.2	32
2	2.1	77	0.9	35	1.48	7.5	31
3	2.3	81	0.95	32	1.50	7.3	32
4	1.7	84	0.8	37	1.46	7.5	30
5	1.6	86	0.75	30	1.52	7.2	31

2.3. Riverbank Collapse Characterisation

When collecting field data, observations were made to get more insights about the erosion that typically affects the riverbanks of the Ningxia–Inner Mongolia catchment. Figure 3 shows an example of the real case study under investigation prior and after the erosion. Observations confirmed that the bank material is cohesive, the riverbank slope is steep (77–86°), and the pattern of bank collapse is considered to be planar, in conformity with previous studies conducted [48,49]. Additionally, while monitoring the site, it was possible to identify different stages of the continuous collapse process characterized as follows: (i) the bank toe erosion initiates (Figure 4a); (ii) tension cracks develop (Figure 4b); (iii) shearing starts and parts of the bank breaks and detaches from the main body (Figure 4c); (iv) bank failure occurs (Figure 4d). When the soil collapses from the bank, part of it is washed away by the water and the remaining part accumulates at the bank toe. The new shape of the bank including the additional material eroded then becomes the initial form of the next bank collapse.

Figure 3. Example of the existing bank slope (**left**) and an example of riverbank collapse along Dengkou reach (**right**).

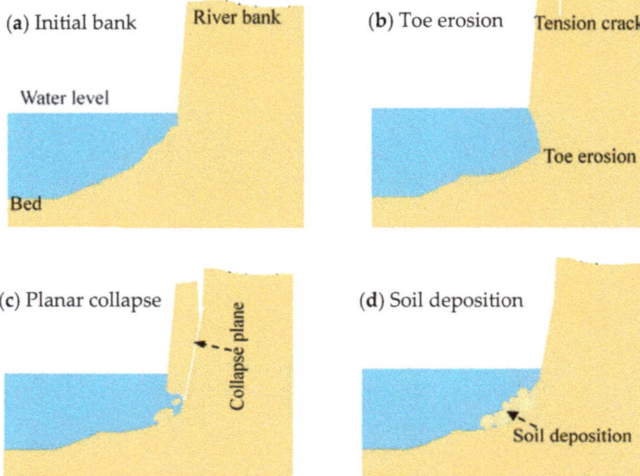

Figure 4. Typical riverbank processes observed at the case study of Ningxia–Inner Mongolia catchment.

2.4. Numerical Modeling

In this study, the BSTEM model, developed by the National Sedimentation Laboratory in Oxford, MS, USA, was used and modified according to the assumptions presented in the introduction.

2.4.1. BSTEM Method

There are two separate modules in the conventional BSTEM model: (i) the toe erosion module and (ii) the stability of the bank module [39]. Equation (1) is commonly applied to predict the width of the bank toe eroded due to the hydraulic conditions impacting on the cohesive riverbank [16,50–52]:

$$B = \kappa(\tau - \tau_c)^a \Delta t \tag{1}$$

where B is the toe erosion width (m); κ is the erodibility coefficient ($m^3\ N^{-1}\ s^{-1}$); τ is the average shear stress (Pa); τ_c is soil's critical shear stress (Pa); a is an exponent usually assumed to be unity, Δt is the time interval (s). κ and τ_c are functions of the soil properties and they are characterized by the following relationship, $\kappa = 0.2\tau_c^{-0.5}$. For non-cohesive soils, τ_c is typically estimated based on the median particle diameter of the soil [53]. Rinaldi [54] noted the difficulty to accurately estimate κ but despite that, other methods provided solutions via a variety of methods to calculate κ and τ_c. One of these methods was developed by Hanson [55] using an in situ jet-test device and his experimental results are currently used universally. In Equation (1), τ can be calculated using Equation (2):

$$\tau = \gamma_w RS \tag{2}$$

where γ_w is the weight of water (N/m^3); R is the hydraulic radius (m); S is the channel slope.

For the specific scenario of planar failure assumed for this study, the riverbank stability analysis involved the computation of a safety factor, identified as the ratio between the resisting and the driving forces applied to the failure zone, which is typically calculated by using Equation (3) [56–59]:

$$F_s = \frac{S_R}{S_D} \tag{3}$$

where F_s is the safety factor with respect to bank failure ($F_s < 1$ indicates the possibility of an eventual collapse); S_R is the shear strength of the soil (kPa); S_D is the driving stress (Pa). S_R can be calculated using Equation (4):

$$S_R = c' + S \tan \phi^b + \Psi \tan \phi' \tag{4}$$

where c' is the effective cohesion (the effective cohesion of the soil is the cohesion of the soil in the anti-shear process) (Pa); S is the normal stress (Pa); ϕ^b is an angle that describes the relationship between the shear strength and the matric suction (°); Ψ is the matric suction which contrasts the pore-water pressure (Pa); ϕ' is the effective internal friction angle (°). S_D can be calculated using Equation (5):

$$S_D = W \sin \beta \tag{5}$$

where W is the weight of the wet soil per unit area of the failure plane (Nm^{-2}); β is the angle of the failure plane (°) [39]. Note that the BSTEM model takes c' into account to represent the cohesion effect. Nevertheless, the influence of cohesive materials on erosion would be considered through other approaches, as well. For example, Dodaro et al. [60,61] modified Shields parameter to predict erosion phenomena at a cohesive sediment mixture.

2.4.2. BSTEM Method Modified

As defined in Section 2.3, the phenomenon of bank collapse is characterized by four different stages. Despite multiple factors influencing the rate of the riverbank collapse, as previously mentioned, continuous collapses are also affected by the intensity of the antecedent collapse magnitude. When the collapse happens, the riverbank affected typically divides into two sections within a moment-frame: one part collapsed is naturally washed away by the river and the remaining part usually deposits at the bank toe. Before the next bank collapse takes place, the additional deposited soil is eroded. Therefore, it is very important to characterize the continuous changing shape of the riverbank to correctly predict the effects of the continuous erosion. Although the original BSTEM model provides some insights into this process and is very helpful for the analysis of the riverbank stability, it should be noted that BSTEM assumes that the falling part which has collapsed from the riverbank does not accumulate on the toe and is immediately removed. Hence it does not accurately represent the reality of the entire erosion process typical on natural rivers. Authors observed multiple shapes and forms during the erosion process following the collapse of material and have tried to characterize each of them, but the literature published to date lacks formulae to accurately estimate and quantify the amount of material deposited. Therefore, to address this gap, three changes were made in the original BSTEM model as follows.

(i) The quantity of soil falling into the river and carried away by the natural streamflow in the form of suspended material can be calculated using Equation (6):

$$M_0 = (S_* - S_v)Hd \tag{6}$$

where M_0 is the weight of soil carried away per unit length (kg/m); S_v is the sediment concentration (kg/m^3), typically obtained by field monitoring; H is the height of the bank (m); d is the width of the flow in the proximity of the bank (m). Because there was a main current which was about 20 m away from the bank, d is assumed as a constant (d = 20). S_* is the capacity of the flow to transport suspended material (kg/m^3) and can be calculated using Equation (7) [62]:

$$S_* = K\left(U_L^3/ghw\right)^m \tag{7}$$

where K and m are parameters that can be obtained from the literature [60]; U_L is flow velocity (m/s); g is gravitational acceleration (m/s^2); h is water depth (m); and w is the velocity of the particles setting (m/s) which can be calculated by using Equation (8) [62]:

$$w = (\gamma_s - \gamma_w)gD^2/25.6\gamma_w \nu \tag{8}$$

where D is the grain size of the material (m); ν is kinematic viscosity coefficient (m^2/s).

(ii) The quantity of deposition soil can be calculated by using Equation (9):

$$M_d = (W_0 - M_0)x \tag{9}$$

where M_d is the quantity of material deposited (kg); W_0 is the quantity of material collapsed (excluding the part suspended in the flow) (kg); x is a ratio between the material deposited and the material collapsed.

(iii) The new shape generated by the deposition of material on the bank toe is assumed to be triangular and the angle away from the riverbank is the angle of repose of sediment. In general, loose sediment grains in water accumulate with an angle close to the sediment angle of repose [63,64].

In the calculation process, the time step is selected as the hour. This decision is justified by the fact that since the flow rate, water depth, flow rate and other data are used as a daily average, the 24 h cycle was considered more appropriate. After the original boundary has passed 24 h, the river bank stability was tested: if the river bank was still stable, then the next set of water depth and flow rate conditions was inserted and the river bank toe erosion module continued to run. When the river bank stability coefficient was less than 1, the collapse occurred and consequently the collapse rate should have been recorded. For this scenario, the amount of collapse was calculated according to the shape of the collapse area identified, and the amount of sediment deposited at the foot of the slope was calculated by using Equations (6)–(9). The shape of the material deposited on the slope was considered to be triangular, and the angle away from the river bank was the angle of repose of the sediment, which was taken as 30 degrees. Based on the collapse rate, the new shape of the river bank was obtained and taken as a new initial condition for the next collapse.

The scheme utilized by the modified BSTEM model can be summarized with the flow chart displayed in Figure 5.

It is complex to quantify in real time the collapse processes along the entire riverbank of the study area due to flow conditions under continuous change, but it is possible to provide an approximation of the riverbank collapse processes by using averaged parameters such as monthly average water depths, flowrates, sediment concentration and average grain size. In addition, there are also parameters that require an accurate calibration during the entire calculation process, such as x. τ_c can be obtained by using values furnished in the literature [65]. By applying trial and error techniques, x = 0.39 corresponds to the optimal numerical value to replicate the realistic distance of collapse measured for this case study and discrepancies obtained between numerical and field measurements are listed in Tables 5–9.

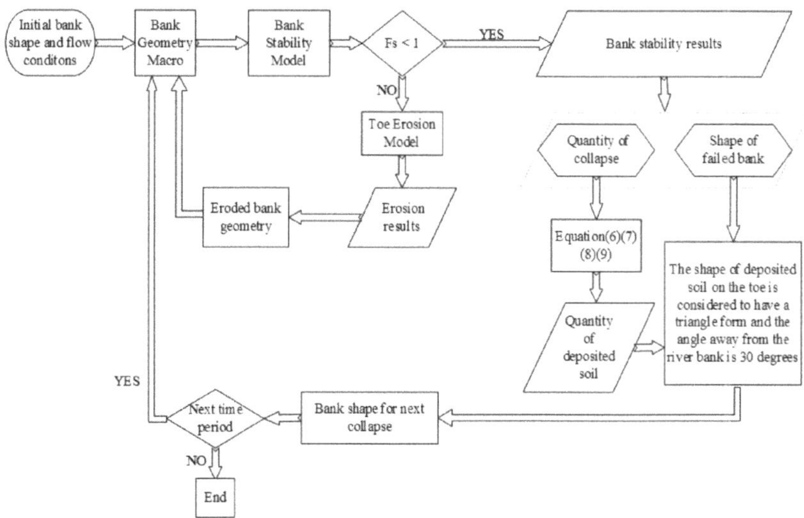

Figure 5. Flow chart of the established model.

Table 5. Numerical results, field measurements and errors for surveying Section 1 of Figure 1.

Time Scale	Calculating Collapse Distance (m)	Monitoring Collapse Distance (m)	Error (%)
21 May–13 July	4.96	4.60	7.83
14 July–9 August	3.24	3.05	6.23
10 August–23 September	5.55	5.25	5.71
23 September–30 September	1.50	1.68	10.71
Total collapse	15.25	14.58	4.60

Table 6. Numerical results, field measurements and errors for surveying Section 2 of Figure 1.

Time Scale	Calculating Collapse Distance (m)	Monitoring Collapse Distance (m)	Error (%)
21 May–13 July	4.99	4.50	10.89
14 July–9 August	3.01	2.80	7.5
10 August–23 September	5.38	5.08	5.91
23 September–30 September	1.13	1.45	22.06
Total collapse	14.51	13.81	5.07

Table 7. Numerical results, field measurements and errors for surveying Section 3 of Figure 1.

Time Scale	Calculating Collapse Distance (m)	Monitoring Collapse Distance (m)	Error (%)
21 May–13 July	5.46	5.80	5.9
14 July–9 August	3.30	3.10	6.5
10 August–23 September	7.15	6.80	5.1
23 September–30 September	0.94	1.10	14.5
Total collapse	16.85	16.80	0.3

Table 8. Numerical results, field measurements and errors for surveying Section 4 of Figure 1.

Time Scale	Calculating Collapse Distance (m)	Monitoring Collapse Distance (m)	Error (%)
21 May–13 July	5.75	5.3	8.5
14 July–9 August	3.72	3.95	5.8
10 August–23 September	6.41	6.2	3.4
23 September–30 September	2.73	2.4	13.8
Total collapse	18.61	17.85	4.3

Table 9. Numerical results, field measurements and errors for surveying Section 5 of Figure 1.

Time Scale	Calculating Collapse Distance (m)	Monitoring Collapse Distance (m)	Error (%)
21 May–13 July	5.93	5.6	5.9
14 July–9 August	3.82	3.6	6.1
10 August–23 September	6.24	6.5	4
23 September–30 September	1.75	1.45	20.7
Total collapse	17.74	17.15	3.4

By analyzing the results displayed in Tables 5–9, it is possible to confirm that the error between numerical and experimental values is between 6% and 20%. There are multiple influencing factors that can affect this comparison and the shape of the bank toe, always changing under the action of water flow which provides the higher impact. As water and sand conditions are continuously varying, the error is considerably higher considering measurements vs numerical results for the smallest time frame monitored (23 September–30 September). On the other hand, for longer time scales, the total error is about 5%. Overall, the numerical results are fairly consistent with the data recorded during the monitoring campaign, considering the significant modifications applied within the numerical model.

3. Results and Discussion

This section provides the further results obtained and their interpretation.

3.1. Collapse Processes Obtained through Numerical Simulations with the Modified BSTEM Model

The riverbank collapses simulated with the modified BSTEM model for the period 21 May–30 September in 2011 are illustrated in Figures 6–10 for surveying river Sections 1–5 respectively. Collapse rates of study sections in different months are listed in Table 10.

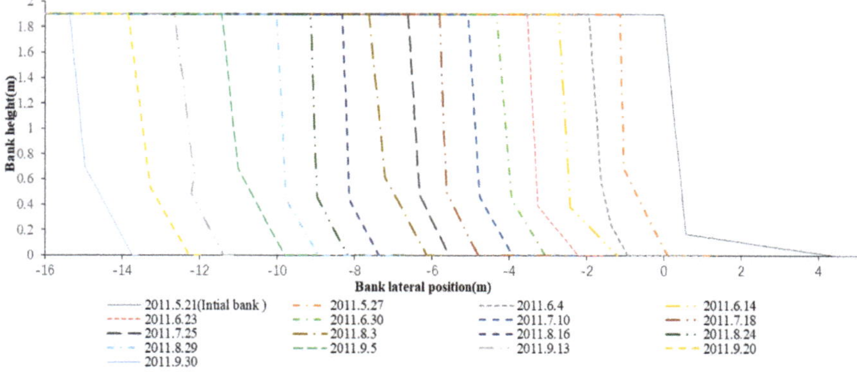

Figure 6. Collapse phases for surveying Section 1.

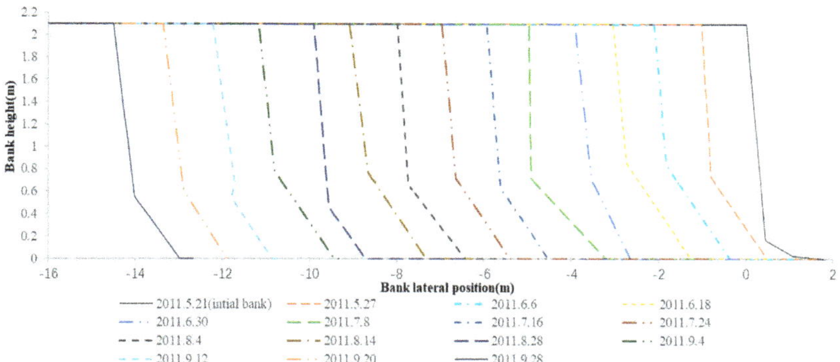

Figure 7. Collapse phases for Section 2.

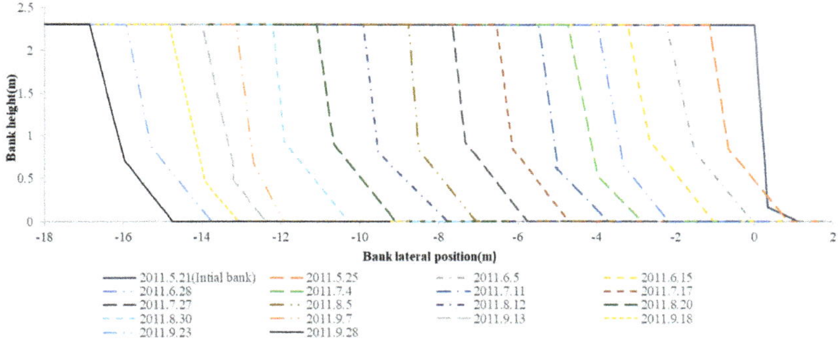

Figure 8. Collapse phases for Section 3.

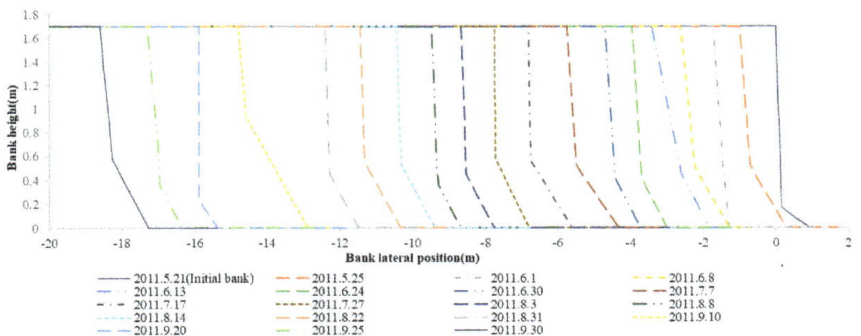

Figure 9. Collapse phases for Section 4.

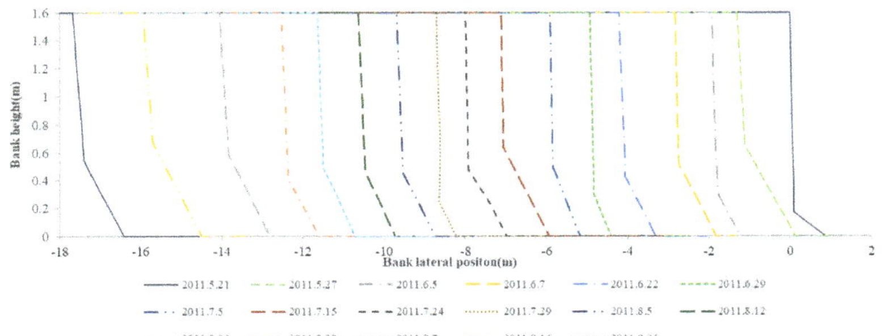

Figure 10. Collapse phases for Section 5.

Table 10. Collapse rates of study sections in different months (Unit: m/d).

Month	Section 1	Section 2	Section 3	Section 4	Section 5
5	0.095	0.094	0.104	0.091	0.118
6	0.106	0.097	0.094	0.123	0.122
7	0.074	0.098	0.119	0.098	0.123
8	0.109	0.095	0.147	0.150	0.124
9	0.178	0.153	0.155	0.207	0.172

To identify hypothetical relationships between hydraulic conditions and collapse rates, different indicators recorded were plotted versus the collapse rates estimated, and factors obtained are displayed in Figure 11.

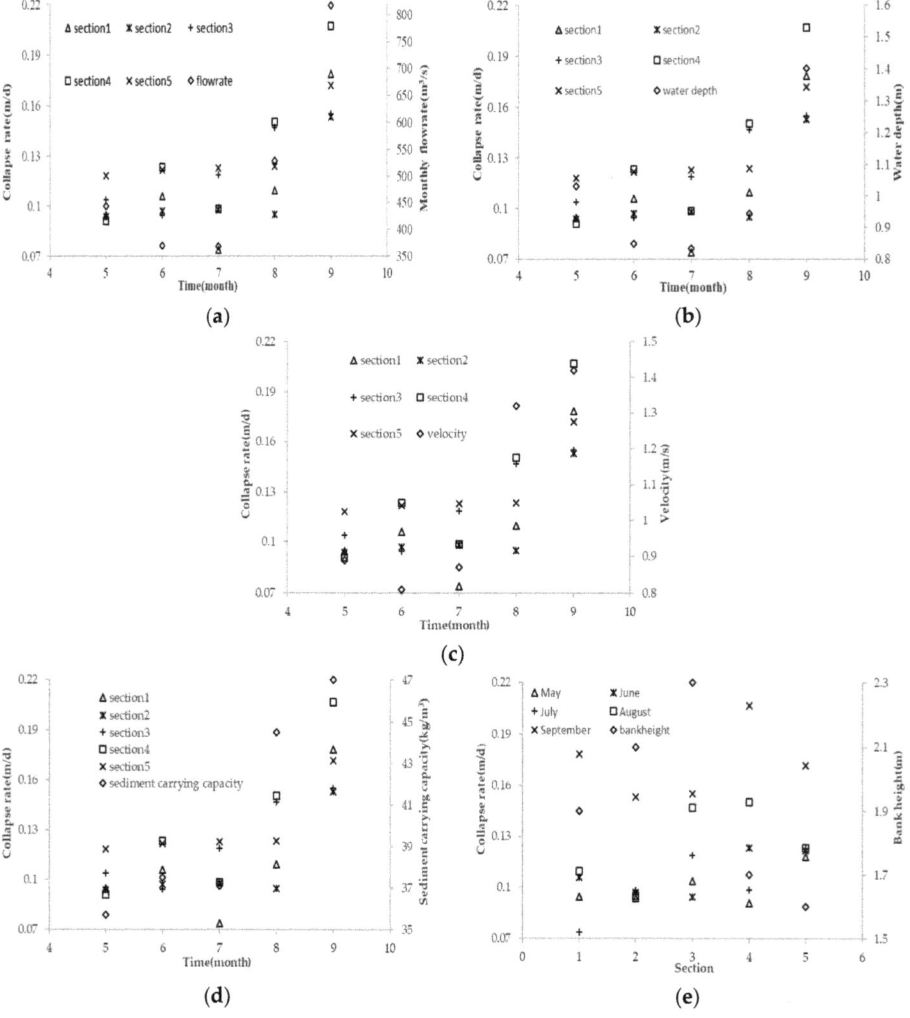

Figure 11. The relationship between flow conditions and collapse rates, the effects of (**a**) flowrate; (**b**) water depth; (**c**) velocity; (**d**) sediment carrying capacity; and (**e**) bank height.

Because the flowrate, water depth, velocity and sediment carrying capacity were of similar magnitudes from June to July, the average collapse rate was used to discuss the effects on collapse rates by influencing factors previously highlighted, and their impacts can be observed in Figure 11.

The authors have attempted to characterize the continuous collapses influenced by multiple factors that can be divided into two categories: autologous and external. The autologous factor considered is the bank height, while external factors include flow rate, water depth, velocity and sediment carrying capacity. Different factors play dissimilar roles on influencing continuous collapse processes. For example, when flow rate, velocity and sediment carrying capacity rise, bank toe erosion can consequently increase due to higher shear stresses. Additionally, when water depth increases, soil shear strength diminishes due to the soil saturation, however the rise of water depth could also improve the bank stability due to the water pressure acting on the bank.

Even if bank collapses are continuously affected by these factors, there are still some interesting regular features observed in this study. For example, in Figure 11a–d, it can be observed that the collapse rate is positively correlated with the flow rate, water depth, velocity and capacity of carrying materials. Although the bank heights are all different in altered sections, collapse rates for the locations considered are higher in the flood season (July to September) than in the dry season (May to June), due to the increase of flow rates, water depth and velocities. Due to the strength of higher flow rates, the flow shear stress is typically greater than the soil shear stress, causing continuous erosion of the bank toe. It should also be noticed then that flow conditions are playing a leading role in the collapse processes and are the main dynamic factors for the quantity of material deposited.

Another interesting phenomenon which can be noticed within the results is that continuous collapse rates are bigger in May than in June, except for Section 3 (Figure 11). During this period, which belongs to the dry season, flow rates, water depth, velocities and sediment carrying capacities are all smaller (Table 3). This is due to the associated phenomena when water depths reduce, with a consequent loss in river bank stability and more intense bank collapses. However, for Section 3, the collapse distance decreases, and the authors believe that this is due to the bank height (2.3 m, the highest among the sections).

3.2. Non Applicability of the Traditional BSTEM Method

In order to further demonstrate that the original BSTEM model is not applicable for the representation of continuous long term bank collapses, numerical and experimental results are listed in Tables 11–15, together with the errors obtained.

Table 11. BSTEM results, field measurements and errors for surveying Section 1.

Time Scale	BSTEM Method Results (m)	Monitoring Collapse Distance (m)	Error (%)
21 May–13 July	7.80	4.60	69.6
14 July–9 August	7.18	3.05	135.4
10 August–23 September	9.98	5.25	89.3
23 September–30 September	2.08	1.68	23.8
Total collapse	27	14.58	85.2

Table 12. BSTEM results, field measurements and errors for surveying Section 2.

Time Scale	BSTEM Method Results (m)	Monitoring Collapse Distance(m)	Error (%)
21 May–13 July	7.70	4.50	71.1
14 July–9 August	5.94	2.80	112.1
10 August–23 September	8.48	5.08	67.2
23 September–30 September	1.73	1.45	19.1
Total collapse	23.84	13.81	72.6

Table 13. BSTEM results, field measurements and errors for surveying Section 3.

Time Scale	BSTEM Method Results (m)	Monitoring Collapse Distance (m)	Error (%)
21 May–13 July	9.63	5.80	66.1
14 July–9 August	6.85	3.10	121.1
10 August–23 September	11.83	6.80	73.9
23 September–30 September	1.36	1.10	24.1
Total collapse	29.68	16.80	76.6

Table 14. BSTEM results, field measurements and errors for surveying Section 4.

Time Scale	BSTEM Method Results(m)	Monitoring Collapse Distance (m)	Error (%)
21 May–13 July	8.9	5.3	68.1
14 July–9 August	9.13	3.95	131.1
10 August–23 September	11.35	6.2	83.2
23 September–30 September	2.9	2.4	21
Total collapse	32.28	17.85	80.8

Table 15. BSTEM results, field measurements and errors for surveying Section 5.

Time Scale	BSTEM Method Results (m)	Monitoring Collapse Distance (m)	Error (%)
21 May–13 July	9.63	5.6	71.9
14 July–9 August	7.85	3.6	118.1
10 August–23 September	12.02	6.5	85
23 September–30 September	1.77	1.45	22
Total collapse	31.27	17.15	82.4

It can be concluded that the errors are all very large for longer monitoring periods. That is due to the hypothesis implemented within the BSTEM method, commonly used to calculate collapse distance, considering no soil deposition at the bank toe after the collapse—an unrealistic condition.

3.3. Discussions

Comparing simulation results using the modified BSTEM model with field data collected, as showed in Tables 5–9, it can be observed that errors range from 0.3% to 5.07%, hence are all acceptable and it is possible to confirm the reliability and accuracy of the modified BSTEM model to calculate continuous riverbank collapse.

Furthermore, additional outcomes have to be highlighted to increase the performance of the numerical model. Firstly, as the initial boundary conditions can have an impact on river bank collapse, supplementary studies have to be conducted to investigate this phenomenon. Secondly, due to the lack of advanced experimental measurement techniques, it is very difficult to get the river bed profile at every time step, and for this study a simplified bank shape was used in the calculations presented. This may be one more reason for the errors calculated. To solve this problem, sonar and electromagnetic wave measurement technologies may be adopted in future studies. Thirdly, in natural rivers, flow conditions such as flow rate and water depth vary continuously. To obtain bank collapse distances, monthly average flow conditions were used in this study, hence more targeted and accurate flow conditions should be used in future studies to improve the results. Finally, riverbed erosion and bank erosion typically occur at the same time and both are influenced by each other. For example, riverbed erosion influences the bank collapse by increasing the bank height and when the bank collapse happens, falling soil directly settles along the area where riverbed erosion happened before depositing at the bank toe.

As all these aspects are crucial, by implementing them within the numerical model it is predicted that results will be more accurate.

4. Conclusions

(1) As confirmed by the observations made during the monitoring of the area, a single bank collapse can be divided into four different stages that can be summarized as: (i) bank toe erosion initiation; (ii) tension crack development; (iii) shearing initiation; and (iv) bank failure occurrence. When a single collapse happens, part of the collapsed soil is washed away by the water and the remaining part accumulates at the bank toe. The new shape of the bank formed after this process along the river becomes the initial form of the next bank collapse. Bank collapse processes in a longer time scale are continuous collapses that are made from single bank collapse during short time scales.

(2) Based on the original BSTEM model, a new modified version was implemented, applying several new assumptions. The original BSTEM assumed that the falling part which had collapsed from the riverbank did not accumulate on the toe and was immediately removed. This does not accurately represent the reality of the entire erosion process typical in natural rivers, and comparisons between experimental and numerical results have shown that, by using the original approach, significant errors will affect the numerical results, as listed in Tables 11–15.

Furthermore, to address this gap, a modified BSTEM was developed to obtain the quantity of collapsed sediment accumulating on the toe. Several parameters were introduced, such as sediment concentration, sediment carrying capacity and sediment setting velocity. The quantity of collapsed sediment accumulated on the toe could have been obtained using Formulas (6)–(9).

Additionally, another main goal modifying the BSTEM model was to obtain the new shape generated by the deposition of material on the bank toe. It should be noted that the boundary conditions play an important role in the calculations of continuous collapse processes for natural rivers. The new shape generated by the deposition of material on the bank toe was the initial boundary condition of next collapse process. Based on this, after observing multiple shapes during the erosion process following the collapse of material, the shape generated by the deposition of material on the bank toe was assumed to be triangular and the angle away from the riverbank was the angle of repose of sediment. After a single collapse, the new shape of the riverbank could have been obtained from the shape of the deposited sediment and the collapsed plain.

(3) Adopting all the previous modifications within the calculation of every single collapse process, the numerical results obtained using the modified BSTEM are consistent with the monitored datasets for continuous processes, as listed in Tables 5–9.

(4) The relationship between collapse rates and influencing factors was discussed and influencing factors were divided into two categories: autologous and external (flow conditions). Different factors play different roles in continuous collapse processes. For example, the flow rate, velocity and sediment carrying capacity rise can increase bank toe erosion by increasing flow shear stress. Water depth rise can reduce soil shear strength by promoting soil saturation, and water depth rise can also increase river bank stability by providing water pressure. Soil characteristics determine the shear strength of each material, and bank height plays an important role in the bank stability analysis. For similar flow conditions, a higher bank with a steeper slope is more unstable. These factors should be considered together to better characterize the collapse mechanisms.

(5) Aspects such as precisely determining the bank shape and quantifying the more singular flow conditions are crucial, and by implementing them within the numerical model it is predicted that results will become more accurate.

Author Contributions: All the authors jointly contributed to this research. A.S. was responsible for the proposition and design of the field monitoring and the modified BSTEM method; M.R. and G.D. analysed the experimental datasets and wrote the paper; and S.W. and F.Z. participated in the experiments.

Funding: The research reported in this manuscript is funded by the National Basic Research Program of China (Grant No. 2011CB403304) and National Natural Science Foundation of China (Grant No. 11372048).

Acknowledgments: The research reported in this manuscript is supported by the National Basic Research Program of China (Grant No. 2011CB403304) and National Natural Science Foundation of China (Grant No. 11372048).

Conflicts of Interest: The authors declare no conflict of interest.

References

1. Nanson, G.C.; Hickin, E.J. A statistical analysis of bank erosion and channel migration in western Canada. *Geol. Soc. Am. Bull.* **1986**, *97*, 497–504. [CrossRef]
2. Shu, A.P.; Li, F.H.; Liu, H.F.; Duan, G.S.; Zhou, X. Characteristics of particle size distributions for the collapsed riverbank along the desert reach of the upper Yellow River. *Int. J. Sediment. Res.* **2016**, *31*, 291–298. [CrossRef]
3. Yao, Z.Y.; Ta, W.Q.; Jia, X.P.; Xiao, J.H. Bank erosion and accretion along the Ningxia–Inner Mongolia reaches of the Yellow River from 1958 to 2008. *Geomorphology* **2011**, *127*, 99–106. [CrossRef]
4. Hou, S. Analysis on recent channel evolution characteristics on Inner-Mongolia reaches of the Yellow River. *Yellow River* **1996**, *18*, 43–44.
5. Hou, S.; Chang, W.; Wang, P. Characteristics and cause of formation of channel atrophy at Inner Mongolia section of the Yellow River. *Yellow River* **2007**, *29*, 25–29.
6. Wu, J.M.; Wang, R.S.; Yao, J.H. Remote sensing monitoring and study on the change of the Yellow River course in Yinchuan plain sector. *Remote Land Resour.* **2006**, *18*, 36–40.
7. Ta, W.Q.; Xiao, H.L.; Dong, Z.B. Long-term morphodynamic changes of a desert reach of the Yellow River following upstream large reservoirs' operation. *Geomorphology* **2008**, *7*, 249–259. [CrossRef]
8. Grabowski, R.C.; Droppo, I.G.; Wharton, G. Erodibility of cohesive sediment: The importance of sediment properties. *Earth-Sci. Rev.* **2011**, *105*, 101–120. [CrossRef]
9. Thorne, C.R. Processes and mechanisms of river bank erosion. In *Gravel-Bed Rivers*; Bathurst, J.C., Thorne, C.R., Eds.; Wiley and Sons: Chichester, UK, 1982; pp. 227–259.
10. ASCE Task Committee on Hydraulics; Bank Mechanics; Modeling of River Width Adjustment. River width adjustment. I: Processes and mechanisms. *J. Hydraul. Eng.* **1998**, *124*, 881–902. [CrossRef]
11. Shields, A.; Ott, W.P.; Uchelen, J.C.V. *Application of Similarity Principles and Turbulence Research to Bed-Load Movement*; Soil Conservation Service Cooperative Laboratory California Institute of Technology: Pasadena, CA, USA, 1936.
12. Buffington, J.M.; Montgomery, D.R. A systematic analysis of eight decades of incipient motion studies, with special reference to gravel-bedded rivers. *Water Resour. Res.* **1997**, *33*, 1993–2029. [CrossRef]
13. Hooke, J.M. An analysis of the processes of riverbank erosion. *J. Hydrol.* **1979**, *42*, 39–62. [CrossRef]
14. Grissinger, E.H. Bank erosion of cohesive materials. In *Gravel-Bed Rivers*; Thorne, C.R., Bathurst, J.C., Eds.; Wiley and Sons: Chichester, UK, 1982; pp. 273–287.
15. Simon, A.; Collison, A.J.C. Pore-water pressure effects on the detachment of cohesive streambeds: Seepage forces and matric suction. *Earth Surf. Process. Landf.* **2001**, *26*, 1421–1442. [CrossRef]
16. Julian, J.P.; Torres, R. Hydraulic erosion of cohesive riverbanks. *Geomorphology* **2006**, *76*, 193–206. [CrossRef]
17. Twidale, C.R. Erosion of an alluvial bank at Birdwood, South Australia. *Z. Geomorphol.* **1964**, *8*, 189–211.
18. Lawler, D.M. The measurement of river bank erosion and lateral channel change: A review. *Earth Surf. Process. Landf.* **1993**, *18*, 777–821. [CrossRef]
19. Rinaldi, M.; Casagli, N. Stability of streambanks formed in partially saturated soils and effects of negative pore water pressures: The Sieve River (Italy). *Geomorphology* **1999**, *26*, 253–277. [CrossRef]
20. Singer, M.J.; Munns, D.N. *Soils: An Introduction*; Prentice Hall Inc.: Upper Saddle River, NJ, USA, 1999; pp. 14–46. ISBN 9780131190191.
21. Couper, P. Effects of silt–clay content on the susceptibility of riverbanks to subaerial erosion. *Geomorphology* **2003**, *56*, 95–108. [CrossRef]
22. Musa, J.J.; Abdulwaheed, S.; Saidu, M. Effect of surface runoff on Nigerian rural roads (a case study of Offa local government area). *AU J. Technol.* **2010**, *13*, 242–248.
23. Wolman, M.G. Factors influencing erosion of a cohesive river bank. *Am. J. Sci.* **1959**, *257*, 204–216. [CrossRef]
24. Knighton, A.D. Riverbank erosion in relation to streamflow conditions, River Bollin-Dean, Cheshire. *East Midl. Geogr.* **1973**, *5*, 416–426.
25. Albidin, R.Z.; Sulaiman, M.S.; Yusoff, N. Erosion risk assessment: A case study of the Langat Riverbank in Malaysia. *Int. Soil Water Conserv. Res.* **2017**, *5*, 26–35. [CrossRef]

26. Yu, M.; Wei, H.; Wu, S. Experimental study on the bank erosion and interaction with near-bank bed evolution due to fluvial hydraulic force. *Int. J. Sediment. Res.* **2015**, *1*, 81–89. [CrossRef]

27. Yang, G.S.; Tuo, W.Q.; Dai, F.N.; Liu, Y.X.; Jing, K.; Li, B.Y.; Zhang, O.Y.; Lu, R.; Hu, L.F.; Tao, Y. Contribution of sand sources to the silting of riverbed in Inner Mongolia section of Huanghe River. *J. Desert Res.* **2003**, *23*, 152–159.

28. Feng, G. The coarse sediment of the Yellow River mainly from wind sand. *J. Soil Water Conserv.* **1992**, *3*, 45–47.

29. Jing, K.; Li, J.; Li, F. Erosion yield from the middle Yellow River basin and tendency prediction. *Acta Geogr. Sin.* **1998**, *53*, 107–115.

30. Jing, K.; Li, J.; Li, F. Study on delimitation of coarse sand area in the middle Yellow River. *J. Soil Water Conserv.* **1997**, *3*, 10–15.

31. Xu, J.H.; Lv, G.C.; Zhang, S.L.; Gan, Z.M. *Definition of the Source Area of Centralized Coarse Sediment and Sediment Yield and Transportation in the Middle Yellow River*; Yellow River Press: Zhengzhou, China, 2000; pp. 11–36, ISBN 9787806214527.

32. Bai, T.; Ma, R.; Ma, X.; Ha, Y.; Huang, Q. Threshold of water and sediment in desert wide valley reaches upper the Yellow River. *J. Desert Res.* **2018**, *38*, 645–650.

33. Yu, M.; Shen, K.; Zhang, J.H. Preliminary investigation on bank properties and sediment source in Yellow River Ningmeng River Reach. *Int. J. Sediment. Res* **2014**, *4*, 39–42.

34. Rijn, L.C.V. *Principles of Sediment Transport in Rivers, Estuaries and Coastal Seas*; Aqua Publications: Delft, The Netherlands, 1993; ISBN 90-800356-2-9.

35. Catano, Y.; Passoni, G.; Pacheco, R.G.; Toro, F.M.; Naranjo, B.J. Comparison between different numerical solutions for the hydraulic dam-break wave. In Proceedings of the Iche 98-International Conference on Hydro-sciences and Engineering, Berlin, Germany, 31 August–3 September 1998.

36. Toro, E.F. *Shock-Capturing Methods for Free-Surface Shallow Flows*; Wiley: New York, NY, USA, 2001; pp. 566–571, ISBN 9780471987666.

37. Cao, Z.; Day, R.; Egashira, S. Coupled and decoupled numerical modeling of flow and morphological evolution in alluvial rivers. *J. Hydraul. Eng.* **2002**, *128*, 306–321. [CrossRef]

38. Hudson, J.; Sweby, P.K. A high-resolution scheme for the equations governing 2D bed-load sediment transport. *Int. J. Numer. Meth.* **2005**, *47*, 1085–1091. [CrossRef]

39. Simon, A.; Curini, A.; Darby, S.E.; Langendoen, E.J. Bank and near-bank processes in an incised channel. *Geomorphology* **2000**, *35*, 193–217. [CrossRef]

40. Zong, Q.; Xia, J.; Deng, C.; Xu, Q.X. Modeling of the composite bank failure process using BSTEM. *J. Sichuan Univ.* **2013**, *45*, 69–78.

41. Simon, A.; Pollen-Bankhead, N.; Mahacek, V.; Langendoen, E. Quantifying reductions of mass-failure frequency and sediment loadings from streambanks using toe protection and other means: Lake Tahoe, United States. *J. Am. Water Resour. Assoc.* **2009**, *45*, 170–186. [CrossRef]

42. Lindow, N.; Fox, G.A.; Evans, R.O. Seepage erosion in layered stream bank material. *Earth Surf. Process. Landf.* **2009**, *34*, 1693–1701. [CrossRef]

43. Wilson, G.V.; Perketi, R.K.; Fox, G.A.; Dabney, S.M.; Shields, F.D.; Cullum, R.F. Soil properties controlling seepage erosion contributions to streambank failure. *Earth Surf. Process. Landf.* **2007**, *32*, 447–459. [CrossRef]

44. Cancienne, R.M.; Fox, G.A.; Simon, A. Influence of seepage undercutting on the stability of root-reinforced streambanks. *Earth Surf. Process. Landf.* **2008**, *33*, 1769–1786. [CrossRef]

45. Xiao, C.; Hao, Y.; Liu, F. Analysis of climate change for Dengkou in recent 52 years. *J. Arid Land Resour. Environ.* **2008**, *22*, 90–93.

46. Kolerski, T.; Shen, H.T. Possible effects of the 1984 St Clair River Ice jam on bed changes. *Can. J. Civ. Eng.* **2015**, *42*, 696–703. [CrossRef]

47. Kolerski, T.; Shen, H.T. St. Clair River ice jam dynamics and possible effect on bed changes. In Proceedings of the 20th IAHR International Symposium on Ice, Lahti, Finland, 14–18 June 2010.

48. Thorne, C.R.; Tovey, N.K. Stability of composite riverbanks. *Earth Surf. Process. Landf.* **1981**, *6*, 469–484. [CrossRef]

49. Shu, A.P.; Gao, J.; Li, F.H. Lateral variation properties in river channels due to bank failure along the desert valley reach of upper Yellow River. *Adv. Water Sci.* **2014**, *25*, 77–82.

50. Osman, A.M.; Thorne, C.R. Riverbank stability analysis. I: Theory. *J. Hydraul. Eng.* **1988**, *114*, 134–150. [CrossRef]

51. Darby, S.E.; Thorne, C.R. Numerical simulation of widening and bed deformation of straight sand-bed rivers. I: Model development. *J. Hydraul. Eng.* **1996**, *122*, 184–193. [CrossRef]

52. Midgley, T.L.; Fox, G.A.; Heeren, D.M. Evaluation of the bank stability and toe erosion model (BSTEM) for predicting lateral retreat on composite streambanks. *Geomorphology* **2012**, *145*, 107–114. [CrossRef]

53. Garcia, M.H. *Sedimentation Engineering*; American Society of Civil Engineers: Reston, VA, USA, 2008; pp. 21–163, ISBN 9780784408148.

54. Rinaldi, M.; Mengoni, B.; Luppi, L.; Darby, S.E.; Mosselman, E. Numerical simulation of hydrodynamics and bank erosion in a river bend. *Water Resour. Res.* **2008**, *44*, 303–312. [CrossRef]

55. Hanson, G.J. Surface erodibility of earthen channels at high stresses. Part II—Developing an in-situ testing device. *Trans. ASAE* **1990**, *33*, 127–131. [CrossRef]

56. Amiritokaldany, E.; Darby, S.E.; Tosswell, P. Bank stability for predicting reach-scale land loss and sediment yield. *J. Am. Water Resour. Assoc.* **2003**, *39*, 897–910. [CrossRef]

57. Langendoen, E.J.; Simon, A. Modeling the evolution of incised streams. II: Streambank erosion. *J. Hydraul. Eng.* **2008**, *134*, 905–915. [CrossRef]

58. Samadi, A.; Amiritokaldany, E.; Darby, S.E. Identifying the effects of parameter uncertainty on the reliability of riverbank stability modelling. *Geomorphology* **2009**, *106*, 219–230. [CrossRef]

59. Taghavi, M.; Dovoudi, M.H.; Amiritokaldany, E.; Darby, S.E. An analytical method to estimate failure plane angle and tension crack depth for use in riverbank stability analyses. *Geomorphology* **2010**, *123*, 74–83. [CrossRef]

60. Dodaro, G.; Tafarojnoruz, A.; Stefanucci, F.; Adduce, C.; Calomino, F.; Gaudio, R.; Sciortino, G. An experimental and numerical study on the spatial and temporal evolution of a scour hole downstream of a rigid bed. In Proceedings of the International Conference on Fluvial Hydraulics, Lausanne, Switzerland, 3–5 September 2014; pp. 1415–1422.

61. Dodaro, G.; Tafarojnoruz, A.; Sciortino, G.; Adduce, C.; Calomion, F.; Gaudio, R. Modified Einstein sediment transport method to simulate the local scour evolution downstream of a rigid bed. *J. Hydraul. Eng.* **2016**, *142*. [CrossRef]

62. Wang, X.K.; Shao, X.J.; Li, D.X. *Fundamental River Mechanics*; China Water and Power Press: Beijing, China, 2002; pp. 54–98, ISBN 7-5084-1093-9.

63. Gaudio, R.; Tafarojnoruz, A.; Calomino, F. Combined flow-altering countermeasures against bridge pier scour. *J. Hydraul. Res.* **2012**, *50*, 35–43. [CrossRef]

64. Tafarojnoruz, A.; Gaudio, R.; Calomino, F. Bridge pier scour mitigation under steady and unsteady flow conditions. *Acta Geophys.* **2012**, *60*, 1076–1097. [CrossRef]

65. Tang, C.B. Laws of sediment incipient motion. *J. Hydraul. Eng.* **1963**, *2*, 3–14.

Article

Numerical Study of the Collapse of Multiple Bubbles and the Energy Conversion during Bubble Collapse

Jing Zhang [1,2,3], Lingxin Zhang [1,2,3,*] and Jian Deng [1,2,3]

1   State Key Laboratory of Fluid Power and Mechatronic Systems, Zhejiang University,
    Hangzhou 310027, China; zj121314@yeah.net (J.Z.); zjudengjian@zju.edu.cn (J.D.)
2   Key Laboratory of Soft Machines and Smart Devices of Zhejiang Province, Zhejiang University,
    Hangzhou 310027, China
3   Department of Mechanics, Zhejiang University, Hangzhou 310027, China
*   Correspondence: zhanglingxin@zju.edu.cn; Tel.: +86-136-5671-5841

Received: 10 December 2018; Accepted: 29 January 2019; Published: 31 January 2019

Abstract: This paper investigates numerically the collapses of both a single cavitation bubble and a cluster consisting of 8 bubbles, concerning mainly on the conversions between different forms of energy. Direct numerical simulation (DNS) with volume of fluid (VOF) method is applied, considering the detailed resolution of the vapor-liquid interfaces. First, for a single bubble near a solid wall, we find that the peak value of the wave energy, or equivalently the energy conversion rate decreases when the distance between the bubble and the wall is reduced. However, for the collapses of multiple bubbles, this relationship between the bubble-wall distance and the conversion rate reverses, implying a distinct physical mechanism. The evolutions of individual bubbles during the collapses of multiple bubbles are examined. We observe that when the bubbles are placed far away from the solid wall, the jetting flows induced by all bubbles point towards the cluster centre, while the focal point shifts towards the solid wall when the cluster is very close to the wall. We note that it is very challenging to consider thermal and acoustic damping mechanisms in the current numerical methods, which might be significant contributions to the energy budget, and we leave it open to the future studies.

Keywords: cavitation bubble, multiple bubble collapse, pressure wave energy, energy conversion rate

---

1. Introduction

As a natural phenomenon, cavitation occurs frequently in a variety of engineering applications. As the local pressure in liquid is lower than its saturated vapor pressure, gas nucleus can develop into cavitation bubbles. When the local pressure is recovered or driven by high pressure waves, the cavitation bubbles collapse, leading to a series of physical scenarios such as micro jetting flows, shock wave emission, heat release etc. In hydraulic systems, cavitation can cause material damage, loss of power, and induce noises. On the other hand, bubble collapse can also be utilized as a tool in the fields of medical treatments and mining, such as drug delivery, ultrasound operation and cavity jet exploits [1]. Numerous experimental and numerical studies have been performed to study the bubble dynamics and the collapses of cavitation bubbles, in order to control or reinforce the effects of cavitation bubble collapse. However, the dynamics of multiple cavitation bubbles and their collapses, particularly the underlying behavior of energy conversion have not been well understood.

Historically, studies on bubble dynamics can date back to 1917, when Reyleigh proposed a theoretical spherical-bubble model [2]. This model was for an ideal spherical bubble, neglecting the effects of viscosity, surface tension, compressibility and mass transfer. It was followed by a series of modifications and perfections [3–8].

In the past few decades, a large amount of works have been undertaken to understand the characteristics of bubble collapse [9–24]. Lauterborn recorded the process of a single bubble collapsing near a rigid wall, in which the shock wave emission was captured by a schlieren system [25–27]. Moreover, the particle image velocimetry (PIV) technique was adopted to monitor the velocity fields by lauterborn & Vogel [28], who captured the micro jetting flows as the bubble collapses asymmetrically. They revealed the relationship between the noises induced by the bubble collapse and the dimensionless distance γ ($\gamma = L/R_{max}$, where L is the distance between the bubble centre and the rigid wall, and R_{max} is the maximum bubble radius). Gonzalez-Avila measured the pressure emission of laser-induced bubble collapse [29], in which the pressure amplitude was found to be up to $1k$ Bar when the bubble was close to the rigid boundary [30]. Merouan [31] measured the temperature variations during the growth and collapse of a single cavitation bubble. More recently, Fortes et al. proposed an analytic model to examine the pressure wave emission during bubble collapse, with the conversion rate of the wave energy found to be strongly influenced by gas content in the bubble [32].

However, not many efforts have been made to study the pressure wave emission from the collapses of multiple bubbles due to the difficulties in experimental setup and the consideration of compressibility of liquid in numerics [33–36]. Generally, a bubble cluster collapses inward, generating a high level impulsive pressure wave, and the symmetries for individual bubbles are lost due to the presence of the rigid wall and their nearby bubbles [37,38]. It has been shown that the shock wave emission focuses inward during the bubble cloud collapse, leading to a very high peak pressure, which can exceed that of the collapse of a single bubble with the equivalent volume [39]. Tiwari et al. [40] simulated the expansion and subsequent collapse of a hemispherical cluster of 50 bubbles adjacent to a plane rigid wall, with the detailed compressible-fluid mechanics of bubble-bubble interactions considered. They found that the peak pressure was associated with the centremost bubble, which caused a corresponding peak pressure on the nearby wall. Despite these advances in the few existing researches, the detailed bubble-scale dynamics of a cluster collapse remain poorly understood.

Here, we investigate the characteristics of the collapses of multiple bubbles by solving directly the Navier-Stokes equations considering the compressiblility of the fluid, with particular concerns on the energy conversion. The volume of fluid method is involved to capture the interfaces between the liquid and the cavitation bubbles. The simulation results are organized into two parts: in the first part, a single bubble is studied as a reference to our further multiple-bubble studies. In the second part, we focus the collapses of multiple bubbles, presenting the unique dynamic behaviors in pressure wave propagation and energy conversion, in contrast to the single bubble situation.

2. Numerical Methods

The numerical work is performed using a finite-volume based code. Assuming that the liquid is compressible and the cavitation bubbles are filled with pure vapour, and considering viscosity and surface tension, the governing equations are formularized as the following

$$\frac{d\rho}{dt} + \rho\nabla\vec{U} = 0, \tag{1}$$

$$\rho\frac{\partial\vec{U}}{\partial t} + \rho\vec{U}\cdot\nabla\vec{U} = \rho\vec{g} - \nabla p + 2\nabla\cdot\left(\mu\overline{\overline{D}}\right) - \frac{2}{3}\nabla\left(\mu\nabla\cdot\vec{U}\right) + \sigma\kappa\vec{N}, \tag{2}$$

$$\frac{d\alpha}{dt} = \frac{\partial\alpha}{\partial t} + \vec{U}\cdot\nabla\alpha = 0, \tag{3}$$

where k is surface curvature, σ is the surface tension coefficient, $\overline{\overline{D}}$ is the strain rate tensor, $\vec{N}$ is the unit normal vector of the interface, α is the volume fraction of the liquid, and ρ and μ are the density and the viscosity respectively of the mixture fluid, which are obtained by weighting of the volume fractions:

$$\rho = \alpha\rho_1 + (1-\alpha)\rho_2, \tag{4}$$

$$\mu = \alpha\mu_1 + (1-\alpha)\mu_2, \tag{5}$$

where subscripts 1 and 2 denote liquid phase (water) and gas phase (vapour), respectively. The densities of each phase are calculated by

$$\rho_1 = \rho_{10} + \psi_1 p, \tag{6}$$

$$\rho_2 = \rho_{20} + \psi_2 p, \tag{7}$$

where ψ_1 and ψ_2 are two constants associated with the respective compressibilities of water and vapour.

In our simulations, the space discretizations are second-order upwind for the convection terms and central differences for the Laplacian terms, respectively. The time discretization is first-order implicit Euler. The pressure-velocity coupling is obtained using the Pressure Implicit Split Operator (PISO) scheme. The preconditioned conjugate gradient (PCG) method is used to treat the pressure equation and the preconditioned biconjugate gradient (PBiCG) method is used for the velocity equations.

The computational domain is a cylinder with a radius of 25 mm, and a height of 50 mm. The grid distributions on some specific sections are shown in Figure 1. The bottom of the cylindrical flow domain is set as a rigid wall, the others are far field boundaries. The initial pressure of the liquid phase is 101,325 Pa, and the pressure of vapor is 3154 Pa. The initial flow is quiescent, and the initial bubbles are spherical, with the radius of 2 mm. The geometrical setup of the multiple-bubble case is shown in Figure 2. The dimensionless wall distance is defined as $\gamma = L/R_{max}$. After carefully carrying out self-consistency tests, we find that the current grid with a cell number of 25,581,192, and an initial maximum time step size of $dt = 1 \times 10^{-7}$ s, are sufficient to assure satisfactory independence of the results with respect to both mesh and time discretizations. We note that the time step is adjustable during the simulations to meet the requirement of local courant number $Co = 0.35$.

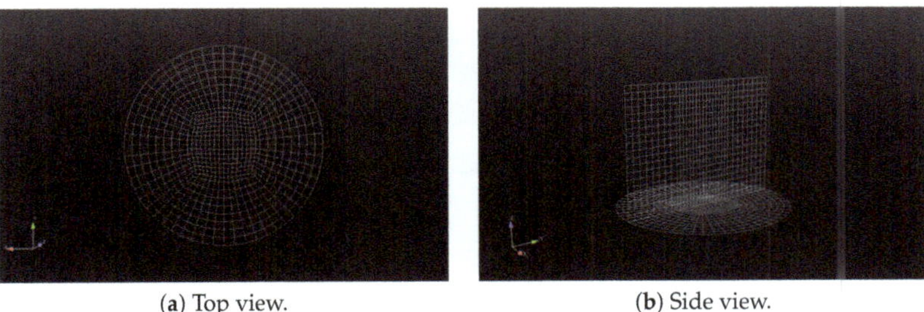

<table>
<tr><td>(a) Top view.</td><td>(b) Side view.</td></tr>
</table>

Figure 1. (a) Top view and (b) side view of the grid distributions within the computational domain. Note that the grid in the central area of the cylindrical domain is refined where the bubbles are placed, and only a coarse grid with 10% of the cells is displayed to diagram the grid distributions.

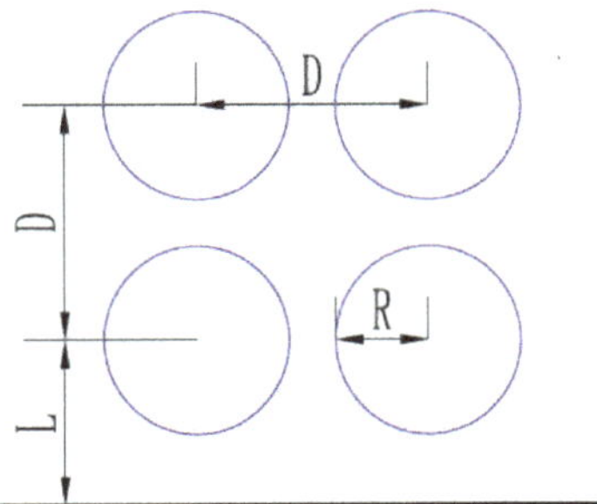

Figure 2. Geometrical description for the simulations of multiple bubbles.

To further validate the spatial resolutions, we carry out simulations with different numbers of cells along the diameter of a single bubble. Figure 3 presents the time histories of the dimensionless radius R^*, where $R^* = R/R_{max}$, and the time is non-dimensionalized according to $T^* = t/t_c$. Here, R_{max} is the maximum or the initial radius of the bubble, and t_c is the collapsing time predicated from the Reyleigh-plesset equation. It is observed that the result converges at 34 cells, which accords well with the theoretical solution. We show in Figure 4a the evolutions of bubble shapes during collapse simulated based on the current resolution of 34 cells. The numerical results are consistent with the previous experimental observation [41].

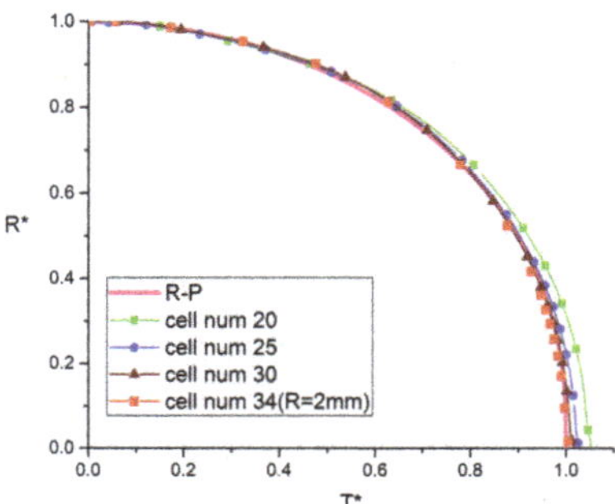

Figure 3. Time histories of radius for a single bubble resolved by different number of cells distributed along the diameter. The reference curve is the theoretical solution of the Rayleigh-Plesset equation.

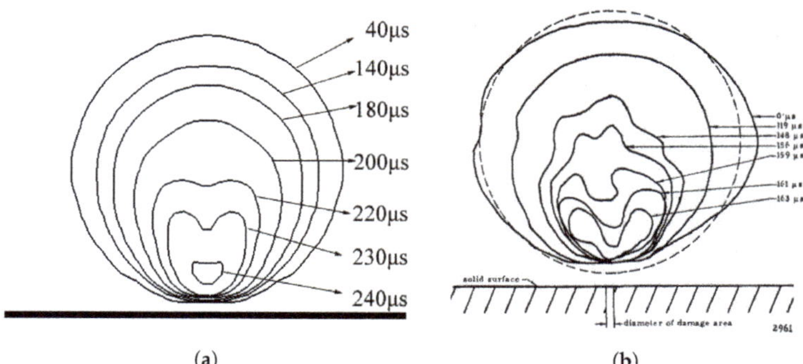

Figure 4. Variations of the bubble shapes during collapse obtained by (**a**) the present numerical simulations and (**b**) previous experiments [41].

In the following sections, according to [42], the potential energy of a bubble is defined as

$$E_{pot} = \frac{4}{3}\pi R^3 \Delta p, \tag{8}$$

where R is the bubble radius and Δp is the pressure difference between two sides of the bubble interface, then the maximum potential energy of a cavitation bubble is defined as $E_{pot-max}$. The total kinetic energy of the flow domain is the integral:

$$E_k = \int \frac{1}{2}\rho U^2 dV. \tag{9}$$

We also define the wave energy emitted by the bubble collapse, following an acoustic approach [28,43], with the following form:

$$E_{wave} = \int \frac{\Delta P^2}{(\rho c)^2} dV,$$ (10)

where c is the sound speed in water, $c = 1500$ m/s in the current simulations. The conversion rate of the wave generation is defined as

$$\eta = \frac{E_{wave}}{E_{pot-max}}$$ (11)

3. Result

3.1. Collapse of a Single Bubble

In this section, we study a single cavitation bubble collapsing near a solid wall. To understand the presence of the solid wall on the non-spherical deformation, and consequently the energy conversion, we carry out a series of simulations by varying the distance between the bubble and the solid wall. The dimensionless distance γ ranges from 1.5 to 4.0.

In Table 1, we present the simulation results. First, we observe that the energy conversion rate reaches its maximum of 29.29% at $\gamma = 12.5$, which is not difficult to understand since the $\gamma = 12.5$ case corresponds to a distance very far from the solid wall, which we believe that the wall effect can be reasonably neglected. The values of maximum kinetic energy for all cases are around 4.0 mJ, with slight variations among different distances. We note that the energies and the conversions at $\gamma = 2.5$ and $\gamma = 4.0$ are very close, exhibiting nontrivial bounding effects of the solid wall.

Table 1. Energies and their conversion rates for various distances for the collapse of a single bubble (note the different units for the two energies).

γ	E_k (mJ)	E_{wave} (µJ)	η (%)
1.5	3.94	352.2	10.73
1.65	3.90	418.1	12.74
1.85	3.97	543.8	16.57
2.0	3.94	564.2	17.19
2.5	4.08	608.8	18.55
3.0	3.95	579.3	17.65
3.5	4.01	630.1	19.20
4.0	4.10	620.6	18.91
12.5	4.36	961.3	29.29

In Figure 5a,b, we present the time histories of different forms of energy at $\gamma = 12.5$ and $\gamma = 1.5$ respectively, to make a direct comparison between a small and a large distances. In Figure 5a, we observe that the potential energy of the bubble decreases during the process of collapse as the bubble shrinks, while the kinetic energy grows from zero, representing the accelerated process of the bubble collapse. As the bubble collapses to a singular point at 199 µs, the kinetic energy reaches its maximum value, while the potential energy drops to nearly zero value. During this whole period, the total energy is approximately equal to the initial bubble potential energy. In other words, the total mechanical energy remains approximately constant, or is conserved. During this process of bubble shrinkage, the potential energy is lost to the re-entrant jetting towards the bubble centre.

After the bubble vanishes, around 200 µs, a pressure wave is emitted and travels outwards, resulting in an intensive wave energy observed in Figure 5a. However, since we consider only limited forms of energy in the simulations and neglect thermal and acoustic damping mechanisms, which might be significant, the total energy is not conserved anymore. We understand that it is very challenging to quantify in detail any switch-over of these dominant damping mechanisms in such violent bubble collapses. We leave it open to the future studies. To help understand this process more intuitively, we present in Figure 6 the flow fields for three time instants demonstrating the alternative dominance of different energy forms. In Figure 6a, at the early stage, the bubble potential energy dominates, while it is converted to kinetic energy when the bubble starts to collapse (see Figure 6b), and in Figure 6c the pressure wave is emitted with only the wave energy prominent in the energy budget.

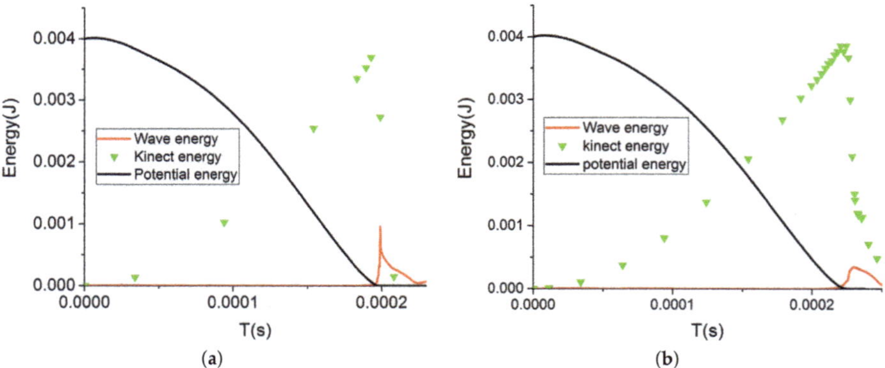

Figure 5. Time histories of different forms of energy in the flow domain during the collapse of a single bubble at (**a**) $\gamma = 12.5$ and (**b**) $\gamma = 1.5$. For both cases, $R_{max} = 2$ mm and $\Delta P = 101{,}325$ Pa.

For comparison, we present the time histories of different forms of energy at $\gamma = 1.5$ in Figure 5b, in which case the initial bubble is very close to the solid wall. Apparently, the changing trend of each form of energy is roughly the same with its large distance counterpart (see Figure 5a). They still differ in some aspects: first, the wave energy appears later than the large distance case due to the confinement of the solid wall, which delays the collapse of the bubble. Second, during the whole process, even after the kinetic energy drops, it is always higher than the wave energy, implying a different physical mechanism. We conjecture that the presence of a rigid boundary can lead to asymmetric collapse, which requires more time. In the process of asymmetric collapse, high speed micro jetting flows are induced towards the rigid boundary, the liquid can thus be pushed away along the wall instead of collapsing towards a singular point. Therefore, it is easy to understand that the kinetic energy dominates even after the bubble collapse.

To get a comprehensive understanding of the distance effects, in Figure 7 we present the variations of wave energy with time for various distances corresponding to Table 1. When the bubble is located far away from the rigid boundary, e.g., $\gamma = 12.5$, the wave energy reaches the maximum value of 0.95 mJ at $t = 1.85$ ms within a very short period of 0.1 ms, and with a very sharp peak. As the distance is reduced, the peak value of wave energy decreases accordingly, and the sharp peak is replaced by a broad distribution of the high wave energy. In specific, the peak value of wave energy for $\gamma = 1.5$ is around 0.3 mJ, only one-third of that for $\gamma = 12.5$. It is interesting to find that there are secondary peaks in the wave energy distributions at $\gamma = 2.5$–4.0. We believe that these secondary peaks are induced by the pressure waves reflected by the solid boundary, which are difficult to be distinguished as we further reduce the distance. The comparison between small and large distances suggests that the near-wall collapses generate less wave energy due to the non-spherical characteristics, but with a longer duration in the computational domain.

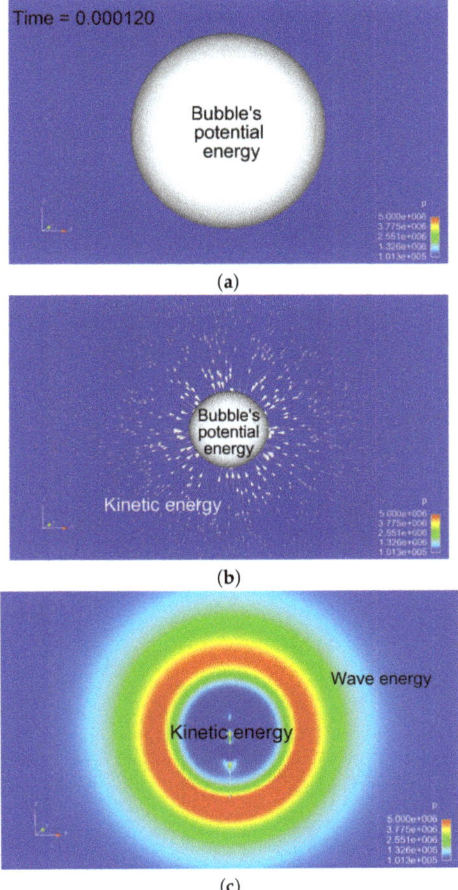

(a)

(b)

(c)

Figure 6. Flow fields with pressure contours for three time instants at $\gamma = 12.5$, demonstrating different dominant forms of energy: (**a**) potential energy, (**b**) kinetic energy and (**c**) wave energy.

Figure 8 shows the variations of kinetic energy with time for various distances. Again, their overall trends are the same, as we have discussed in Figure 5, and they are consistent with the wave energy variations shown in Figure 7. There are also some slight differences among various distances. In short, the wall effects on kinetic energy can be concluded as: first, the drop of kinetic energy is delayed as the bubble is placed closer to the wall. Second, the sharp peak is replaced by broader distributions of the kinetic energy as the bubble stays closer to the wall. Both conclusions are consistent with the wave energy variations, as shown in Figure 7.

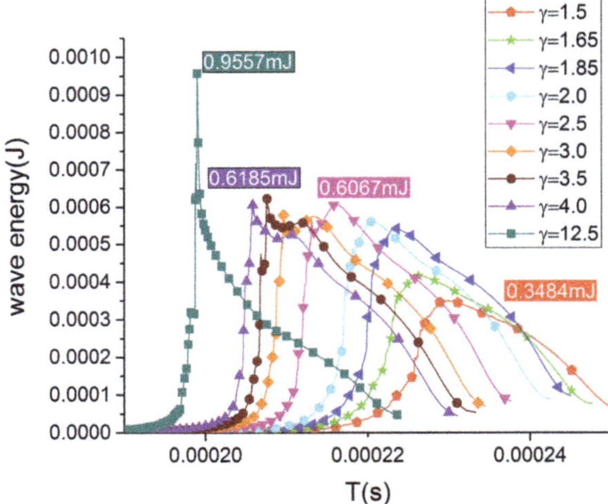

Figure 7. Wave energy variations in the flow fields at different values of γ.

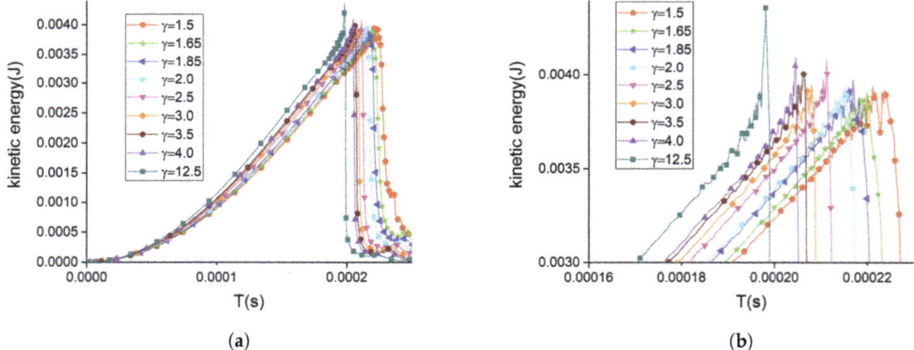

Figure 8. (**a**) Kinetic energy variations in the flow fields at different values of γ, and the enlarged views around the peak regions are shown in (**b**).

3.2. Collapses of Multiple Bubbles

The asymmetric bubble collapse can be brought by placing a solid wall, as we have studied in the last section for a single bubble, it can also be caused by the presence of nearby bubbles. In this section, we study the collapses of multiple bubbles, or more specifically, 8 bubbles, which are placed in a cubic region with two layers, and 4 bubbles on each layer.

In Table 2, we present the peak values of both the kinetic and wave energies, as well as their conversion rates. The value of E_k slightly grows with the distance γ, while η decreases as γ increases. In specific, the energy conversion rate at $\gamma = 1.5$ is approximately twice of that at $\gamma = 11.25$.

To examine the dynamic evolutions of individual bubbles, in Figures 9 and 10, we present the instantaneous shapes of the bubbles in three subsequent time instants. As shown in Figures 9a and 10a for the initial bubble configurations, the two cases represent respectively the small ($\gamma = 1.5$) and large ($\gamma = 11.5$) distances of the bubbles away from the solid wall. For the small distance, as shown in Figure 9, the wall effect leads to asymmetric collapses of the bubble cluster, with the bubbles in the upper layer collapsing faster than that in the lower layer, due to the solid wall below the lower layer. While at $\gamma = 12.5$, as shown in Figure 10, the asymmetry is brought only by the presence of nearby bubbles, therefore, the collapses and the evolutions of bubbles are symmetric with respect to the symmetry plane between the two layers, which is a straightforward demonstration that the solid wall does not affect the bubble collapse markedly when they are spaced sufficiently far away. The two cases differ in the focal point. For the large distance case, the jetting flows from all bubbles point to the centre of the bubble cluster, while for the small distance case, the focal point shifts towards the solid wall.

Table 2. Energies and their conversion rates for multiple bubbles.

γ	E_k (mJ)	E_{wave} (mJ)	η (%)
1.5	27.96	2.97	11.31
2.0	28.65	2.96	11.27
3.75	29.91	2.09	7.96
6.25	29.28	2.03	7.73
11.25	30.04	1.40	5.33

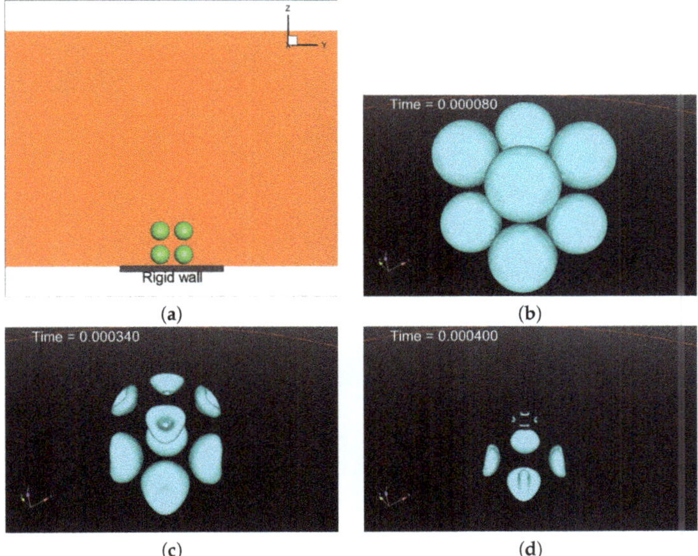

Figure 9. Deformations of multiple bubbles at $\gamma = 1.5$ for (**a**) the initial shapes (side view), (**b**) T = 80 µs, (**c**) T = 340 µs and (**d**) T = 400 µs.

Figure 11a presents the potential energy variations of multiple bubbles. The drops of potential energy slow down at the small distances, which is apparent, because the collapses of bubbles have been delayed due to the confinement of the solid wall. Comparing to the single bubble, for the multiple bubbles, the solid wall plays a similar role, as the bubbles are very close to the wall. However, the trend of wave energy conversion is quite different with that of the single bubble, which might be raised by the strongly nonlinear interactions between the bubbles. Similar comparisons can be made in the kinetic energy variations, as shown in Figures 8b and 11b.

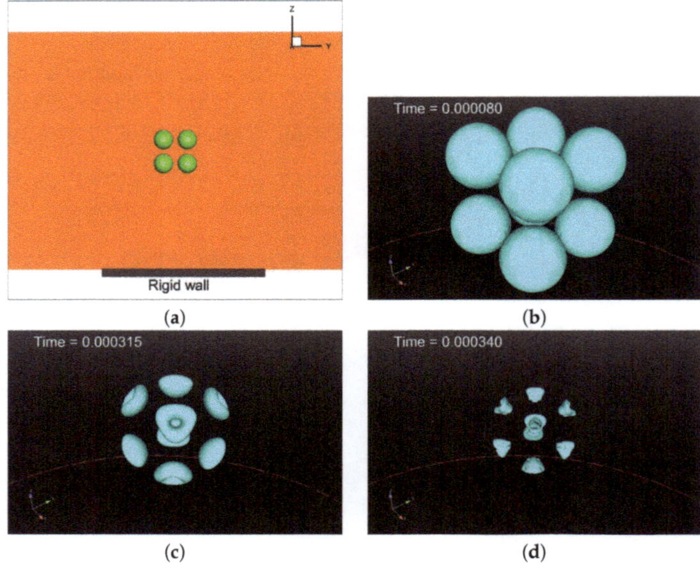

Figure 10. Deformations of multiple bubbles at γ = 11.5 for (**a**) the initial shapes (side view), (**b**) T = 80 µs, (**c**) T = 315 µs and (**d**) T = 340 µs.

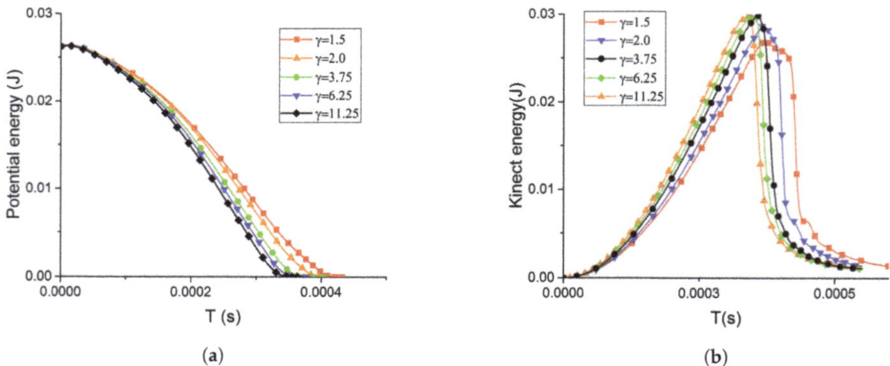

Figure 11. (**a**) Potential energy and (**b**) kinetic energy variations in the flow fields for the collapses of multiple bubbles.

The most remarkable difference between the single bubble and the multiple bubbles lies in the wave energy. Figure 12 presents the wave energy variations in the simulations of multiple bubbles. In contrast to that of the single bubble (see Figure 7), here, the ascending and descending branches of the wave energy variations are nearly symmetric with respect to the peak location. The peak values of wave energy for the small distance cases are higher than that of the large distance cases, resulting in the same behaviour for the energy conversion rate. This relationship between the peak wave energy is opposite to that of the single bubble.

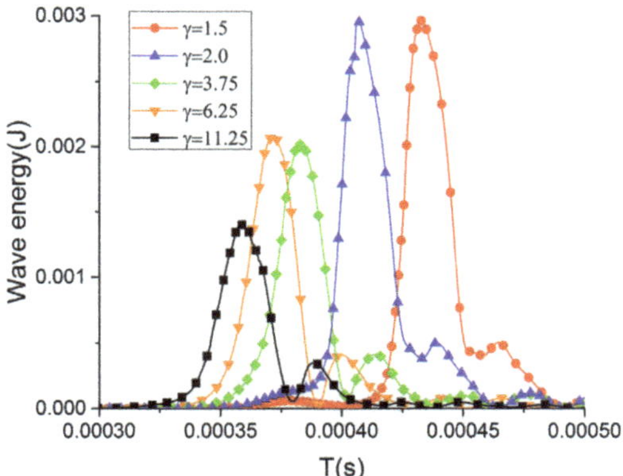

Figure 12. Variations of wave energy in the flow fields during the collapsed of multiple bubbles.

4. Conclusions

In the present work, we carry out numerical simulations to study the collapses for both a single bubble and a bubble cluster with 8 bubbles. Different forms of energy are evaluated to understand the underlying physical mechanisms of bubble collapse.

The first part concentrates on a single bubble collapsing near a solid wall. The intuitive physical rule tells that the energy should be conserved, which has indeed been observed during the first stage of collapse. As the bubble shrinks, the potential energy determined by the vapour volume converts into kinetic energy. Further, as the bubble vanishes, a part of kinetic energy starts to convert into other energy forms such as wave energy, or is lost due to the thermal and acoustic damping mechanisms, which have unfortunately not been considered in the current simulations. We report that a rigid boundary can affect the process of bubble collapse by deforming the bubble into non-spherical shapes, delaying the collapse, and consequently decreasing the conversion rate.

For a cluster consisting of 8 bubbles, their collapses are more complicated than the single bubble case, because the individual bubbles can be affected by both the solid wall and the surrounding bubbles, and both can break the symmetry of their geometric configurations. By examining the evolutions of individual bubbles during the collapse, we observe that when the bubbles are far away from the solid wall, the jetting flows from all bubbles point towards the cluster centre, while the focal point shifts towards the solid wall as the cluster is located very close to the wall. Moreover, we find that the collapse differentiates from the single bubble case mainly in the relationship between the bubble-wall distance and the wave energy, or equivalently the energy conversion rate.

Author Contributions: Conceptualization, J.Z. and L.Z.; Methodology, L.Z. and J.Z.; Investigation, J.Z.; Data Curation, J.Z.; Writing-Original Draft Preparation, J.Z. and L.Z.; Writing-Review and Editing, J.D.; Visualization, J.Z.; Funding Acquisition, L.Z. and J.D.

Funding: This research was funded by the National Natural Science Foundation of China (No. 11772298, No. 11272284) and the State Key Program of National Natural Science of China (No. 11332009).

Conflicts of Interest: The authors declare no conflict of interest.

References

1. Ohl, S.W.; Klaseboer, E.; Khoo, B.C. Bubbles with shock waves and ultrasound: A review. *Interface Focus* **2015**, *5*, 20150019. [CrossRef] [PubMed]
2. Lordrayleigh, O.M.F.R.S. VIII. On the pressure developed in a liquid during the collapse of a spherical cavity. *Philos. Mag.* **1917**, *34*, 94–98.
3. Plesset, M.S. The Dynamics of Cavitation Bubbles. *J. Appl. Mech.* **1949**, *16*, 277–282. [CrossRef]

4. Nigmatulin, R.; Khabeev, N. Dynamics of vapor-gas bubbles. *Fluid Dyn.* **1976**, *11*, 867–871. [CrossRef]

5. Nigmatulin, R.; Khabeev, N. Heat exchange between a gas bubble and a liquid. *Fluid Dyn.* **1974**, *9*, 759–764. [CrossRef]

6. Ida, M.; Naoe, T.; Futakawa, M. Suppression of cavitation inception by gas bubble injection: A numerical study focusing on bubble-bubble interaction. *Phys. Rev. E* **2007**, *76*, 046309. [CrossRef] [PubMed]

7. Fuster, D.; Colonius, T. Modelling bubble clusters in compressible liquids. *J. Fluid Mech.* **2011**, *688*, 352–389. [CrossRef]

8. Chahine, G.L. Numerical simulation of bubble flow interactions. *J. Hydrodyn. Ser. B* **2009**, *21*, 316–332. [CrossRef]

9. Alloncle, A.; Dufresne, D.; Autric, M. Visualisation of laser-induced vapor bubbles and pressure waves. In *Bubble Dynamics and Interface Phenomena*; Springer: Dordrecht, The Netherlands, 1994; pp. 365–371.

10. Brujan, E.A.; Keen, G.S.; Vogel, A.; Blake, J.R. The final stage of the collapse of a cavitation bubble close to a rigid boundary. *Ultrason. Sonochem.* **2011**, *18*, 59–64. [CrossRef]

11. Dular, M.; Coutier-Delgosha, O. Thermodynamic effects during growth and collapse of a single cavitation bubble. *J. Fluid Mech.* **2013**, *736*, 44–66. [CrossRef]

12. Field, J.E. *Experimental Studies of Bubble Collapse*; Springer: Dordrecht, The Netherlands, 1994; pp. 17–31

13. Fujikawa, S.; Akamatsu, T. Effects of the non-equilibrium condensation of vapour on the pressure wave produced by the collapse of a bubble in a liquid. *J. Fluid Mech.* **1980**, *97*, 481–512. [CrossRef]

14. Goh, B.H.T.; Ohl, S.W.; Klaseboer, E.; Khoo, B.C. Jet orientation of a collapsing bubble near a solid wall with an attached air bubble. *Phys. Fluids* **2014**, *26*, 221–240. [CrossRef]

15. Isselin, J.C.; Alloncle, A.P.; Autric, M. On laser induced single bubble near a solid boundary: Contribution to the understanding of erosion phenomena. *J. Appl. Phys.* **1998**, *84*, 5766–5771. [CrossRef]

16. Kapahi, A.; Hsiao, C.T.; Chahine, G.L. *Shock-Induced Bubble Collapse versus Rayleigh Collapse*; IOP Publishing: Bristol, UK, 2015; p. 012128.

17. Philipp, A.; Lauterborn, W. Cavitation erosion by single laser-produced bubbles. *J. Fluid Mech.* **2000**, *361*, 75–116. [CrossRef]

18. Shaw, S.J.; Schiffers, W.P.; Gentry, T.P.; Emmony, D.C. The interaction of a laser-generated cavity with a solid boundary. *J. Acoust. Soc. Am.* **2000**, *107*, 3065. [CrossRef] [PubMed]

19. Shutler, N.D. A Photographic Study of the Dynamics and Damage Capabilities of Bubbles Collapsing Near Solid Boundaries. *J. Fluids Eng.* **1965**, *87*, 511. [CrossRef]

20. Smith, W.R.; Wang, Q.X. Viscous decay of nonlinear oscillations of a spherical bubble at large Reynolds number. *Phys. Fluids* **2017**, *29*, 082112. [CrossRef]

21. Sugimoto, Y.; Yamanishi, Y.; Sato, K.; Moriyama, M. Measurement of bubble behavior and impact on solid wall induced by fiber-holmium: YAG laser. *J. Flow Control Meas. Vis.* **2015**, *3*, 135–143. [CrossRef]

22. Supponen, O.; Kobel, P.; Obreschkow, D.; Farhat, M. The inner world of a collapsing bubble. *Phys. Fluids* **2015**, *27*, 94–98. [CrossRef]

23. Zhang, S.; Duncan, J.H.; Chahine, G.L. The final stage of the collapse of a cavitation bubble near a rigid wall. *J. Fluid Mech.* **2006**, *257*, 147–181. [CrossRef]

24. Wang, Z.; Pecha, R.; Gompf, B.; Eisenmenger, W. Single bubble sonoluminescence: Investigations of the emitted pressure wave with a fiber optic probe hydrophone. *Phys. Rev. E* **1999**, *59*, 1777. [CrossRef]

25. Lauterborn, W.; Vogel, A. *Shock Wave Emission by Laser Generated Bubbles*; Springer: Berlin, Heidelberg, 2013; pp. 67–103.

26. Lauterborn, W. Experimental investigation of cavitation-bubble collapse in the neighbourhood of a solid boundary. *J. Fluid Mech.* **1975**, *72*, 391–399. [CrossRef]

27. Lauterborn, W.; Kurz, T. Physics of bubble oscillations. *Rep. Prog. Phys.* **2010**, *73*, 106501. [CrossRef]

28. Vogel, A.; Lauterborn, W.; Timm, R. Optical and acoustic investigations of the dynamics of laser-produced cavitation bubbles near a solid boundary. *J. Fluid Mech.* **2006**, *206*, 299–338. [CrossRef]

29. Staudenraus, J.; Eisenmenger, W. Fibre-optic probe hydrophone for ultrasonic and shock-wave measurements in water. *Ultrasonics* **1993**, *31*, 267–273. [CrossRef]

30. Gonzalez Avila, S.R. The acoustic pressure generated by the non-spherical collapse of laser-induced cavitation bubbles near a rigid boundary. In Proceedings of the 10th International Symposium on Cavitation, Baltimore, MD, USA, 14–16 May 2018.

31. Merouan Hamdi, O.C.D. Measurements of the temperature variations during the growth and collapse of cavitation bubbles. In Proceedings of the 10th International Symposium on Cavitation, Baltimore, MD, USA, 14–16 May 2018.
32. Fortes-Patella, R.; Challier, G.; Reboud, J.L.; Archer, A. Energy Balance in Cavitation Erosion: From Bubble Collapse to Indentation of Material Surface. *J. Fluids Eng.* **2014**, *135*, 011303. [CrossRef]
33. Bui, T.T.; Ong, E.T.; Khoo, B.C.; Klaseboer, E.; Hung, K.C. A fast algorithm for modeling multiple bubbles dynamics. *J. Comput. Phys.* **2006**, *216*, 430–453. [CrossRef]
34. Han, B.; Köhler, K.; Jungnickel, K.; Mettin, R.; Lauterborn, W.; Vogel, A. Dynamics of laser-induced bubble pairs. *J. Fluid Mech.* **2015**, *771*, 706–742. [CrossRef]
35. Bremond, N.; Arora, M.; Ohl, C.D.; Lohse, D. Controlled multibubble surface cavitation. *Phys. Rev. Lett.* **2006**, *96*, 224501. [CrossRef]
36. Wang, Q. Multi-oscillations of a bubble in a compressible liquid near a rigid boundary. *J. Fluid Mech.* **2014**, *745*, 509–536. [CrossRef]
37. Chahine, G.L.; Duraiswami, R. Dynamical Interactions in a Multi-Bubble Cloud. *ASME J. Fluids Eng.* **1992**, *114*, 680–686. [CrossRef]
38. Tomita, Y.; Robinson, P.B.; Tong, R.P.; Blake, J.R. Growth and collapse of cavitation bubbles near a curved rigid boundary. *J. Fluid Mech.* **2002**, *466*, 259–283. [CrossRef]
39. Brujan, E.A.; Ikeda, T.; Matsumoto, Y. Shock wave emission from a cloud of bubbles. *Soft Matter* **2012**, *8*, 5777–5783. [CrossRef]
40. Tiwari, A.; Pantano, C.; Freund, J.B. Growth-and-collapse dynamics of small bubble clusters near a wall. *J. Fluid Mech.* **2015**, *775*, 1–23. [CrossRef]
41. Kling, C.L.; Hammitt, F.G. A photographic study of spark-induced cavitation bubble collapse. *J. Borderl. Stud.* **1970**, *94*, 75–90. [CrossRef]
42. Vogel, A.; Lauterborn, W. Time-resolved particle image velocimetry used in the investigation of cavitation bubble dynamics. *Appl. Opt.* **1988**, *27*, 1869–1876. [CrossRef]
43. Ward, B.; Emmony, D.C. The Energies and Pressures of Acoustic Transients Associated with Optical Cavitation in Water. *Opt. Acta Int. J. Opt.* **1990**, *37*, 803–811. [CrossRef]

Article

Numerical Analysis on the Hydrodynamic Performance of an Artificially Ventilated Surface-Piercing Propeller

Dongmei Yang *, Zhen Ren, Zhiqun Guo and Zeyang Gao

College of Shipbuilding Engineering, Harbin Engineering University, Harbin 150001, China; forfanqiang@126.com (Z.R.); guozhiqun@hrbeu.edu.cn (Z.G.); gaozeyangpc@hotmail.com (Z.G.)
* Correspondence: yangdongmei@hrbeu.edu.cn; Tel.: +86-594-8258-8360

Received: 11 September 2018; Accepted: 18 October 2018; Published: 23 October 2018

Abstract: When operated under large water immersion, surface piercing propellers are prone to be in heavy load conditions. To improve the hydrodynamic performance of the surface piercing propellers, engineers usually artificially ventilate the blades by equipping a vent pipe in front of the propeller disc. In this paper, the influence of artificial ventilation on the hydrodynamic performance of surface piercing propellers under full immersion conditions was investigated using the Computational Fluid Dynamics (CFD) method. The numerical results suggest that the effect of artificial ventilation on the pressure distribution on the blades decreases along the radial direction. And at low advancing speed, the thrust, torque as well as the efficiency of the propeller are smaller than those without ventilation. However, with the increase of the advancing speed, the efficiency of the propeller rapidly increases and can be greater than the without-ventilation case. The numerical results demonstrates the effectiveness of the artificial ventilation approach for improving the hydrodynamic performance of the surface piercing propellers for high speed planning crafts.

Keywords: surface-piercing propeller; artificial ventilation; hydrodynamic performance; numerical simulation

---

1. Introduction

Surface-piercing propellers (SPPs) are also known as Surface penetrating propellers. The SPP is so named because part of the propeller is above the water surface and the rest is under the water during normal operation. Compared with the conventional propellers, the SPPs mainly have the following three advantages (Ding et al. [1]): (1) the resistance of the appendages, such as the paddle shaft and the shaft bracket, is minimized; (2) propeller diameter is no longer limited by some parameters such as soak depth and stern frame; and (3) the cavitation erosion on the blade surface is substantially reduced. Due to these advantages, the SPPs become an optimal choice for high-speed boats and some shallow draft ships, and thus have really good application prospects.

However, in some cases the SPPs might be operated under the large water immersion, e.g., the boats are at the bow-up status, which makes the SPPs be in heavy load conditions. One of the practical approaches to solve this problem is setting an aeration pipe in front of SPPs to ventilate the blades. This approach has been adopted in some actual boats and achieved significant results. Nonetheless, few attentions have been paid on how the ventilation of blades improves the hydrodynamic characteristics of SPPs. To this end, this paper is dedicated to investigate the influence of the ventilation of blades on the hydrodynamic performance of SPPs using numerical methods.

In some simplified studies, the surface-piercing process of an SPP blade can be deemed as the water-entry process of a 2D profile, which is easy to be understood. Zhao et al. [2] used the non-linear boundary element method to investigate the characteristics of the sprays generated by the water-entry of wedges with different bottom-angles. It was found that the increase of the wetted area directly leads to a great slamming pressure on the wetted surface. And when the slope angle of the bottom is larger than 30°, there is no typical slamming pressure concentration phenomenon due to the smaller wetted surface. Young et al. [3], in his doctoral dissertation, carried out similar studies on the water-entry of wedges, and analyzed the pressure distribution on the wetted surface of a flat plate under different entry angles.

Yari et al. [4] conducted a preliminary study on a wedge entering the water under different inclination angles. The obtained pressure distribution, free surface deformation, and so on agree well with the experimental data. Yu et al. [5] carried out a numerical study on the water entry problem of a 2D cross-section profile, and found that the cup shape of the trailing edge can enhance the hydrodynamic load on the edge, while reducing the lift-to-drag ratio as well as the efficiency of the propeller.

Ghassemi et al. [6] investigated the hydrodynamics of SPPs named SPP-1 and SPP-2 under full and half immersed conditions using a boundary element method (BEM). The numerical results agree well with experimental ones. In the study, they found that the Weber number has significant influence on the ventilation status of the SPP. Kinnas et al. [7] studied the bubble flow around the hydrofoil and SPPs using a BEM, in which they investigated the hydrodynamic effects of the bubble flow on the hydrofoil and SPPs with various blade profiles. Young et al. [8,9] carried out a series of studies on large-scale SPPs and super-vacuum paddles also by using the BEM, which contributed to the knowledge of the hydrodynamics of these propellers.

With the development of the computer technology, the CFD methods become popular and play the role of benchmark in the numerical studies on surface-piercing propellers. Young et al. [3] predicted the hydrodynamic performance of large-scale surface paddles using a coupled boundary element method–finite element method (BEM–FEM), and found that the numerical results compared well with those obtained by the Reynolds Averaged Navier–Stokes (RANS) solver (Fluent software), in which the Reynolds stress tensor was modeled using the SST form of the $k - \omega$ turbulence model. Shi et al. [10] investigated the wake of surface-piercing propellers, as well as the pulsating pressure on the paddles and the deformation of the free surface after the paddles. Alimirzazadeh et al. [11] employed the OpenFOAM software to study the hydrodynamic performance of SPP-841B surface paddle under different sway angles and immersion depths. In the study, the $k - \omega$ based Shear Stress Transport (SST) model with automatic wall functions (mixed formulation) was employed. The numerical results agree with experimental ones well under different immersion depths, while poorly under different sway angles. They found that increasing the yaw angle would decrease the thrust coefficient and torque coefficient, while the efficiency was improved. In addition, the maximum efficiency point has not been achieved at the zero shaft yaw angle, due to the implementation of SPPs.

Obviously, in the existing works the most attention has been paid to investigating the hydrodynamic performance of SPPs under the natural operation conditions, while very little attention has been paid on the artificial ventilated SPPs. It is believed that the hydrodynamic characteristics of the artificial ventilated SPPs should be significantly different from the without-ventilation case. Thereby, we were motivated to study the SPPs under artificial ventilation conditions. In this paper, a right-handed three-blade paddle was proposed for the investigation, and an aeration pipe was installed in front of the paddle disk for the ventilation purpose. The numerical simulation was carried out based on the commercial CFD software Star-CCM+ [12]. The rotation of the blades was realized by using the overlapped mesh technique.

2. SPP Model and Numerical Setup

The submergence ratio of SPPs is defined as $It = h/D$, where h denotes the immersion depth of the propeller disc, D is the diameter of the propeller, as shown in Figure 1.

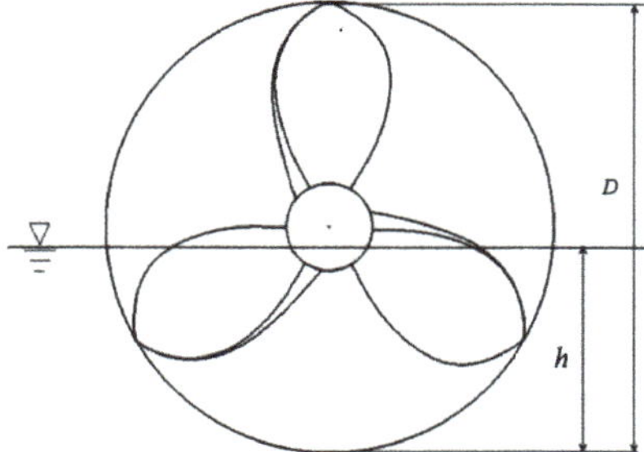

Figure 1. The submergence ratio of surface-piercing propellers (SPPs), which is defined as the ratio of the immersion depth h to the propeller diameter D.

Generally, the SPPs work on the water-air interface, and the blades rush into and out of the water in a staggered manner. In this process, the air is sucked into the water forming an air cavity, which connects the atmosphere and the paddle. This phenomenon is known as the ventilation of SPPs.

When the submergence ratio is sufficient large, the torque on the paddle will dramatically increase. For the adjustable SPPs, the torque can be reduced by decreasing the immersion depth. For the fixed SPPs, however, it is commonly difficult to adjust the immersion depth. Although the paddle may pump some air into water when it rotates, the torque can still be very large. For the purpose of torque reduction, an aeration pipe is generally installed in front of the paddle to ventilate the blades and thus reduce the torque.

2.1. SPP Model

The SPP proposed for the calculation is the SPP-1 type surface-piercing propeller model given in Table 1 [6], with the rotation direction being right-handed and the profile being S-C type. The diameter of the aeration pipe is 50 mm, and the distance from the outlet of the aeration pipe to the paddle surface is 100 mm. The sketch of the propeller and the vent pipe are shown in Figure 2.

Table 1. SPP-1 propeller geometric parameters [6].

Parameters	
Diameter D	0.2
Tilt angle/°	10
Number of leaves Z	3
Pitch ratio P/D	1.6
Disk ratio AE/AO	0.5
Hub diameter ratio d/D	0.2
Side angle/°	0
Profile	S-C

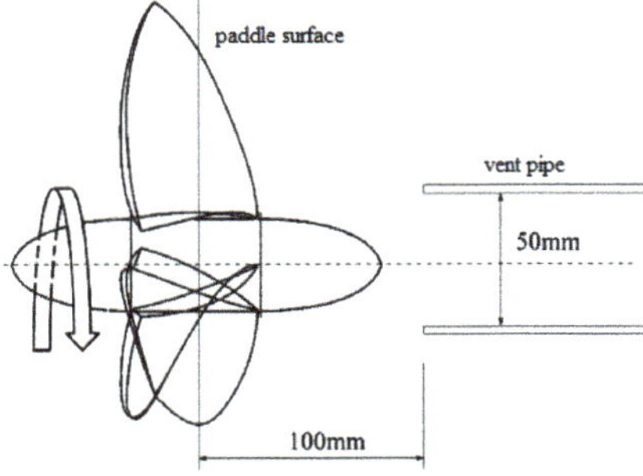

Figure 2. The sketch of the propeller and the vent pipe.

2.2. Control Equation

Under the full immersion condition, ventilation through the aeration pipe allows the paddle always being in an unsteady air-water two-phase flow field. The VOF (volume of fluid) model [13] was selected for capturing the interface between water and air. Air and water in the flow field are deemed incompressible.

Mass conservation equation can be written as:

$$\frac{\partial p}{\partial t} + \nabla(\rho \vec{\mu}) = S_m \tag{1}$$

The above equation is a general expression of the mass conservation equation, where the term S_m can be any user-defined source term added to the continuity term.

The momentum equations are given as:

$$\frac{\partial(\rho \mu_i)}{\partial t} + \frac{\partial(\rho \mu_i \mu_j)}{\partial x_j} = -\frac{\partial p}{\partial x_i} + \frac{\partial}{\partial x_j}\left(\mu \frac{\partial \mu_i}{\partial x_j} - \rho \overline{\mu_i' \mu_j'}\right) + S_i \tag{2}$$

where μ_i is the velocity component in the xi direction ($i, j = 1, 2, 3$) of the Cartesian coordinate system, p the fluid pressure, ρ the fluid density, μ the dynamic viscosity coefficient, t the time, and S_i the volume force.

In this paper, the $k - \omega$ SST [14,15] was employed as the turbulence model, which could take the transport characteristics of turbulent shear forces into account, and make possible to obtain more accurately results of the flow separation in the counter pressure gradient region [16].

2.3. Computational Domain and Grid Division

In order to make the flow field around the propeller be fully developed, and the numerical results be accurate and reliable, the size of the flow domain in the simulation should be large enough. In practice, the entire flow domain is divided into two parts: a cylindrical rotating zone containing the paddle and an outer cylindrical stationary zone. The outer diameter of the stationary zone is 5.0*D*, the inlet is 3.5*D* from the paddle plane, and the outlet is 7.0*D* from the paddle plane. The diameter of the rotation domain is 1.2*D*, and both end surfaces of the cylinder are 0.45*D* from the paddle surface. Figure 3 shows an oblique view of the propeller's computational domain.

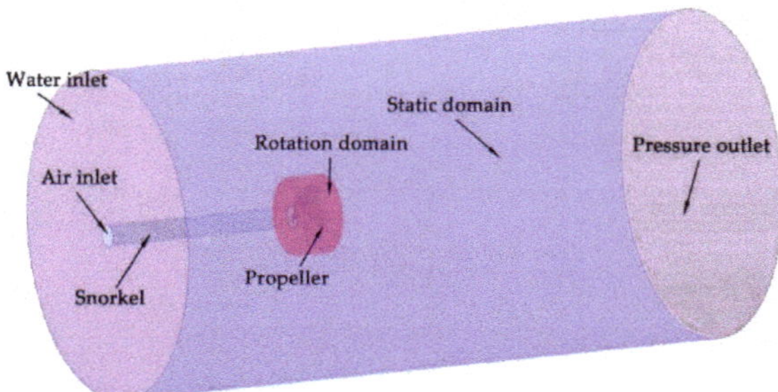

Figure 3. Oblique view of propeller and computational field.

To simulate the rotation of the paddle in the flow field, the overlapped mesh method was employed. At the overlapping domain between the stationary and the rotational domains, the information of the flow field is exchanged through the overlapping grids. The mesh in the overlapping domain should be the same size as much as possible. We set the polyhedral mesh in the inner rotational domain, while the cutting body mesh in the outer stationary domain. Moreover, local mesh densification was performed around the overlapping domain. Finally, the total number of grids is 3.5 million, in which the rotation domain is 1.7 million, and the stationary domain is 1.8 million. The min and max size of computational grids are 7.65×10^{-4} mm and 11 mm, respectively. The surface mesh is shown in Figure 4.

Figure 4. Mesh on the propeller and aeration pipe.

2.4. Grid Convergence

Let $Y+$ be the non-dimensional wall distance defined as $Y+ \equiv u_*y/\nu$, where u_* is the friction velocity at the nearest wall, y the distance to the nearest wall, and ν the local kinematic viscosity of the fluid. In this sub-section the sensitivity of numerical results with respect to $Y+$ is studied. Generally, the range of dimensionless wall distance $Y+ \approx 30 \sim 100$ is acceptable for blade surfaces. Table 2 lists the error of the thrust coefficient K_t and torque coefficient $10K_q$ compared with the test value with different $Y+$ value. It can be seen that three errors reach minimum when the $Y+$ is equal to 60. Therefore, we set $Y+ = 60$ in the meshing setup.

Table 2. Results for the $Y+$ for selection.

$Y+$	K_t	$10K_q$	ErrorK_t	Error$10K_q$
Exp.	0.3093	0.8579		
30	0.2759	0.7758	-10.81%	-9.57%
40	0.2791	0.7840	-9.77%	-8.61%
50	0.2825	0.7956	-8.69%	-7.26%
60	0.2857	0.8016	-7.65%	-6.56%
70	0.2836	0.7977	-8.33%	-7.01%
80	0.2800	0.7843	-9.48%	-8.57%
90	0.2786	0.7779	-9.95%	-9.32%

Grid convergence study is important in the mesh generation process. Here four grid numbers, 1.77 million (G1), 2.5 million (G2), 3.5 million (G3), and 5 million (G4) are selected for the convergence study. The grid number and simulation results are given in Table 3. From these results, it is clearly noticed that with the growth of grid number, the results get closer to the experimental ones, i.e., the grid convergence can be guaranteed in these numbers. However, the greater the grid number, the lower computational efficiency for the numerical simulation. It is found that the accuracy was only slightly improved in the G4 case at the expense of a lot of calculation time. Hence, when determining the grid number, one should make a compromise between accuracy and efficiency. In this paper, the fine mesh of G3 was adopted in the present simulations, which achieves both computational efficiency and accuracy.

Table 3. Results from the grid convergence.

Grid	Control Volumes	K_t	εK_t	$10K_q$	εK_q
G1	1,770,000	0.2756	10.89%	0.7761	9.53%
G2	2,500,000	0.2806	9.28%	0.7874	8.21%
G3	3,500,000	0.2857	7.63%	0.8016	6.65%
G4	5,000,000	0.2884	6.77%	0.8070	5.93%

2.5. Boundary and Initial Conditions

Due to the air-suck phenomenon, the flow around the surface-piercing paddle is usually an air-water two-phase unsteady one. The density of water and air in the simulation is 997.56 kg/m^3 and 1.18 kg/m^3, respectively. The dynamic viscosity of water and air is 8.89 $\times$ 10^{-4} Pa·s and 1.86 $\times$ 10^{-5} Pa·s, respectively. The VOF approach was employed to track the air-water interface. The principle of this approach is to determine the interface by counting the volume ratio function of the fluid in each cell rather than the motion of the particle on the free surface.

The fluid density can be expressed as:

$$m\frac{dv}{dt} = \sum \alpha_m \rho_m \tag{3}$$

where α_m is the volume fraction of the mth type of fluid in each cell, which satisfies the following condition:

$$\sum_{m=1}^{n} \alpha_m = 1 \tag{4}$$

In this paper, velocity inlet was divided into air velocity inlet and water velocity inlet. Both inlets have the same velocity vector. The volume fraction of the overall inlet was set as the mixture of air and water fractions, where the volume fraction of air at the air velocity inlet was set to 1, and the water volume fraction was set to 0. The outlet is set as a pressure outlet and the standard atmospheric pressure is made as the reference pressure. The outer wall of the stationary zone is set as a plane of symmetry, and the surfaces of the surface-piercing paddle and the vent pipe are set to be non-slip and non-penetrable. The outer boundary of the rotation domain is set to overlapped grids.

The second-order schemes were applied for spatial discretization and linear interpolation, as well as the time integration. The time step is 6.94 $\times$ 10^{-5} s, during which the propeller rotates 1° under the given rotational speed. As suggested by Blocken and Gualtieri [17], this time step makes the condition CFL $\leq$ 1 satisfied. The numerical calculation was performed on a PC with an Intel Xeon CPU X5690 (6 cores, 3.46 GHz), and one case simulation approximately costs 40 h.

2.6. Validation of the Numerical Setup

To verify the aforementioned numerical setup, the hydrodynamic performance of a propeller in the open water was investigated and compared with experimental results from Ghassemi et al. [6]. The main parameters of the SPP-1 propeller are listed in Table 1. The hydrodynamic characteristics are plotted using solid lines (see Figure 5). Let T, Q, n, and V_A be the thrust, torque, rotational speed, and forward speed of the propeller, respectively. The relevant parameters for the propeller can be nondimensionalized as follows.

Speed coefficient:

$$J = \frac{V_A}{nD} \tag{5}$$

Thrust coefficient:

$$K_t = \frac{T}{\rho n^2 D^4} \tag{6}$$

Torque coefficient:

$$m\frac{dv}{dt} = f \tag{7}$$

Promote efficiency:

$$\eta = \frac{K_t}{K_q}\frac{J}{2\pi} \tag{8}$$

In the numerical and experimental setup, the submergence ratio for the propeller is $It = 1/3$, and the rotational speed is $n = 2400$ rpm. The speed coefficient J ranges from 0.85 to 1.45, which was obtained by varying the magnitude of the incoming flow velocity V_A.

By using the CFD solver, we obtained the thrust coefficient K_t, the torque coefficient $10K_q$ of the propeller, and the open water efficiency η. $10K_q$ is used due to the fact that K_q is one order smaller than K_t and η. Figure 5 compares the CFD results with the experimental ones given in Table 1 [6].

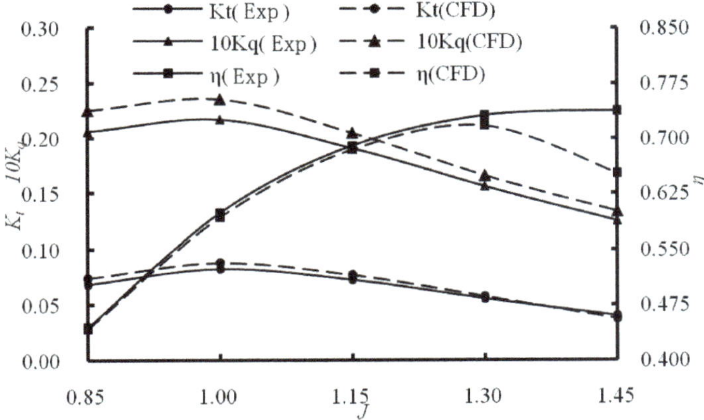

Figure 5. Comparison of CFD and experimental results at $It = 1/3$.

As shown in Figure 5, the thrust coefficient K_t obtained by the CFD is slightly larger than the experimental one when $J = 0.85$–1.30, and then the discrepancy gradually decreases with the growth of J. The torque coefficient $10K_q$ has a similar trend but the discrepancy between CFD and experimental results is slightly larger than that in the K_t case. In contrast, the open water efficiency η obtained by the CFD is always slightly smaller than the experimental one when $0.85 < J < 1.30$, but the error rapidly grows with J when $1.30 < J < 1.45$ and reaches its maximum 12% at $J = 1.45$. This study demonstrates that the numerical setup in the CFD solver is effective and can provide the required precision for the hydrodynamic simulation of the propeller with the existence of air-water interface.

3. Numerical Results of the Ventilated SPP

3.1. Calculation of Hydrodynamic Coefficient of Wigley Ship

Figure 6 depicts the wake of the ventilated SPP at $J = 1.15$, from which it can be seen that the majority of the air stream flow bypasses the hub, while the rest of the flow is swung away by the rotating blades. Figure 6 displays a snapshot of the air volume fraction (green color) flowing around the rotating propeller. One can find that the trailing edges are enclosed by air bubbles due to the relatively low pressure around them. The air bubbles on the trailing edges generally stretch from the root to the tip of the blades, though on the leading edges the bubbles may only concentrate around the root domain of the blades. In the wake, three helical air bubbles are forming and rotating about the centric air stream. In the near propeller F field, the helical air bubbles and the centric air stream are connected to each other. With the flow moving downstream, the helical air bubbles gradually separate from the centric air stream, and then both helical air bubbles and centric air stream keep breaking into smaller bubbles until all of them disappear in the far field. Through setting up the interface and defining two-phase flow displayable distribution, it also shows the air and water distribution on the two cross sections $x/D = -0.5$ and $x/D = -1.0$ after the paddle plane. The red and blue colors refer to air and water, respectively. It can be seen that the air-content ratio gradually decreases along the downstream direction of the wake.

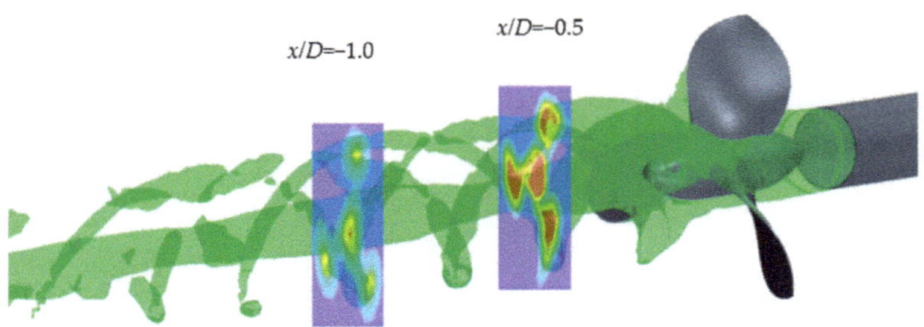

Figure 6. The wake after the ventilated propeller at $J = 1.15$ when $x/D = -1.0$ (left section) and $x/D = -0.5$ (right section).

3.2. Effect of Ventilation on the Hydrodynamic Performance of the SPP

Table 4 compares the numerical results for the ventilated SPP with the unventilated one under full submergence condition.

Table 4. Numerical results for the SPP under full submergence conditions.

Speed Coefficient J		Unventilated SPP			Ventilated SPP		
		K_t	$10K_q$	η	K_t	$10K_q$	η
0.85	Value	0.3763	0.9679	0.5259	0.3199	0.8452	0.5120
	Variation/%				−14.99	−12.68	−2.64
1.00	Value	0.3093	0.8579	0.5738	0.2749	0.7441	0.5880
	Variation/%				−11.12	−13.26	2.47
1.15	Value	0.2495	0.7502	0.6087	0.2308	0.6462	0.6537
	Variation/%				−7.49	−13.86	7.39
1.30	Value	0.1897	0.6426	0.6108	0.1867	0.5389	0.7168
	Variation/%				−1.58	−16.14	17.35

From Table 4, it can be seen that the thrust coefficient K_t and torque coefficient $10K_q$ of the propeller decrease after ventilation. With the growth of the speed coefficient J (from 0.85 to 1.30), the influence of ventilation on the thrust coefficient K_t gradually diminishes, while the effect on the torque coefficient $10K_q$ is reinforced. On the other hand, the efficiency η of the ventilated propeller is slightly less than the unventilated one under low speed coefficient J. However, with the growth of the speed coefficient J, the efficiency η of the ventilated propeller quickly increases as compared to the unventilated one.

As one sees, the thrust T and the torque Q decrease after ventilation. This is mainly due to the fact that the air bubbles attached to the surfaces of the blades after ventilation. However, with the growth of the speed factor, the effect of the ventilation on the thrust coefficient K_t is getting weaker, while the effect on the torque coefficient $10K_q$ becomes stronger, which makes the efficiency η of the ventilated propeller greater than the unventilated one.

3.3. Effect of Ventilation on the Pressure Distribution

Figure 7 portrays the pressure distribution cloud on the blades of the unventilated SPP at $J = 0.85$. Figure 7a,b show the pressure distribution on the pressure surface and the suction surface, respectively. The arrows in the figures indicate the rotational direction of the propeller. From Figure 7a, one notes that on the pressure surface the high pressure is mainly concentrated on the region near the leading edge of the blades. Obviously, the closer the leading edge, the higher the pressure. While the closer the trailing edge, the lower the pressure. On the thick trailing edge, the pressure may even be negative. In contrast, as shown in Figure 7b, the lower pressure mainly locates on the region near the leading edge of the blades.

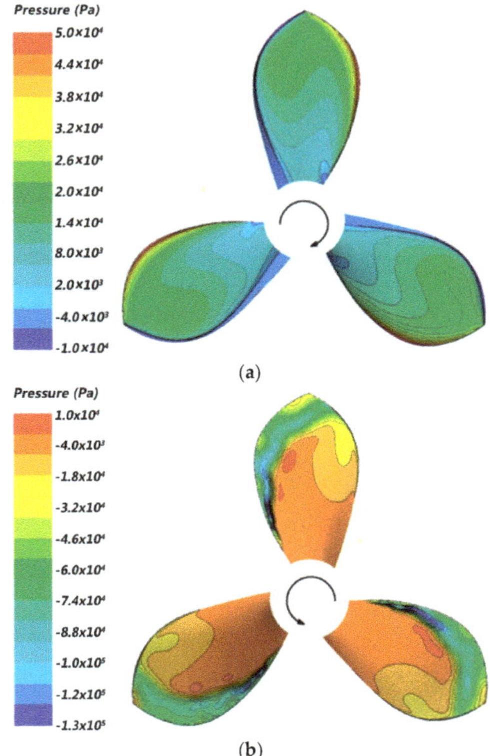

Figure 7. The pressure distribution on the blades at $J = 0.85$ under unventilated condition. (**a**) The pressure distribution on the pressure surface; (**b**) The pressure distribution on the suction surface.

As shown in Figure 7b, there exists a pressure jump near the leading edge, which is caused by the geometric characteristics of the SPP, i.e., the relative large pitch of the SPP induces a vortex formed after the leading edge, as shown in Figure 8. Figure 8a,b depict a non-dimensional velocity distribution cloud at the $0.5R$ section and a non-dimensional tangential velocity vector near the leading edge, respectively. The flow velocity is nondimensionalized with respect to the forward speed of the propeller V_A. From Figure 8a,b, one can find that there exists a high flow velocity zone on the suction surface near the leading edge. The local high velocity produces a sudden pressure drop on the suction surface. Moreover, it can be seen from Figure 8a that around the root of the thick trailing edge, the flow velocity is also very high, which makes the pressure in this zone be low and thus the air bubbles can adhere to this zone.

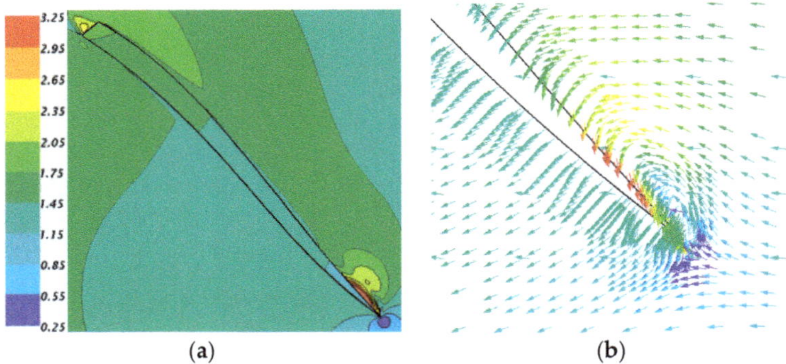

(a)   (b)

Figure 8. Non-dimensional velocity distribution cloud at 0.5*R* section. (**a**) The non-dimensional velocity cloud; (**b**) The non-dimensional velocity vector cloud near the leading edge.

Figure 9 shows the pressure distribution on the surfaces of the ventilated SPP. Figure 9a,b are the pressure distribution on the pressure surface and the suction surface, respectively. Comparing Figure 9a with Figure 7a, it can be found that the pressure on the trailing edge of the pressure surface increases after ventilating due to the fact that air bubbles adhere to it. On the other hand, comparing Figure 9b with Figure 7b, it can be found that the area of the high pressure zone on the suction surface is significantly increased. This is because of the adsorption of air on the suction surface after ventilation. The increase in the area of the high pressure zone on the suction surface results in a reduction in the thrust and torque of the SPP.

Figure 10 plots the pressure distribution at $J = 0.85$ on various profiles (0.24*R*, 0.50*R* and 0.70*R*) of the ventilated/unventilated SPP. C_p is the pressure coefficient. Blue and red lines refer to the pressure distribution profiles of the unventilated and ventilated SPP, respectively. From Figure 10a one can see that on the profile near the root (0.24*R*), when there is no ventilation, the difference between the pressure on the pressure surface and the one on the suction surface is very large, especially near the leading edge. After ventilating, the pressure on the suction surface of the blade significantly increases and evenly distributes along the entire chord direction. The pressure almost remains unchanged because the air bubbles completely cover the root region of the blades, and in the vicinity of the leading edge ($x/C \leq 0.1$), the pressure on the pressure surface and the suction surface are almost the same. This is because both the pressure surface and suction surface near the leading edge are covered by the air bubble. The same phenomenon can be observed around the trailing edge.

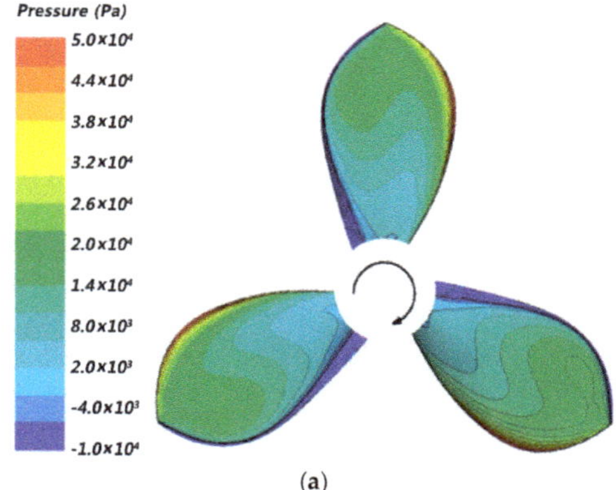

(a)

Figure 9. *Cont.*

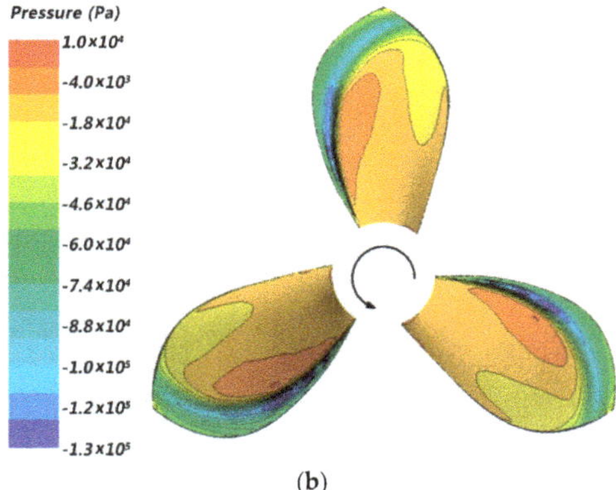

(**b**)

Figure 9. The pressure distribution on the blades at $J = 0.85$ under ventilated condition. (**a**) The pressure distribution on the pressure surface; (**b**) The pressure distribution on the suction surface.

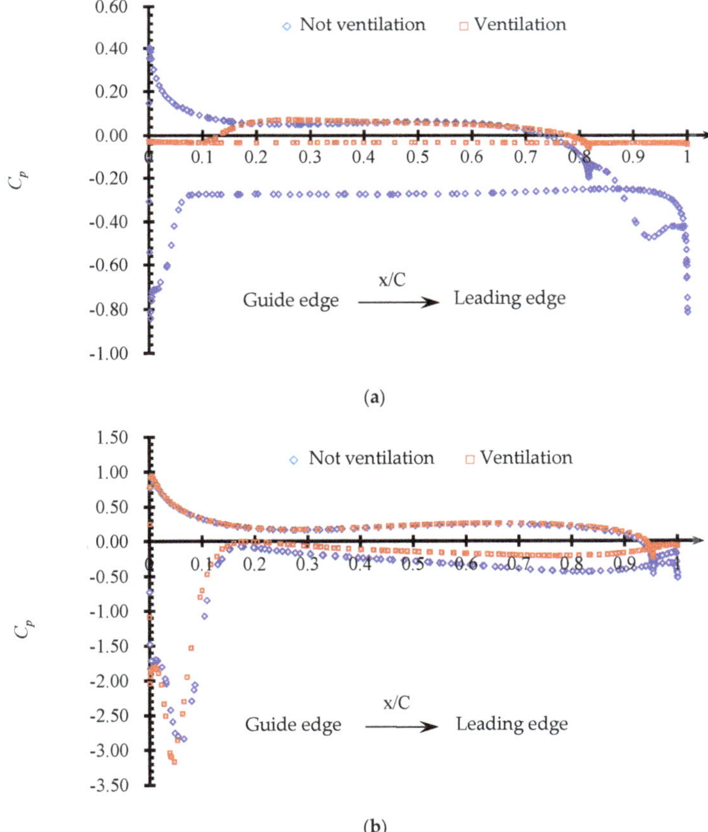

Figure 10. *Cont.*

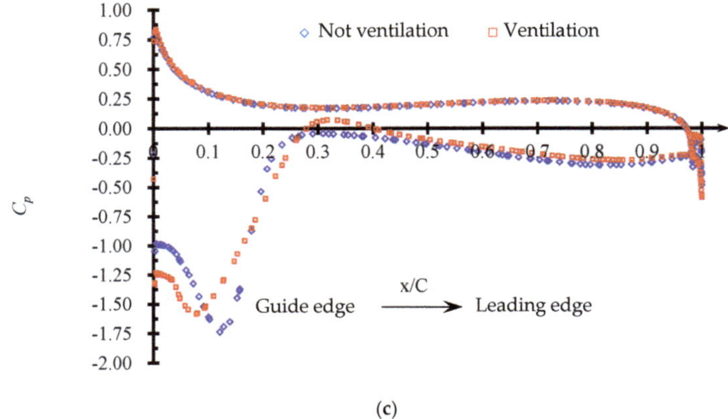

(c)

Figure 10. The pressure distribution profiles at $J = 0.85$. (**a**) The pressure distribution profile at $0.24R$; (**b**) The pressure distribution profile at $0.50R$; (**c**) The pressure distribution profile at $0.70R$.

As shown in Figure 10b, on the profile $0.50R$, the pressures on the pressure surface of the ventilated and unventilated SPP are almost the same. And only on the suction surface the pressure from the ventilated SPP is slightly bigger than the unventilated one due to the ventilation.

On the profile closer to the tip of the blade ($0.70R$), see Figure 10c, the pressure distribution on the pressure surface of the ventilated and unventilated SPP are completely the same. Only in the vicinity of the guide edge were the pressures on the suction surface of the ventilated and unventilated SPP slightly different, which is due to the non-uniformity of the flow field.

Comparing Figure 10a–c, one observes that the effect of ventilation on the pressure distribution of the SPP is limited to a zone in the vicinity of the blade root. In this zone, the ventilation has significant effect on the pressure. And along the outward radius direction, the effect of the ventilation gradually diminishes.

4. Conclusions

In this paper, the effect of artificial ventilation on the hydrodynamic performance of a surface-piercing propeller (SPP) was investigated using the CFD technique. The present work is different from previous works as follows. In previous works, such as Ghassemi et al. [6,7], Young et al. [9–11], and Alimirzazadeh et al. [13], the SPPs work under half immersed conditions with the propeller naturally ventilated. In such conditions, all blades experience periodic water-entry and water-exit processes. While in the present work, the SPP works under full immersion conditions with artificial ventilation through a vent pipe in front of the propeller disc. In such condition, all blades are partially surrounded by air bubbles, in which the phenomenon as well as the hydrodynamic performance of the SPP are significantly different from those given in literatures. The main conclusions are summarized as follows:

(1)   The turbulence model SST and the overlapped mesh method can make faithful simulation for the ventilated SPP in unsteady flow field, and precisely forecast its hydrodynamic performance.

(2)   In the wake of the ventilated SPP, there are three helical air bubbles that rotate about the centric air stream. With the flow moving downstream, the helical air bubbles and the centric air stream gradually break into smaller bubbles until all of them disappear in the far field.

(3)   The effect of ventilation on the pressure distribution of the SPP is limited to a zone in the vicinity of the blade root (less than $0.5R$). In this zone, the ventilation has significant effect on the pressure, and along the outward radius direction, the effect gradually diminishes.

(4)   The thrust coefficient (K_t) and torque coefficient ($10K_q$) of the propeller decrease after ventilation. However, with the growth of the forward speed of the SPP, the influence of the ventilation on

the thrust coefficient is smaller than on the torque coefficient, which makes the efficiency (η) of the ventilated SPP, be significantly greater than that of unventilated SPP. The numerical results demonstrate the effectiveness of the ventilation approach for improving the hydrodynamic performance of the SPPs for high speed planning crafts.

Author Contributions: D.Y. performed the numerical calculations; Z.R. and Z.G. (Zhiqun Guo) completed the geometric modeling and mesh generation; Z.G. (Zhiqun Guo) and Z.G. (Zeyang Gao) analyzed the post-processing data; D.Y. wrote the paper.

Funding: This project is supported by the National Natural Science Foundation of China (Grant No. 51509053, No. 51579056, and No. 51579051).

Conflicts of Interest: The authors declare no conflict of interest. The founding sponsors had no role in the design of the study; in the collection, analyses, or interpretation of data; in the writing of the manuscript, and in the decision to publish the results.

References

1. Ding, E.B.; Tang, D.H.; Zhou, W.X. A review of research on semi-submerged propellers. *J. Ship Mech.* **2002**, *6*, 75–84.
2. Zhao, R.; Faltinsen, O. Water entry of two-dimensional bodies. *J. Fluid Mech.* **1993**, *246*, 593–612. [CrossRef]
3. Young, Y.L.; Savander, B.R. Numerical analysis of large-scale surface-piercing propellers. *Ocean Eng.* **2011**, *38*, 1368–1381. [CrossRef]
4. Yari, E.; Ghassemi, H. Numerical analysis of surface piercing propeller in unsteady conditions and cupped effect on ventilation pattern of blade cross-section. *J. Mar. Sci. Technol.* **2016**, *21*, 501–516. [CrossRef]
5. Yu, Y.Q.; Yu, J.X.; Ding, E.B. Numerical study of two-dimensional "cup" shaped supercavitation cross-section inflow. *J. Ship Mech.* **2008**, *12*, 539–544.
6. Ghassemi, H.; Shademani, R. Hydrodynamic characteristics of the surface-piercing propellers for the planing craft. *J. Mar. Sci.* **2009**, *8*, 267–274.
7. Kinnas, S.A.; Young, Y.L. Modeling of cavitating or ventilated flows using BEM. *Int. J. Numer. Methods Heat Fluid Flow* **2003**, *13*, 672–697. [CrossRef]
8. Young, Y.L.; Kinnas, S.A. Numerical analysis of surface piercing Propellers. In *2003 Propeller and Shaft Symposium*; Society of Naval Architects and Marine Engineers: Virginia Beach, VA, USA, 2003.
9. Young, Y.L.; Kinnas, S.A. Analysis of supercavitating and surface piercing propeller flows via BEM. *J. Comput. Mech.* **2003**, *32*, 269–280. [CrossRef]
10. Shi, Y.X.; Zhang, L.X.; Shao, X.M. Numerical simulation of hydrodynamic performance of semi-submerged propeller. *Mechatron. Eng.* **2014**, *31*, 985–990.
11. Alimirzazadeh, S.; Roshan, S.Z.; Seif, M.S. Unsteady RANS simulation of a surface piercing propeller in oblique flow. *Appl. Ocean Res.* **2016**, *56*, 79–91. [CrossRef]
12. CD-adapco Group. *STAR-CCM+ User Guide Version 9.06*; CD-adapco Group: Melville, NY, USA, 2014.
13. Zhang, J.; Fang, J.; Fan, B.Q. A review of theories and applications of VOF methods. *Adv. Sci. Technol. Water Resour.* **2005**, *4*, 67–70.
14. Menter, F.R. Zonal two-equation k-ω turbulence model for aerodynamics flows. In Proceedings of the 23rd Fluid Dynamics, Plasmadynamics, and Lasers Conference, Orlando, FL, USA, 6–9 July 1993; AIAA Paper 93-2906. The American Institute of Aeronautics and Astronautics (AIAA): Reston, VA, USA, 1993.
15. Calomino, F.; Alfonsi, G.; Gaudio, R.; D'lppplito, A.; Lauria, A.; Tafarojnoruz, A.; Artese, S. Experimental and Numerical Study of Free-Surface Flows in a Corrugated Pipe. *Water* **2018**, *10*, 638. [CrossRef]
16. Huang, S.; Wang, C.; Wang, S.Y. Application and comparison of different turbulence models in the calculation of propeller hydrodynamic performance. *J. Harbin Eng. Univ.* **2009**, *5*, 481–485.
17. Blocken, B.; Gualtieri, C. Ten iterative steps for model development and evaluation applied to Computational Fluid Dynamics for Environmental Fluid Mechanics. *Environ. Model. Softw.* **2012**, *33*, 1–22. [CrossRef]

Article

Study of Cavitation Bubble Collapse near a Wall by the Modified Lattice Boltzmann Method

Yunfei Mao, Yong Peng * and Jianmin Zhang

State Key Laboratory of Hydraulics and Mountain River Engineering, Sichuan University, Chengdu 610065, China; maoyunfeigy@163.com (Y.M.); zhangjianmin@scu.edu.cn (J.Z.)
* Correspondence: pengyongscu@foxmail.com; Tel.: +86-187-8023-3156

Received: 2 September 2018; Accepted: 3 October 2018; Published: 12 October 2018

Abstract: In this paper, an improved lattice Boltzmann Shan-Chen model coupled with Carnahan-Starling equation of state (C-S EOS) and the exact differential method (EDM) force scheme is used to simulate the cavitation bubble collapse in the near-wall region. First, the collapse of a single cavitation bubble in the near-wall region was simulated; the results were in good agreement with the physical experiment and the stability of the model was verified. Then the simulated model was used to simulate the collapse of two cavitation bubbles in the near-wall region. The main connection between the two cavitation bubble centre lines and the wall surface had a 45° angle and parallel and the evolution law of cavitation bubbles in the near-wall region is obtained. Finally, the effects of a single cavitation bubble and double cavitation bubble on the wall surface in the near-wall region are compared, which can be used to study the method to reduce the influence of cavitation on solid materials in practical engineering. The cavitation bubble collapse process under a two-dimensional pressure field is visualized, and the flow field is used to describe the morphological changes of cavitation bubble collapse in the near-wall region. The improved lattice Boltzmann Method (LBM) Shan-Chen model has many advantages in simulating cavitation problems, and will provide a reference for further simulations.

Keywords: collapse near a wall; double cavitation bubble; tilt distribution cavitation; parallel cavitation; pseudopotential lattice Boltzmann model

---

1. Introduction

When a liquid is heated at a constant temperature or depressurized by static or dynamic methods at a constant temperature, steam bubbles or vapour-filled cavitation bubbles appear and develop over time. Cavitating water in a low-pressure zone involves a large amount of vapour forming a two-phase flow, and when the water flows through a region with a higher pressure downstream, the cavitation bubbles collapse under the effect of the pressure or temperature. In the collapse event, the cavitation phenomenon includes the nascent development and collapse of cavitation bubbles, which is an unsteady compressible multiphase turbulent flow phenomenon involving mass transfer between the gas and liquid phases. During the collapse of the cavitation bubble, high pressure of up to thousands of atmospheres is generated. When the collapse of the cavitation bubble occurs within a certain distance from a solid sidewall, the sidewall is subjected to continuous impact, causing fracture or fatigue damage of the material, which leads to erosion and cavitation. Corrosion is associated with a series of problems such as mechanical efficiency reduction and equipment damage in devices. Therefore, the formation and collapse of cavitation bubbles near a wall has been a focus of cavitation research.

Cavitation phenomena were first observed experimentally using a flow field. Kling and Hammitt [1] and Lauterborn [2,3] used high-speed photography techniques to study spark- and laser-induced cavitation bubble collapse processes. Lauterborn [3] studied the collapse of cavitation bubbles near a wall. Some scholars have also conducted in-depth physical research on cavitation [4–13]. However, although the bubble collapse process can be observed through this test method, data such as pressure field and velocity field are not available. With the development of numerical simulation technology, more methods have emerged to simulate the bubble collapse process, such as the interface tracking method, volume-of-fluid (VOF) method, and level set method. LBM is another new and efficient numerical simulation technology that has been developed as a digital model. Relative to traditional computational fluid dynamics (CFD), LBM has substantial advantages. Sukop and Or [14] first simulated the expansion and collapse of cavitation bubbles using the lattice Boltzmann Shan-Chen model; Chen et al. [15] used the lattice Boltzmann method to simulate two-dimensional cavitation within static and shear flows. The Rayleigh-Plesset equation results were compared with their simulation results. Shan et al. [16] and Zhou et al. [17] used LBM to simulate the growth and collapse of a single bubble under a two-dimensional pressure field and the sound field at a plane rigid wall. Kucera and Blake [18] used the mirror image method to simulate cavitation erosion of a plane rigid interface upon single bubble collapse near the interface and simulated bubbles near a rigid interface at different angles but did not consider the influence of cavitation on the flow field. Li et al. [19] used shadow photography and light deflection to investigate the motion of a single cavitation bubble core and the collapse time under different cone angles. Zhang et al. [20] investigated the three-dimensional (3D) cavitation bubble phenomenon at a low liquid pressure and successfully reproduced bubble growth in low-pressure water. Mishra et al. [21] investigated the coupling between the hydrodynamics of a collapsing cavity and the resulting solute chemical species introduced by cavitation based on the Shan-Chen multiphase model. Other scholars have carried out some numerical simulations on cavitation [22–33].

Previous simulations of cavitation have focused on the relationship between bubble collapse and the position of a single cavitation bubble relative to the boundary or the collapse of multiple cavitation bubbles, but no simulation of the density field or pressure field has been performed. In this paper, based on the Shan-Chen lattice Boltzmann method, a C-S state equation and an exact difference method that can accurately derive the external force term are coupled. The non-equilibrium extrapolation format and pressure boundary are used to simulate the collapse of a bubble in the near-wall region. First, the collapse evolution law of a single cavitation bubble is obtained, which is consistent with the physical experiment. On this basis, the collapse evolution law of two cavitation bubbles in the near-wall region is studied, and the collapse evolution law and flow field changes of two cavitation bubbles at different pressures are obtained.

2. Basic Principle of the LBM

LBM is considered to be one of the most effective methods to solve multiphase flow problems. Commonly used two-dimensional models are D2Q7, D2Q9, etc., where D is the dimension of space and Q is the number of discrete velocities. In this paper, the two-dimensional case is simulated, and in order to ensure the calculation accuracy, nine discrete velocities are used, so the D2Q9 model is used for simulation. The specific arrangement is shown in Figure 1.

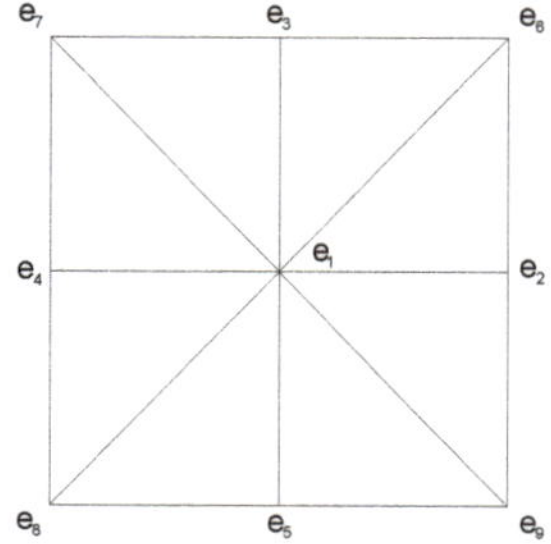

Figure 1. D2Q9 model.

In the Boltzmann model, discrete particle distribution function f_i is used to replace the fluid particle distribution function:

$$f_i(x + ce_i\Delta t, t + \Delta t) = f(x, t) - \frac{1}{\tau}[f_i(x, t) - f_i^{eq}(x, y)] + F_i(x, t) \tag{1}$$

where $f_i(x, t)$ is a single particle density distribution function, $f_i^{eq}(x, t)$ is the equilibrium particle distribution function, τ is the relaxation time, and the kinematic viscosity is $\nu = c_s^2(\tau - 0.5)\Delta t$. The discrete speed can be defined as

$$[e_1, e_2, e_3, e_4, e_5, e_6, e_7, e_8, e_9] = c \begin{bmatrix} 0 & 1 & 0 & -1 & 0 & -1 & -1 & 1 \\ 0 & 0 & 1 & 0 & -1 & 1 & -1 & -1 \end{bmatrix}, \tag{2}$$

where $c = \Delta x / \Delta t$ is the grid velocity and Δx and Δt are the grid step and time step, respectively. The lattice sound velocity is $c_s = c/\sqrt{3}$. In the D2Q9 model, the equilibrium distribution function can be expressed as

$$f_i^{eq}(x, t) = w_i \rho [1 + \frac{e_i u}{c_s^2} + \frac{(e_i u)^2}{2c_s^4} - \frac{u^2}{2c_s^2}], \tag{3}$$

where u is the fluid velocity and w_i is the weighting factor of the equilibrium distribution function, where $w_1 = 4/9$, $w_i = 1/9(i = 2, 3, 4, 5)$, $w_i = 1/9(i = 6, 7, 8, 9)$. The discrete velocity weight coefficient is related to the discrete strategy. In the discretized velocity space, the macroscopic velocity and density of the fluid can be expressed as

$$\rho = \sum_i f_i(x, t) \tag{4}$$

$$\rho u = \sum_i f_i(x, t) e_i. \tag{5}$$

The single-component multiphase flow interaction force $F_i(x, t)$ is modelled as follows:

$$F(X, t) = -G\psi(X, t)\sum_{i=2}^{9} \xi\psi(X + e_i\Delta t, t)e_i, \tag{6}$$

where G is the interaction strength and $\zeta_i = 1/9(i = 2, 3, 4, 5)$, $\zeta_i = 1/36(i = 6, 7, 8, 9)$. According to Yuan and Schaefer (2006) [34], the interaction potential can be expressed as $\psi = \sqrt{2(p - \rho c_s)/Gc_s^2}$, where $c_s = \sqrt{RT}$ is the lattice sound velocity. Through this interaction potential, different equations of state can be applied.

The C-S state equation can be expressed as follows:

$$p = \rho RT \frac{1 + b\rho/4 + (b\rho/4)^2 - (b\rho/4)^3}{(1 - b\rho/4)^3} - a\rho^2, \tag{7}$$

where $a = 0.4963 R^2 T_c^2 / p_c$ and $b = 0.18727 RT_c / p_c$. The critical parameters can be expressed as follows:

$$T_c = \frac{0.3773a}{bR} \tag{8}$$

$$p_c = \frac{0.0706}{b^2} \tag{9}$$

The forces of the fluid act all over the objects placed in it, and how to integrate these forces into the LBM is very important. An EDM with second-order accuracy is suitable. The Boltzmann equation is discretized in the velocity space, and the term in accordance with the LB equation is derived by the EDM. The external force term in Equation (1) can be expressed as

$$F_i = f_i^{eq}(\rho, u + F\Delta t/\rho) - f_i^{eq}(\rho, u),$$ (10)

where F is the total interaction force between the fluid and the solid.

3. Physical Model

In order to obtain the evolution of cavitation bubbles in the near-wall region, the physical models are shown in Figure 2.

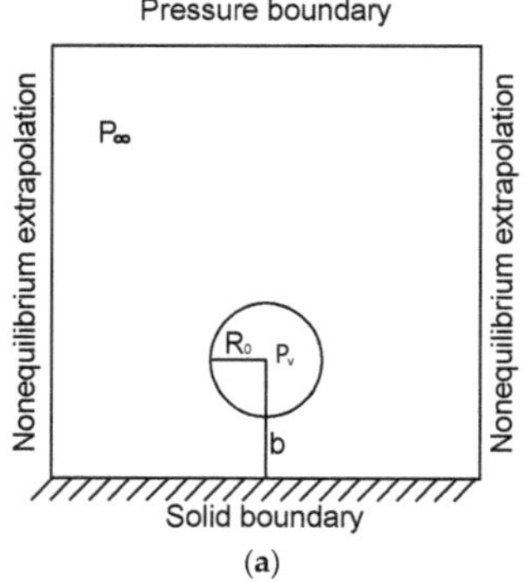

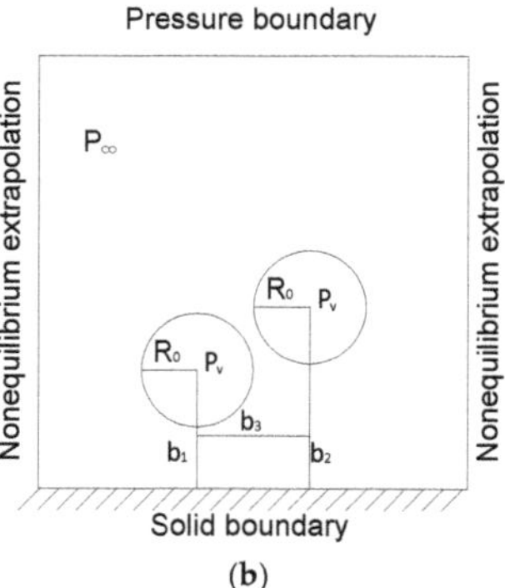

Figure 2. Physical model. (**a**) (single bubble model); (**b**) (double bubbles model) (R_0—bubble initial radius; b_i—distance; P_v—vapour pressure in bubble; P_∞—ambient pressure).

The figure above shows the initial layout of the simulation. Here, (a) is the physical model of the study of the evolution of a single cavitation bubble in the near-wall region. R_0 is the radius of the cavitation bubble, b is the distance from the centre of the cavitation bubble, P_v is the pressure in the cavitation bubble, and P_∞ is the pressure outside the cavitation bubble. The left and right boundaries are infinite areas, the upper boundary is the pressure inlet, and the lower boundary is the rigid wall. (b) is the physical model for the study of the evolution of the double cavitation bubble in the near-wall region. b_1 is the distance from the centre of the left cavitation bubble to the rigid wall, b_2 is the distance from the centre of the right cavitation bubble to the rigid wall, and b_3 is the horizontal distance of the centre of the two cavitation bubbles. The other settings are the same as those of the physical model of a single cavitation bubble in the study of the collapse of the near-wall region.

4. Simulation Content and Parameter Initialization Settings

In this paper, the numerical simulation of cavitation collapse in the near-wall region mainly includes three parts: First, the evolution law of the collapse of a single cavitation bubble in the near-wall region is obtained by numerical simulation, and the bubble deformation is compared with the results of classical experiments to verify the model. Second, the pressure field of the bubble is simulated and theoretically study the mechanism of the erosion of the bubble on the solid sidewall. Third, the previously validated model is used to study the evolution of the double cavitation bubble in the near-wall region. In this paper, if there is no special explanation, the unit in the text always uses the grid unit, the length unit is lu, the time unit is ts, the mass unit is mu, and the temperature unit is tu; thus, the density unit is $mu \cdot lu^{-3}$, and the pressure unit is $mu \cdot lu^{-1} ts^{-2}$. The simulated initial layout is the same as the physical model. In the simulation, $T/T_c = 0.689$ is used to simulate and the equilibrium pressure $p = 0.0028\ mu \cdot lu^{-1} ts^{-2}$ is gotten by the equal area rule is. In the C-S state of equation, $a = 1$,

$b = 4$ and $R = 1$ [34] are adopted. The initial temperature is set to a specific temperature, the velocity is zero, and the density field is initialized as follows:

$$\rho(x, y) = \left(\rho_{liquid} + \rho_{gas}\right)/2 + \left(\rho_{liquid} - \rho_{gas}\right)/2 \cdot [\tan h((2(\sqrt{(x - x_1)^2 + (y - y_1)^2} - R_0)/W)], \quad (11)$$

where x_1 and y_1 is the location of the middle of the bubble at the initial moment, the hyperbolic tangent function $tanh = (e^x - e^{-x})/(e^x + e^{-x})$ and the phase interface width is $W = 4$. Since the cavitation bubble radius is approximately ten times the width of the phase interface, it can obtain better numerical stability of the model, but it is essential to ensure that the bubble collapse is not affected by other boundaries and that the calculation cost is reduced. Therefore, our simulation calculation area is 401×401.

Through the C-S state equation, the P-V curve of the gas-liquid isotherm curve can be obtained, and then the Maxwell construction should be used, which can be stated as

$$\int_{V_{m,l}}^{V_{m,g}} P dV_m = p_0\left(V_{m,g} - V_{m,l}\right), \quad (12)$$

where P is the pressure in the EOS and p_0 is a constant pressure. When the equation is established, p_0 is the equilibrium pressure, and Vm,g and Vm,l are the physical quantities that characterize the equilibrium gas pressure and the equilibrium liquid pressure, respectively. In addition, the coexistence densities ρ_v and ρ_l of gas and liquid, respectively, can also be determined by phase separation simulation with a slight random disturbance of the initial density. In the calculation process, ρ_l need to be slightly adjusted to ensure ρ_l has the same density as the pressure boundary so that an additional pressure difference between the inside and the outside of the bubble is obtained after the fluid balance in the entire calculation domain.

The collapse of cavitation bubbles under the influence of a single wall surface is studied, so infinite areas are needed on both the left and right sides to ensure the computing domain is unaffected by the boundary fluctuations. Under the premise of the minimum calculation area, the unbalanced extrapolation format is well suited for our needs. Therefore, the non-equilibrium extrapolation format is used for the left and right borders. The bottom boundary uses a standard bounce-back format. In addition, a pressure boundary condition is applied at the inlet, and the liquid pressure in the calculation zone is equal to the pressure boundary pressure.

5. Study of the Evolution of a Single Cavitation Bubble

Detailed experimental data have been obtained for the evolution of cavitation bubble collapse at different distances from the wall [35]. However, traditional experimental data and numerical simulation have great limitations. For example: Plesset and Chapman [36] made six assumptions, including negligible surface tension, constant vapour pressure and ambient pressure, an incompressible liquid, non-viscous flow, and no permanent gas, which are challenging to satisfy in experiments and LBM simulations, resulting in inapplicability to specific practical problems. In this paper, using the improved Shan-Chen model, numerical simulation data of cavitation bubble collapse at different locations are obtained at a specific temperature. This part of the study mainly consists of two parts: the first part verifies the simulation results by comparison with physical experiments, and the second part obtains the evolution law of the collapse of a single cavitation bubble in detail. The following is a comparative analysis of density field images of LBM numerical simulations and experimental images of cavitation bubbles collapsing at two different positions with the same radius and pressure. Through comparison with the experimental data of Philipp [35], the LBM calculation results which are shown in Figures 3 and 4 are found to be consistent with the qualitative analysis of the experimental data which are shown in Figures 5 and 6.

Figure 3 shows the change of density field during the collapse of cavitation bubbles. In this simulation, a dimensionless quantity $\lambda = b/R_0$ was introduced, which is the amount that characterizes the distance from the centre of the bubble to the wall. The figures below show cases with $\lambda = 1.6$,

$R_0 = 80$ lu and $\lambda = 2.5$, $R_0 = 80$ lu. The cavitation bubble is initially a circle. The bubble size and the thickness of the gas-liquid boundary layer are controlled by Equation (11). Since the pressure difference exists inside and outside the bubble, the bubble is deformed by extrusion. Due to the influence of the bottom rigid sidewall, the longitudinal flow is blocked, and a negative pressure forms under the bubble to induce longitudinal expansion of the bubble. Due to the shrinkage and deformation of the bubbles, the volume is decreasing, and the surrounding liquid fills the space created by the bubbles, resulting in a decrease in the density and pressure around them. Then, the pressure of the upper pressure boundary is first transmitted to the upper surface of the bubble, and a high-pressure zone is formed in the upper part of the bubble that acts together with the low-pressure zone of the cavitation bubble to form a depression at the upper portion ($t = 470$). Due to the rebound effect of the liquid and the relatively high speed of movement of the upper portion of the bubble, a relatively large conical high-pressure region is formed in the upper portion, which is crucial in the subsequent deformation. Over time, the depression continues to expand ($t = 530$), and with the influence of the surrounding high pressure, the bubble gradually shrinks, assuming a crescent shape ($t = 570$). When the sag causes the upper surface of the bubble to touch the lower surface, a large pressure difference directly breaks down the empty bubble, forming a micro shock wave, which has a destructive effect on the wall surface ($t = 588$). At the same time, a complex sound field is generated, which causes additional damage to the rigid wall. By comparison, the morphology of cavitation bubble collapse differs for different dimensionless parameters λ and ΔP is confirmed. In our simulation, a crescent-shaped bubble is formed at $\lambda = 1.6$ and is then broken down to form two bubbles, which generates a micro-shock; however, at $\lambda = 2.5$, the bubble is squashed directly. A crescent-shaped bubble is formed, but the bubble does not break in the middle and finally collapses in the form of a small bubble. The calculation results are consistent with the results of Philipp [35]. When the bubble is too far from the wall surface, the sidewall has a small retarding effect on the bubble, and a strong negative pressure is unlikely to form under the bubble, so when the bubble collapses, the bubbles are gradually crushed and collapsed, and the impact of the formed pressure on the wall surface is also alleviated.

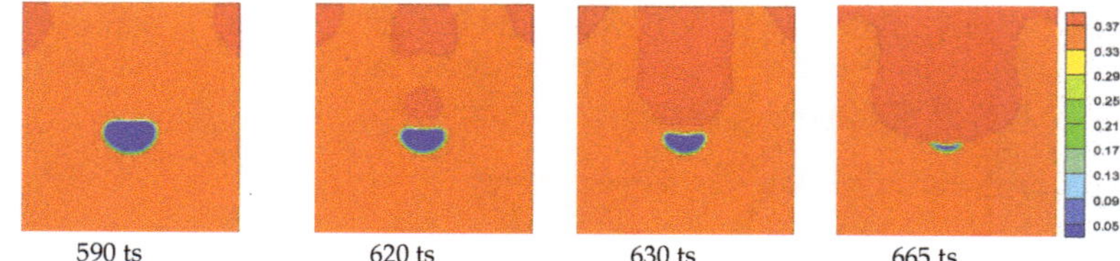

Figure 3. Density field of the LBM simulation: $\lambda = 2.5$, $R_0 = 80$ lu.

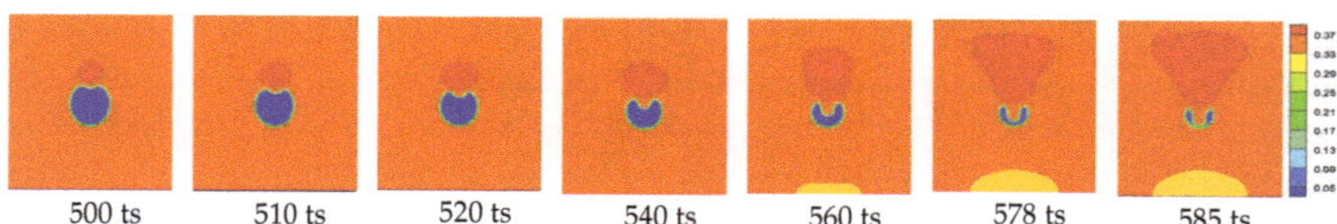

Figure 4. Density field of the LBM simulation: $\lambda = 1.6$, $R_0 = 80$ lu.

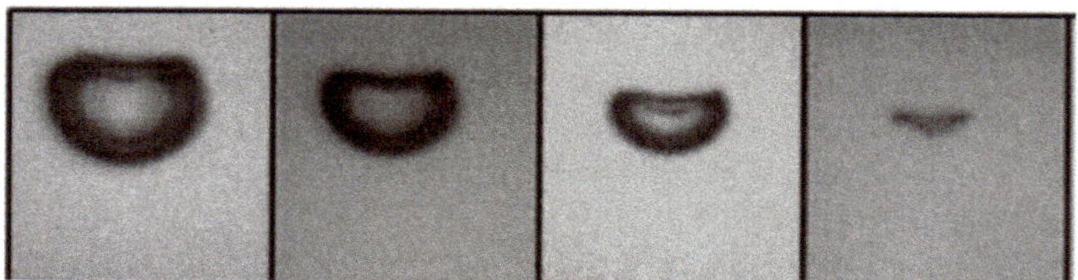

Figure 5. Experimental images: $\lambda = 2.5$, $R_0 = 1.45 \times 10^{-3}$ m.

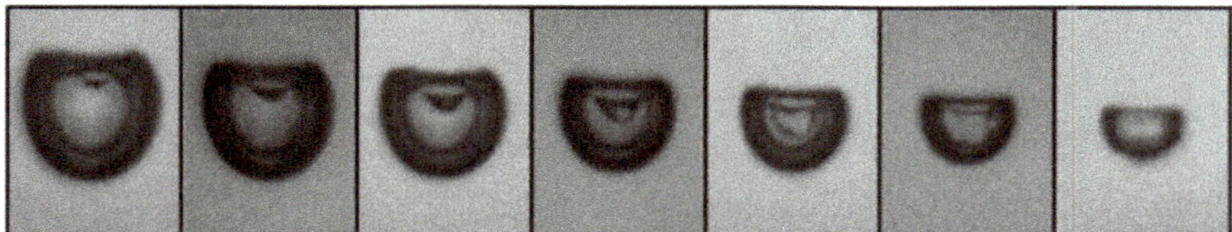

Figure 6. Experimental images: $\lambda = 1.6$, $R_0 = 1.45 \times 10^{-3}$ m.

6. Study of the Evolution of a Double Cavitation Bubble

In the previous section, the collapse evolution of a single cavitation bubble in the near-wall region was discussed, which was found to be in good agreement with the physical experiment, thus verifying the stability of the numerical simulation model. However, in actual engineering, cavitation bubbles do not appear alone. Cavitation clouds typically form in cavitation-concentrated areas, and hundreds of cavitation bubbles interact and collapse under the extra pressure; thus, the erosion of the wall is more complex. This research area merits further study and is instructive for the possibility of reducing cavitation damage. In our study, using the previously validated mathematical model, the collapse evolution law by simulating the collapse of two cavitation bubbles in the near-wall region under pressure induction was obtained. Taking two of these cases as examples, the evolution of the two cavitation bubbles in the near-wall region under pressure induction was analysed. For convenience of description, the left and right bubbles are designated left bubble (LB) and right bubble (RB), respectively. The spatial direction is set as shown in Figure 1. For example, the upper right is the e_6 direction. Similar to the previous simulation, the dimensionless quantities λ_1, λ_2, and λ_3 is introduced, where $\lambda_1 = b_1/R_0$, which is the amount that characterizes the distance of the LB centre from the wall; $\lambda_2 = b_2/R_0$, which is the amount that characterizes the distance of the LB centre from the wall surface; and $\lambda_3 = b_3/R_0$, which is the amount that represents the horizontal distance between the LB and RB centres.

6.1. Case 1: Numerical Simulation of Tilt Distribution Cavitation with Two Cavitation Bubbles

When $\lambda_1 = 1.2$, $\lambda_2 = 2.7$, and $\lambda_3 = 1.5$, the line connecting the centres of the two cavitation bubbles is at an angle of $45°$ to the wall surface. The density field and the pressure field coupling with the velocity field of the cavitation bubble collapse process are shown in Figures 7 and 8, respectively.

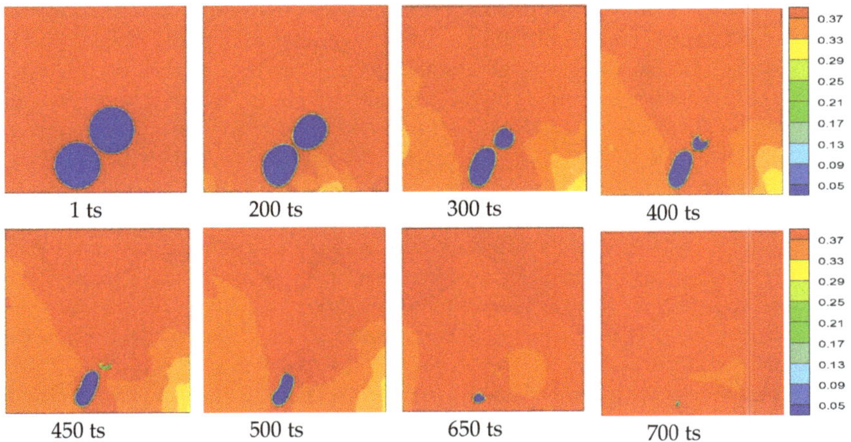

Figure 7. Density field of cavitation bubble collapse.

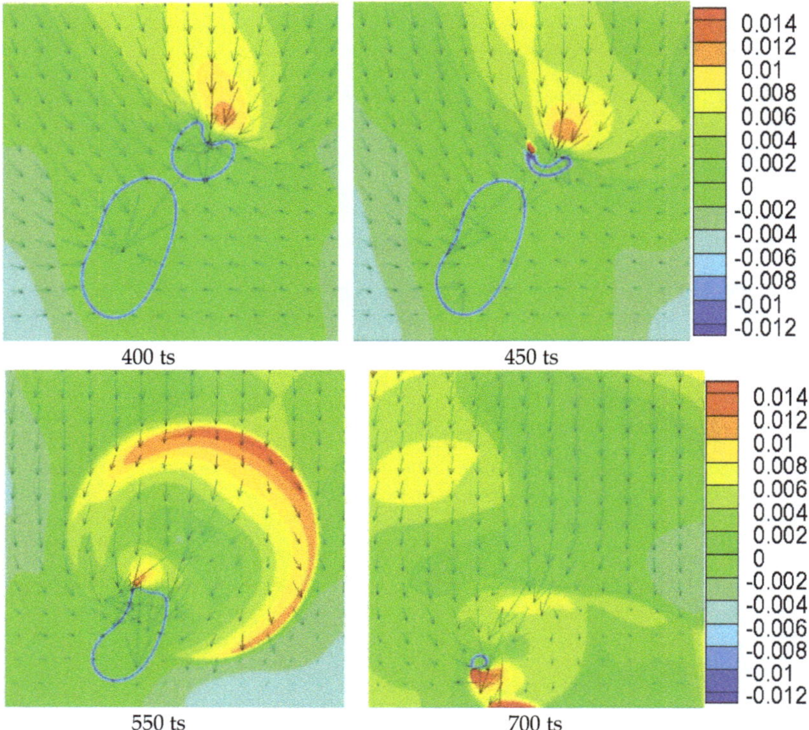

400 ts   450 ts

550 ts   700 ts

Figure 8. Pressure field and velocity field of cavitation bubble collapse.

The cavitation bubble distribution at the initial moment is as shown by $t = 1$ ts, and the specific parameter settings are the same as the previous settings. The collapse process of each cavitation bubble was analysed. When the simulation proceeds to $t = 200$ ts, the change of LB is similar to the collapse of a single cavitation bubble. Due to the blockage of the wall surface, a low-pressure zone is formed in the bottom region, and LB exhibits an elongated deformation state in the e_5 direction. The e_6 direction of the LB and the e_8 direction of the RB are attracted to each other, and the opposite direction is deformed. If the gas density and the liquid density are in equilibrium at this time, the two cavitation bubbles would attract each other and eventually merge into a single bubble. However, since the liquid has been pressurized during the simulated initialization, the additional generated pressure prevents the emergence of the two bubbles. As the simulation progresses, the deformation is further aggravated, and the bubbles decrease under the action of additional pressure. Between them, the change of LB is the most easily detected. When the time reaches $t = 350$ ts, the LB is elongated and deformed due to the low pressure resulting from the blockage of the wall surface; the e_6 direction of RB appears the same as the change of a single cavitation bubble in the near-wall region, and a depression occurs. The change becomes more apparent at 400 ts. Around the shrinking bubble, the liquid fills the newly available space, and a slightly lower pressure appears around the cavitation bubble. The transition can be observed in the velocity field. Around the bubble, the velocity direction tends to support cavitation. Note that in the figure of the pressure field and the velocity field, for the sake of convenience, only the velocity of the liquid is shown and we do not plot the velocity inside the bubble. The LB exhibits an elliptical shape that exhibits an inclination in the e_6–e_8 direction under the action of the sidewall, the attraction of RB and the external extreme pressure. The RB is also elliptical due to the attraction of the two bubbles, but the deformation is not as strong as with the bubble on the left. At this time, relative with the collapse rule of a single cavitation bubble in the near-wall region, LB is equivalent to rigidity avoidance for RB, so a low-pressure zone is generated between the two bubbles, and high pressure is generated by the upper pressure boundary. A pressure difference is generated in the e_6–e_8 direction of the RB caused the deformation to continue to produce. Inspection of the velocity field reveals that the maximum velocity occurs in the same direction of the RB. At t = 450 ts, the cavitation bubble is crescent-shaped. The bubble is strongly compressed from the upper part of the depression until the entire bubble is completely collapsed from the upper part of the bubble, which occurs because the flow field is not completely symmetrical on both sides of the e_6–e_8 direction. The upper part is first crushed by the pressure transferred from the pressure boundary. The other reason is that the most attractive part of LB is located at the nearest position of the two bubbles. The velocity field indicates that the maximum velocity direction has been deflected, and RB begins to collapse from top to bottom under the pressure difference, producing a huge jet and complex sound field. The huge pressure generated by the collapse acts in conjunction with other factors to promote the continued collapse of LB. At t = 550 ts, the pressure field and velocity field images indicate that the flow field changes due to the collapse pressure of RB, and the micro-jet generated by RB in the e_6 direction collides with the downward flowing liquid to generate a high voltage and noise, and energy begins to be consumed; furthermore, the micro-jet in the e_7 direction overlaps with the original flow field, and the flow state becomes complex. As the simulation proceeds, the flow field exhibits vortices, and LB begins to collapse. First, the cavitation bubble e_6 direction begins to shrink under the influence of the pressure of RB, which causes a protrusion in the upper portion of the cavitation bubble. At this time, the energy generated by the RB collapse is insufficient to continue to compress LB. Under the joint action of the upper pressure boundary pressure and the high pressure generated by the RB collapse, a high-pressure region (t = 550 ts) is formed at the upper convex portion, and collapse from the upper portion of the cavitation bubble is induced. As RB is destroyed, the e_9 direction finally collapses, and a jet from the e_7 to the e_9 direction is generated in the flow field, so that the entire flow field exhibits vortices. Finally, LB collapses under the combined action of various factors, and the generated pressure impacts the wall surface to avoid impact. The pressure field diagram of t = 700 ts indicates that a high pressure is generated on the wall surface. Note that after RB collapse, LB collapse occurs under the influence of various factors, and the mechanism of action is relatively complex. The LB high-pressure zone rotates around the cavitation bubble. This visualization is a powerful way to illustrate the complex vortices and other phenomena in the flow field, involving other physical quantities. The parameters of this study cannot describe the more detailed characteristics. The specific mechanism needs further study.

6.2. Case 2: Numerical Simulation of Two Parallel Cavitation Bubbles

When λ_1 = 1.2, λ_2 = 1.2, and λ_3 = 2.4, that is, the line connecting the centres of the two cavitation bubbles is parallel to the wall surface, and the density field of the cavitation bubble collapse process is shown in Figure 9 as well as the pressure field and the velocity field shown in Figure 10.

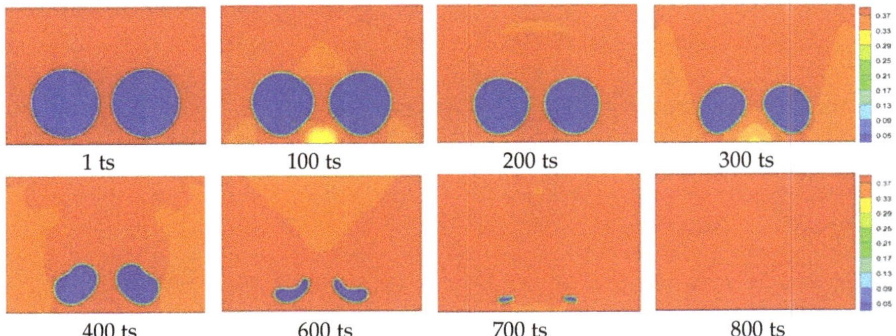

Figure 9. Density field of cavitation bubble collapse.

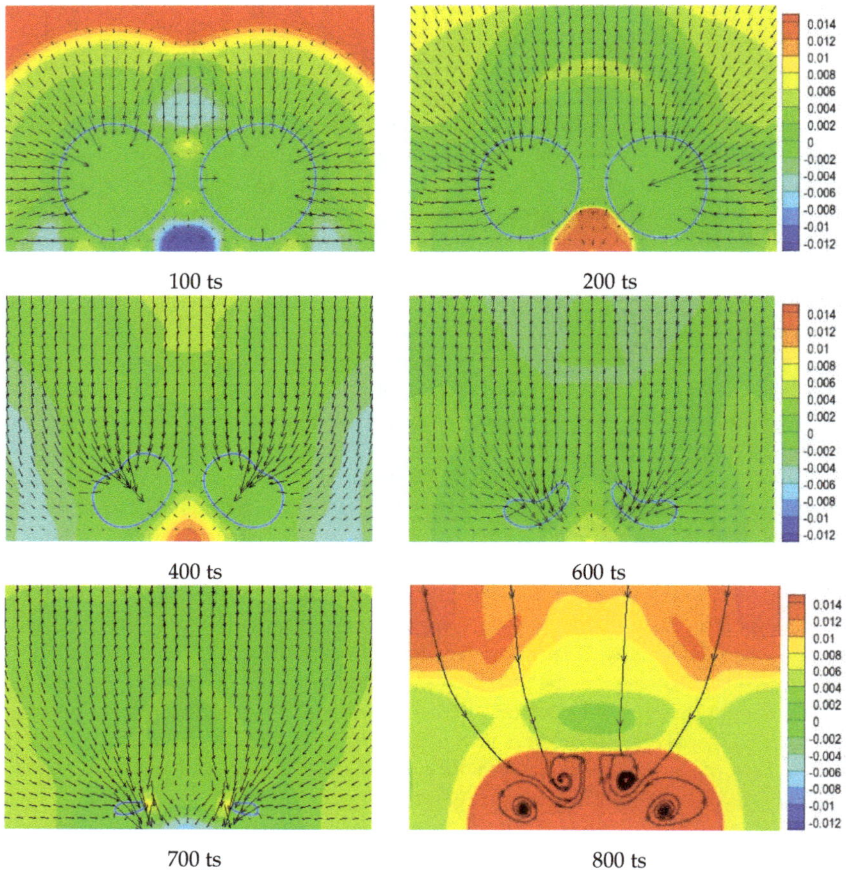

100 ts   200 ts

400 ts   600 ts

700 ts   800 ts

Figure 10. Pressure field and velocity field of cavitation bubble collapse (800 ts is the pressure field and streamline diagram).

The figure shows the change in the two cavitation bubbles in the near-wall region and the initial boundary and density settings are the same as described above. Since the changes of the two bubbles are basically the same, the LB is used as an example for analysis. At $t = 100$ ts, low pressure is created at the bottom of the cavitation bubbles due to the retardation of the rigid wall. At the same time, the two cavitation bubbles are attracted to each other. The cavitation bubbles that are not attracted to the rigid wall and the cavitation bubbles begin to shrink under the action of the additional pressure, and the new space that is generated from the reduced area was filled by the surrounding liquid. Through the analysis of the pressure field, the conclusion that the pressure decreases in the e_9 direction of LB can be obtained. The bottom region of the symmetry axis is defined as the pressure change region, which is the region where the pressure change is most obvious during the cavitation process, except for the pressure region generated by the collapse of the cavitation bubble itself. In the simulation, under the condition of the specific additional pressure and the relative position of the cavitation bubble, a new cavitation bubble is generated in the pressure change zone, but it will collapse quickly. Although the new cavitation bubbles are produced for a short period of time, the effect on the flow field is very important and will be explained in detail in the analysis of the maximum wall pressure behind. When the simulation is carried out to $t = 200$ ts, for LB, the high pressure generated by the collapse of the new cavitation bubble in the pressure change zone continues to act on the e_9 direction of the cavitation bubble, showing a slightly lifted shape. As the simulation progresses to t = 400 ts, the deformation continues; the deformation of the cavitation bubbles is already evident, and the cavitation bubbles are inclined at an inclination angle of 45°, and the cavitation bubbles become crescent-shaped. This is mainly due to the interaction between the cavitation bubbles, the interaction between the cavitation bubbles and the wall surface, and the collapse of the new cavitation bubbles in the pressure change zone. In addition, the velocity field indicates that the velocity of the LB in the e_7 direction is the largest. At t = 600 ts, the LB appears asymmetric on the axis in the e_7–e_9 direction. The upper part is strongly influenced by the pressure boundary and the lower part is strongly influenced by the rigid wall. Therefore, the position where the pressure difference between the upper side and the lower side of the LB is the largest is located at the centre, so that the cavitation bubbles collapse from the upper portion until they completely collapse. For each bubble, a complex eddy current phenomenon occurs in the flow field because the collapse does not occur instantaneously but spreads from the top to the adjacent region. The 800 ts streamline diagram clearly shows the eddy currents in the flow.

7. Maximum Wall Pressure

The maximum wall pressure of a single cavitation bubble and two parallel distributed cavitation bubbles collapsed in the near-wall region was identified in this study. The initial conditions are set as follows: For the case of a single cavitation bubble, the value of λ (from 1.05–2.0) with a series of gradients is used to simulate the maximum wall pressure resulting from single cavitation bubble collapse under different initial conditions. Since the maximum pressure of a single cavitation bubble on the wall is only likely to occur below the cavitation bubble (e_5 direction), the pressure of the wall below the cavitation bubble is recorded to find the maximum value. For the case of a single cavitation bubble, a lambda value (from 1.05–2.0) with a series of gradients is used to simulate the maximum wall pressure resulting from a single cavitation bubble collapse under different initial conditions. Since the maximum pressure of a single cavitation bubble on the wall is only likely to occur below the cavitation bubble (e_5 direction), the pressure of the wall below the cavitation bubble is recorded to find the maximum value. For the case of two cavitation bubbles, λ_3 = 2.4 is used and, likewise, the value of λ (from 1.05 to 2.0) with a series of gradients was used to simulate the maximum wall pressure resulting from the collapse of two cavitation bubbles under different initial conditions, where $\lambda = \lambda_1 = \lambda_2$. Since the cavitation bubble collapse process is complicated, the maximum wall pressure is not generated at the same position for the collapse process under different initial conditions. The wall pressure below the cavitation bubble (e_5 direction) and the pressure change zone is recorded to find the maximum value The comparison of the maximum wall pressure under different initial conditions is shown in Figure 11.

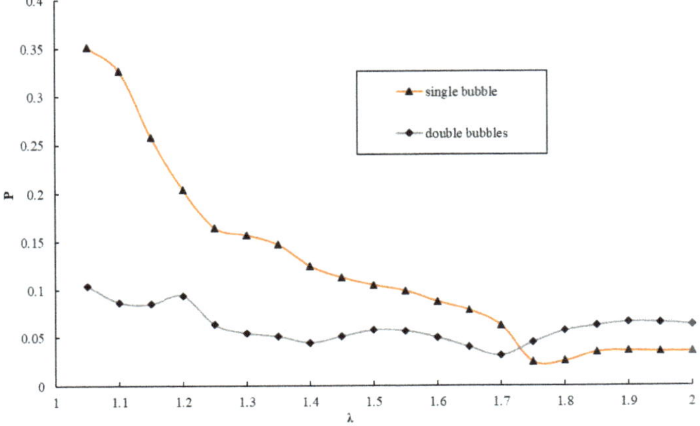

Figure 11. Comparison of the maximum wall pressure under different initial conditions.

Some conclusions are given by analysis:

- The maximum wall pressure generated by the collapse of a single cavitation bubble in the near-wall region decreases with increasing distance from the wall surface, and it first drops sharply, then slowly decreases and finally becomes relatively stable. This is because the closer to the wall, the smaller the thickness of the fluid that the micro shock wave generated by cavitation passes to the wall and the smaller the blockage effect of the flow field, the greater the wall pressure generated.

- It is complex for the case where a double parallel cavitation bubble collapses in the near-wall region because the maximum wall pressure produces a different position under different initial conditions. When $\lambda = 1.05$–1.10, the wall below the cavitation bubble is subjected to the maximum wall pressure, because the cavitation bubble is closer to the wall surface, and the generated micro-shock is transmitted to the wall surface almost unimpeded and the cavitation bubble of the pressure change region is not fully developed, which is not enough to generate create a huge pressure with the lifting force to bubble. When $\lambda = 1.15$–1.25, the maximum wall pressure occurs in the pressure change zone where new cavitation bubbles are generated and collapse rapidly, resulting in a large wall pressure exceeding the maximum wall pressure at other locations. When $\lambda = 1.40$–2.0, the area where the maximum wall pressure is generated is the pressure change zone, but the pressure generation at this time is not when the new cavitation bubble collapses but the pressure after the collapse of the two cavitation bubbles overlaps each other on the wall. The change in wall pressure depends mainly on the size of the new cavitation bubble induced and the pressure generated by the collapse and the lifting force of the bubbles.

- When $\lambda < 1.7$, the wall pressure generated by the collapse of a single cavitation bubble is relatively large. This is because, for the case where the double cavitation bubble collapses in the near-wall region, the pressure generated by the collapse of the new cavitation bubble induced in the pressure change region has an effect of lifting force on the cavitation bubble. Thereby, the bottom pressure of the cavitation bubble collapse is increased, and the blockage effect of the wall surface is weakened, which is equivalent to an increase of λ, and the resulting wall pressure is relatively small. When $\lambda > 1.7$, the wall pressure of the pressure change zone is formed by the superposition of pressure generated by the collapse of two cavitation bubbles, so the generated wall pressure is greater than the pressure generated by the collapse of a single cavitation bubble.

In summary, the maximum wall pressure generated by the collapse of the two cavitation bubbles in the near-wall region is smaller than the single cavitation bubble at $\lambda < 1.7$; the closer the bubble is to the wall, the greater the influence on the wall surface. When $\lambda > 1.7$, the distance between the cavitation bubble pair and the wall surface has little effect on the maximum wall pressure. This is a point worthy of further study with different initial conditions. It can be used as a reference for reducing cavitation damage in practical engineering.

8. Conclusions

Based on the improved lattice Boltzmann Shan-Chen model coupled with C-S EOS, the collapse phenomenon of cavitation bubbles in the near-wall region is studied. In this paper, the collapse of a single bubble in the near-wall region is simulated and compared with the physical experiment; the results are in good agreement. Then, using the verified mathematical model, the simulation of the collapse of two cavitation bubbles in the near-wall region under different pressure-induced conditions is carried out, and the collapse evolution law is obtained. And analysed the effect of wall pressure on different initial conditions. By analysing in detail the density field changes and pressure field changes during bubble collapse, the following conclusions can be drawn:

1. For the case where a single cavitation bubble collapses in the near-wall region, when $\lambda = 1.6$, a crescent-shaped bubble is formed that is broken down to form two bubbles, and a micro-shock is generated; when $\lambda = 2.5$, the bubbles are crushed to form crescent-shaped bubbles, but the

bubbles do not break in the middle but rather ultimately collapse in the form of a small bubble. The shape of the cavitation bubble is related to the distance of the cavitation bubble from the rigid wall.

2. For the numerical simulation of tilted distribution cavitation with two cavitation bubbles, in the early stage of simulation, the collapse behaviour is similar to that of the single-bubble case. Subsequently, the e_6 direction of RB has a concave deformation, which is very important in the collapse of the two bubbles. The velocity field indicates that the maximum velocity appears in the depression of RB. The tremendous pressure generated by this velocity directly penetrates the bubble; thereafter, the bubble is crescent-shaped until collapsing completely. Due to the asymmetry of the collapse, the flow field becomes complex after RB collapses and particularly so after LB collapses, and vortices appear in the flow field.

3. For the numerical simulation of parallel distribution cavitation of two bubbles, the bubble collapses under the blocking effect of the rigid wall and the attraction of the two bubbles, and the two bubbles collapse simultaneously. Therefore, the mutual influence during the collapse process is smaller than that of Case 1; in addition, the erosion effect of the bubble collapse on the wall surface is the result of superimposing the pressure fields formed by the collapse of the two bubbles. However, for each bubble, since the collapse does not occur instantaneously but is collapsed from the upper part of the closest position of the two cavitation bubbles, a complex vortex phenomenon occurs in the flow field.

4. By comparing the maximum wall pressure generated by cavitation under different initial conditions, the factors affecting the maximum wall pressure are obtained. For a single cavitation bubble, the distance from the wall is the most important factor. For two cavitation bubbles, the lifting effect of the new induced cavitation bubble collapse is the most important factor.

The results indicate that the improved LBM Shan-Chen model has many advantages in simulating cavitation problems, providing a reference for further simulation.

All the abbreviations are explained in the Appendix A (Table A1).

Author Contributions: Conceptualization, Y.P. and J.Z.; Methodology, Y.P. and J.Z.; Software, Y.M.; Investigation, Y.M.; Data Curation, Y.M.; Writing-Original Draft Preparation, Y.P. and J.Z.; Writing-Review & Editing, Y.M.; Visualization, Y.M.; Funding Acquisition, Y.P.

Funding: This research was funded by the National Natural Science Foundation of China (51579166) and the National Key Research and Development Program of China (2016YFC0401705).

Conflicts of Interest: The authors declare no conflict of interest.

Appendix A

Table A1. All variables with definitions.

f_i	**Single Particle Density Distribution Function**
f_i^{eq}	equilibrium particle distribution function
τ	relaxation time
ν	kinematic viscosity
c	grid velocity
Δx	grid step
Δt	time step
c_s	lattice sound velocity
ω_i	weighting factor
u	fluid velocity
F_i	interaction force
ψ	interaction potential
T_c	critical temperature
P_c	critical pressure
F	total interaction force

Table A1. *Cont.*

f_i	Single Particle Density Distribution Function
P_v	pressure of the cavitation bubble
P_∞	the pressure outside the cavitation bubble
R_0	radius of the cavitation bubble
b (b_1, b_2, b_3)	distance between corresponding points
T	lattice temperature
a (in Equation (11))	parameter of the C-S state of equation
b (in Equation (11))	parameter of the C-S state of equation
R (in Equation (11))	parameter of the C-S state of equation
$V_{m,g}$	the equilibrium gas pressure
$V_{m,l}$	the equilibrium liquid pressure
$\lambda(\lambda_1, \lambda_2, \lambda_3)$	dimensionless value that characterizes the distance

References

1. Kling, C.L.; Hammitt, F.G. A photographic study of spark-induced cavitation bubble collapse. *J. Basic Eng.* **1972**, *94*, 825–832. [CrossRef]
2. Lauterborn, W. High-speed photography of laser-induced breakdown in liquids. *Appl. Phys. Lett.* **1977**, *31*, 663–664. [CrossRef]
3. Lauterborn, W.; Bolle, H. Experimental investigations of cavitation-bubble collapse in the neighbourhood of a solid boundary. *J. Fluid Mech.* **1975**, *72*, 391–399. [CrossRef]
4. Reuter, F. Electrochemical wall shear rate microscopy of collapsing bubbles. *Phys. Rev. Fluids* **2018**, *3*, 063601. [CrossRef]
5. Watanabe, R.; Yanagisawa, K.; Yamagata, T.; Fujisawa, N. Simultaneous shadowgraph imaging and acceleration pulse measurement of cavitating jet. *Wear* **2016**, *358–359*, 72–79. [CrossRef]
6. Reuter, F.; Cairos, C.; Mettin, R. Vortex dynamics of collapsing bubbles: Impact on the boundary layer measured by chronoamperometry. *Ultrason. Sonochem.* **2016**, *33*, 170–181. [CrossRef] [PubMed]
7. Cui, P.; Zhang, A.M.; Wang, S.P.; Khoo, B.C. Ice breaking by a collapsing bubble. *J. Fluid Mech.* **2018**, *841*, 287–309. [CrossRef]
8. Rossello, J.M.; Urteaga, R.; Bonetto, F.J. A novel water hammer device designed to produce controlled bubble collapses. *Exp. Therm. Fluid Sci.* **2018**, *92*, 46–55. [CrossRef]
9. Pal, A.; Joseph, E.; Vadakkumbatt, V.; Yadav, N.; Srinivasan, V.; Maris, H.J.; Ghosh, A. Collapse of Vapor-Filled Bubbles in Liquid Helium. *J. Low Temp. Phys.* **2017**, *188*, 101–111. [CrossRef]
10. Daou, M.M.; Igualada, E.; Dutilleul, H.; Citerne, J.-M.; Rodriguez-Rodriguez, J.; Zaleski, S.; Fuster, D. Investigation of the collapse of bubbles after the impact of a piston on a liquid free surface. *AIChE J.* **2017**, *63*, 2483–2495. [CrossRef]
11. Ma, X.J.; Huang, B.A.; Zhao, X.; Wang, Y.; Chang, Q.; Qui, S.; Fu, X.; Wang, G. Comparisons of spark-charge bubble dynamics near the elastic and rigid boundaries. *Ultrason. Sonochem.* **2018**, *43*, 80–90. [CrossRef] [PubMed]
12. Gong, S.W.; Ohl, S.W.; Klaseboer, E.; Khoo, B.C. Interaction of a spark-generated bubble with a two-layered composite beam. *J. Fluids Struct.* **2018**, *76*, 336–348. [CrossRef]
13. Oh, J.; Yoo, Y.; Seung, S.; Kwak, H.-Y. Laser-induced bubble formation on a micro gold particle levitated in water under ultrasonic field. *Exp. Therm. Fluid Sci.* **2018**, *93*, 285–291. [CrossRef]
14. Sukop, M.C.; Or, D. Lattice Boltzmann method for homogeneous and heterogeneous cavitation. *Phys. Rev. E* **2005**, *71*. [CrossRef] [PubMed]
15. Chen, X.P.; Zhong, C.W.; Yuan, X.L. Lattice Boltzmann simulation of cavitating bubble growth with large density ratio. *Comput. Math. Appli.* **2012**, *61*, 3577–3584. [CrossRef]
16. Shan, M.L.; Zhu, C.P.; Zhou, X.; Yin, C.; Han, Q.B. Investigation of cavitation bubble collapse near rigid boundary by lattice Boltzmann method. *J. Hydrodyn.* **2016**, *28*, 442–450. [CrossRef]
17. Zhou, X.; Shan, M.L.; Zhu, C.P.; Chen, B.Y.; Yin, C.; Ren, Q.G.; Han, Q.B.; Tang, Y.B. Simulation of Acoustic Cavitation Bubble Motion by Lattice Boltzmann Method. *Appl. Mech. Mater.* **2014**, *580–583*, 3098–3105. [CrossRef]

18. Kucera, A.; Blake, J.R. Approximate methods for modeling cavitation bubbles near boundaries. *Bull. Austral. Math. Soc.* **1990**, *41*, 1–44. [CrossRef]

19. Li, B.B.; Zhang, H.C.; Han, B.; Chen, J.; Ni, X.W.; Lu, J. Investigation of the collapse of laser-induced bubble near a cone boundary. *Acta Phys. Sin.* **2012**, *61*, 21.

20. Zhang, X.M.; Zhou, C.Y.; Islam, S.; Liu, J.Q. Three-dimensional cavitation simulation using lattice Boltzmann method. *Acta Phys. Sin.* **2009**, *58*, 8406–8414.

21. Mishra, S.K.; Deymier, P.A.; Muralidharan, K.; Frantziskonis, G.; Pannala, S.; Simunovic, S. Modeling the coupling of reaction kinetics and hydrodynamics in a collapsing cavity. *Ultrason. Sonochem.* **2010**, *17*, 258–265. [CrossRef] [PubMed]

22. Schanz, D.; Metten, B.; Kurz, T.; Lauterborn, W. Molecular dynamics simulations of cavitation bubble collapse and sonoluminescence. *New J. Phys.* **2012**, *14*. [CrossRef]

23. Du, T.; Wang, Y.; Huang, C.; Liao, L. A numerical model for cloud cavitation based on bubble cluster. *Theor. Appl. Mech. Lett.* **2017**, *7*, 231–234. [CrossRef]

24. Wang, Q.; Yao, W.; Quan, X.; Cheng, P. Validation of a dynamic model for vapor bubble growth and collapse under microgravity conditions. *Int. Commun. Heat Mass Transf.* **2018**, *95*, 63–73. [CrossRef]

25. Ogloblina, D.; Schmidt, S.J.; Adams, N.A. Simulation and analysis of collapsing vapor-bubble clusters with special emphasis on potentially erosive impact loads at walls. *EPJ Web Conf.* **2018**, *180*, 9. [CrossRef]

26. Chen, Y.; Lu, C.J.; Chen, X.; Li, J.; Gong, Z.X. Numerical investigation of the time-resolved bubble cluster dynamics by using the interface capturing method of multiphase flow approach. *J. Hydrodynam.* **2017**, *29*, 485–494. [CrossRef]

27. Ming, L.; Zhi, N.; Chunhua, S. Numerical Simulation of Cavitation Bubble Collapse within a Droplet. *Comput. Fluids* **2017**, *152*, 157–163. [CrossRef]

28. Peng, K.; Tian, S.; Li, G.; Huang, Z.; Yang, R.; Guo, Z. Bubble dynamics characteristics and influencing factors on the cavitation collapse intensity for self-resonating cavitating jets. *Petrol. Explor. Dev.* **2018**, *45*, 343–350. [CrossRef]

29. Xue, H.H.; Shan, F.; Guo, X.S.; Tu, J.; Zhang, D. Cavitation Bubble Collapse near a Curved Wall by the Multiple-Relaxation-Time Shan-Chen Lattice Boltzmann Model. *Chin. Phys. Lett.* **2017**, *34*, 83–87. [CrossRef]

30. Tagawa, Y.; Peters, I.R. Bubble Collapse and Jet Formation in Corner Geometries. *Phys. Rev. Fluids* **2018**, *2*. [CrossRef]

31. Goncalves, E.; Zeidan, D. Numerical study of bubble collapse near a wall. *AIP Conf. Proc.* **2018**, *1978*. [CrossRef]

32. Peng, Y.; Mao, Y.F.; Wang, B. Study on C–S and P–R EOS in pseudo-potential lattice Boltzmann model for two-phase flows. *Int. J. Mod. Phys. C* **2017**, *28*. [CrossRef]

33. Peng, Y.; Wang, B.; Mao, Y. Study on Force Schemes in Pseudopotential Lattice Boltzmann Model for Two-Phase Flows. *Math. Probl. Eng.* **2018**. [CrossRef]

34. Yuan, P.; Schaefer, L. Equations of state in a lattice Boltzmann model. *Phys. Fluids* **2006**, *18*. [CrossRef]

35. Philipp, A.; Lauterborn, W. Cavitation erosion by single laser-produced bubbles. *J. Fluid Mech.* **2000**, *361*, 75–116. [CrossRef]

36. Plesset, M.S.; Chapman, R.B. Collapse of an initially spherical vapour cavity in the neighbourhood of a solid boundary. *J. Fluid Mech.* **1970**, *47*, 283–290. [CrossRef]

Article

Three-Dimensional Aerators: Characteristics of the Air Bubbles

Shuai Li [1,2], Jianmin Zhang [3,*], Xiaoqing Chen [1,2] and Jiangang Chen [1,2]

[1]   Key Laboratory of Mountain Hazards and Earth Surface Processes, Chinese Academy of Sciences, Chengdu 610041, China; lishuai@imde.ac.cn (S.L.); xqchen@imde.ac.cn (X.C.); chenjg@imde.ac.cn (J.C.)

[2]   Institute of Mountain Hazards and Environment, Chinese Academy of Sciences and Ministry of Water Conservancy, Chengdu 610041, China

[3]   State Key Laboratory of Hydraulics and Mountain River Engineering, Sichuan University, Chengdu 610065, China

*   Correspondence: zhangjianmin@scu.edu.cn; Tel.: +86-139-8187-8609

Received: 26 August 2018; Accepted: 29 September 2018; Published: 12 October 2018

Abstract: Three-dimensional aerators are often used in hydraulic structures to prevent cavitation damage via enhanced air entrainment. However, the mechanisms of aeration and bubble dispersion along the developing shear flow region on such aerators remain unclear. A double-tip conductivity probe is employed in present experimental study to investigate the air concentration, bubble count rate, and bubble size downstream of a three-dimensional aerator involving various approach-flow features and geometric parameters. The results show that the cross-sectional distribution of the air bubble frequency is in accordance with the Gaussian distribution, and the relationship between the air concentration and bubble frequency obeys a quasi-parabolic law. The air bubble frequency reaches an apex at an air concentration (C) of approximately 50% and decreases to zero as $C = 0\%$ and $C = 100\%$. The relative location of the air-bubble frequency apex is 0.210, 0.326 and 0.283 times the thickness of the layers at the upper, lower and side nappes, respectively. The air bubble chord length decreases gradually from the air water interface to the core area. The air concentration increases exponentially with the bubble chord length. The air bubble frequency distributions can be fit well using a "modified" gamma distribution function.

Keywords: three-dimensional aerator; air concentration; air bubble frequency; air bubble chord length

---

1. Introduction

Cavitation erosion caused by high velocity flows is a common phenomenon in spillways or chutes with high-head dams. A commonly adopted counter-measure to reduce or prevent cavitation damage is aeration [1–6]. Three-dimensional aerators are used to enhance air entertainment into water and to prevent cavitation erosion on spillways. Full interfacial aeration is commonly observed at the free surface of the upper, lower and side nappe along the free jet (Figure 1). The effect of air entrainment is based on the jet disintegration process due to turbulence and secondary interactions with the surrounding atmosphere.

Figure 1. Air water flow in three-dimensional aerators: (**a**) Three Gorges Project; (**b**) Xiang Jia-Ba Project.

Air entrainment at the interfacial area is a fundamental parameter in numerical models for simulating two-phase flows. Likewise, it is also an essential consideration when designing optimum aerators. During the last few decades, numerous studies have been conducted to investigate the air concentration and free-surface aeration on an aerator. A series of experiments were conducted to study the free-surface aeration and diffusion characteristics of the bottom aeration devices [7–11]. Results indicate that the quantities and geometric scales of air bubbles are important parameters in characterizing the air concentration in the interfacial area. Chanson [12] described the flow structure of the developing aerated region in a flat chute and found that the maximum mean air content in different cross-sections was 12%. Toombes [13] showed that the maximum bubble count rates typically correspond to air concentration between 40% and 60%. The distribution of velocity indices at the different nappes, which stays at the downstream of an expanding chute aerator, were found to be different [14] and a solution was also found to improve both the bottom and lateral cavities' length [15]. The longitudinal position of the upper trajectory apex was found to be further downstream than that of the lower trajectory apex [16]. Kramer and Hager [17] found that both the entrainment rate and dimensionless bubble count rate increased with jet length. The mean air concentration is affected by aerators, but the air concentration at the bottom is limited [18]. Furthermore, Toombes and Chanson [19] studied flows past a backward-facing step, where the void fraction distribution, bubble count rate, local air distribution and water chord size were indicated. The results showed that the relationship between bubble count rate and void fraction can be defined by a quasi-parabolic function, while probability density functions of local chord size exhibited a quasi-log-normal shape. Some researchers [20–24] also established analytical equations to predict the air concentration distributions at the aerators' downstream. Numerical models such as the lattice Boltzmann method were found to successfully simulate two-phase flows and free-surface flows [25–27]. These models may be used to complement physical models and experimental measurements after careful validation and verification of model parameters such as the void fraction distribution, velocity, and air-bubble chord sizes. Chanson [28] reviewed the recent progress in turbulent free-surface flows and the mechanics of aerated flows and highlighted that physical modelling tests are efficient methods to validate phenomenological, theoretical and numerical models.

Although numerous studies regarding air–water flows have been conducted over the years, only a few researchers, however, have systematically investigated the air bubble count rate and bubble size of high velocity air–water flows downstream of the pressure outlet joined by a three-dimensional aerator (sudden vertical drop and lateral enlargement) in the tunnel. There is also a lack of knowledge regarding the influence of the bubble count rate and bubble size on the air concentration in two-phase flows, which is necessary for improving turbulence models. Thus, further insight is needed regarding the details of air entrainment in high-speed flows and the related core area features. In this study, a series of experimental investigations were systematically conducted to further the knowledge of air water flows among the upper, lower and side nappes downstream of a three-dimensional aerator (Figure 2). The experimental data and equations obtained from this study provide novel information and a quantitative description of the aeration flow structure at the high-speed interfacial area.

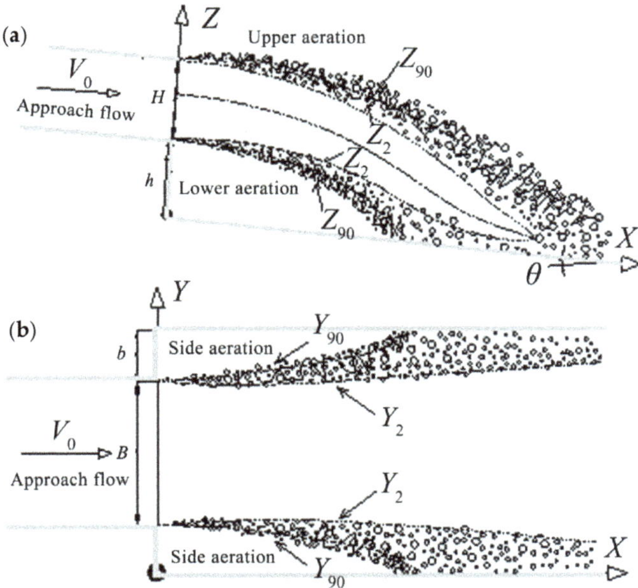

Figure 2. Schematic with the relevant parameters: (a) side view and (b) plan view.

2. Experimental Setup

Experiments were performed at the State Key Laboratory of Hydraulic and Mountain River Engineering, Sichuan University, China. The experimental setup (Figure 3) includes an upstream reservoir, a pressure section, a sudden vertical drop and lateral enlargement aerator, a free flow section, a tail-water section, and an underground reservoir. Water is discharged (can be up to 0.5 m^3/s) by a continuous and stable water supply system (2.5 m wide, 3.5 m long, and 6.3 m high). The components of the pressure inlet were assembled with a smooth steel plate. The pressure section is a rectangular pipe (0.25 m wide, 0.15 high and 2.0 m long) with a variable inclination angle varied with the downstream chute. The sudden fall-expansion aerator (height h × width b) was installed at the end of the pressure section. The free-flow section was fabricated from polymethyl methacrylate (PMMA) with a roughness height of 0.008 mm. The upper, lower, and side nappes were exposed to atmospheric pressure. The constant water levels corresponding to a particular head were maintained in the upstream reservoir.

Experiments were carried out for mean velocities (V_0) of 6 m/s, 7 m/s, 8 m/s, and 9 m/s; chute slopes of 0%, 10%, and 25%; and aerator sizes (h, b) of 0.025m, 0.045m, and 0.065 m. The test cases are summarized in Table 1 ($0.9 < q_w < 1.35$, $0.89 < R_e < 1.34$). It should be noted that some combinations (series 7–10, and 13 in Table 1) pertain to cases involving three-dimensional (vertical drop and lateral enlargement) aerators. The other cases involve only either vertical drop aerators or lateral enlargement aerators. Upstream flows were supercritical ($4.95 < Fr < 7.42$) for all of the investigated flow conditions. Downstream of the aerator, a free jet, with the use of a ventilated air cavity below and/or beside it, was formed.

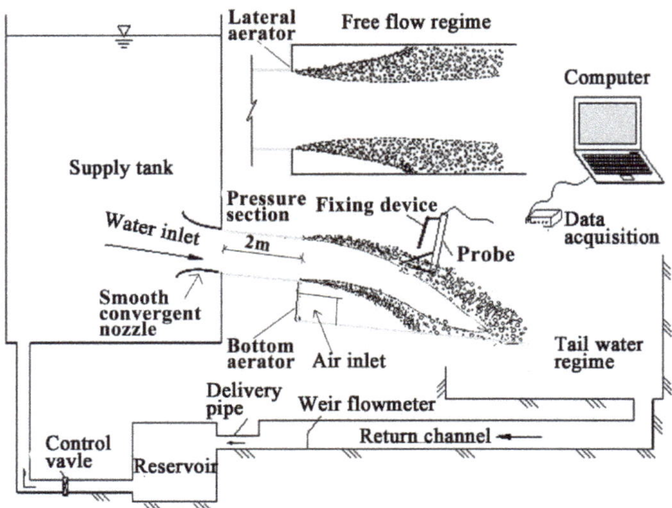

Figure 3. Sketch of the experimental model and test equipment.

Table 1. Summary of the operating conditions for the experiments.

Series	(h, b)/m	θ
1	(0.025, 0.000)	10%
2	(0.045, 0.000)	10%
3	(0.065, 0.000)	10%
4	(0.000, 0.025)	10%
5	(0.000, 0.045)	10%
6	(0.000, 0.065)	10%
7	(0.025, 0.025)	10%
8	(0.045, 0.045)	10%
9	(0.065, 0.065)	10%
10	(0.045, 0.045)	0%
11	(0.000, 0.045)	0%
12	(0.045, 0.000)	0%
13	(0.045, 0.045)	25%
14	(0.000, 0.045)	25%
15	(0.045, 0.000)	25%

The air concentration, bubble count rate, and chord length were recorded using a CQY—Z8a measurement instrument [14,29,30] with a double-tip conductivity probe (Figure 4). Different voltage indices between air and water were measured by the platinum tip. The probe consisted of two identical tips, including an external stainless steel electrode with 0.7-mm-diameter and an internal concentric platinum electrode with 0.1-mm-diameter. The tips are aligned in the flow direction and the distance between the two tips is 12.89 mm. Both tips were connected to an electronics device with a response time less than 10 μs. The vertical translation of the probe was dominated by a fixed device with an accuracy of 1 mm. The probe measurements were taken at regular intervals of 5 mm along the vertical direction from the chute bottom to the free surface (or side nappe surface) and 100 mm along the flow direction. At each location, signals were recorded at a scan rate (f) of 100 kHz per channel for a scan period (T) of 10 s. The scan period was based on the findings of Toombes [13] which indicated that a scanning period of 10 s was long enough to provide a reasonable representation of the flow characteristics while maintaining realistic time and data storage constraints.

Figure 4 presents a typical aeration structure detected by a probe at a fixed location along the aeration flow. Assuming each segment is either air or water, the air concentration can be presented as the probability of any discrete element being air. The air concentration is computed as the encountering air's probability at the leading tip of the probe.

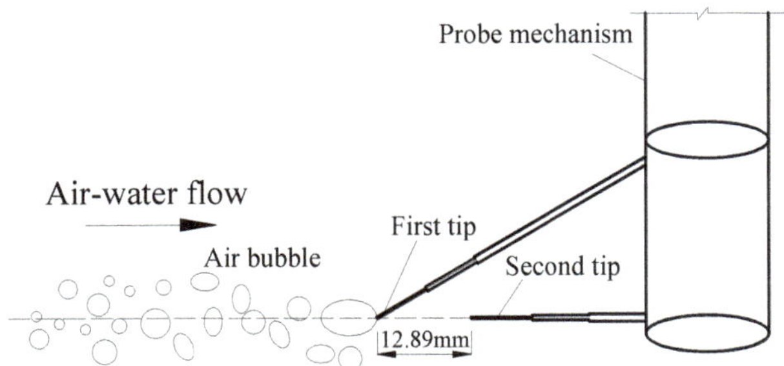

Figure 4. Sketch of the double-tip conductive probe.

The number of data (N) can be obtained by multiplying the frequency (f) of the measurement instrument with the sampling time T, i.e., $N = f \times T$. Those data were classified according to two categories, air and water signals. The air concentration (C) was computed as the encountering air's probability at the leading tip of the probe, which can be expressed as [26]

$$C = \frac{\sum\limits_{i=1}^{N} R_i}{N} \tag{1}$$

where R_i represents the ratio of bubbles measured all throughout the whole measurement time. When the probe tip is completely immersed in air bubbles $R_i = 1$, otherwise $R_i = 0$.

The length of bubble chord is defined as the straight distance between the two intersections of the interface [12]. Note that an air bubble is defined as a volume of air (i.e., air entity), which can be detected by the leading tip of the probe between two continuous air–water interface events. The bubble chord length is calculated by

$$d_i = \frac{v \times n_i}{f} \tag{2}$$

where n_i is the number of detected bubbles recorded; and v is the bubble velocity equal to the local mean aeration velocity (i.e., no slip between the air and water phases).

3. Results

The air bubble frequency and chord length are two important parameters used to characterize the air concentration of aerated flows. We first describe air bubble frequency among the upper, lower and side air–water mixed layers in the experimental flows with various conditions. In this section, the relationships between air concentration with air bubble frequency, and with the relative location at which the maximum air bubble frequency are also discussed. Next, we identify the effects of initial flow velocity on air bubble chord length. In addition, the relationships between the air concentration and air bubble chord length, and between air bubble count rate and air bubble size are also evaluated.

3.1. Air Bubble Frequency

The air bubble frequency characterizes the flow fragmentation, which is proportional to the specific area of the interface between two phases. The air bubble frequency, or air bubble count rate, is a function of bubbles' shape and size, surface tension, and shear forces of the fluid. Predicting how air concentration affects the bubble frequency is a complex problem [19]. A simplified analogy would be to consider on aeration flows past a fixed probe, as a series of discrete one-dimensional air and water elements (Figure 4). Here, the air bubble frequency (*Fa*) is defined as the number of air-structures (*Na*) per unit time (*t*) detected by the leading tip of the probe sensor (i.e., $Fa = Na/t$).

Typical air bubble frequency distributions at each cross-section along the upper, lower, and side nappes are presented in Figure 5. The air bubble frequency distributions exhibit unitary self-similarity along the thickness of the mixed layer and the horizontal distance. The trend of the cross-sectional distribution tends to initially increase and then decrease from the aeration interface to the inside of the water following a Gaussian distribution that is defined by the following functions:

$$\frac{F_a}{(F_a)_{\max}} = \begin{cases} \dfrac{1}{\sqrt{2\pi}\sigma_0} e^{-\dfrac{\left(\frac{Z-Z_2}{Z_{90}-Z_2}-\mu_0\right)^2}{2\sigma^2}} & \text{the upper nappe} \\[2em] \dfrac{1}{\sqrt{2\pi}\sigma_0} e^{-\dfrac{\left(\frac{Z-Z_{90}}{Z_2-Z_{90}}-\mu_0\right)^2}{2\sigma^2}} & \text{the lower nappe} \\[2em] \dfrac{1}{\sqrt{2\pi}\sigma_0} e^{-\dfrac{\left(\frac{Y-Y_{90}}{Y_2-Y_{90}}-\mu_0\right)^2}{2\sigma^2}} & \text{the side nappe} \end{cases} \tag{3}$$

where, $F_a/(F_a)_{\max}$ is the dimensionless air bubble frequency and $(F_a)_{\max}$ is the maximum bubble frequency in the cross-section; The characteristic value of μ_0 reflects the depth corresponding to the air bubble frequency; σ_0 represents the degree of deviation between the relative depth and its mean value.

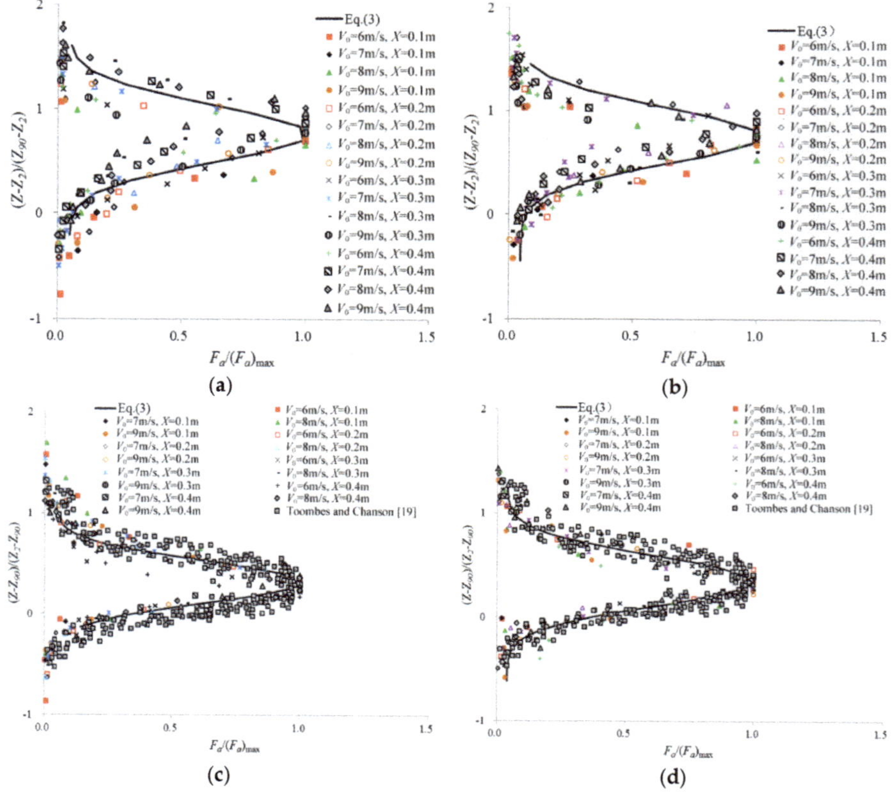

Figure 5. *Cont.*

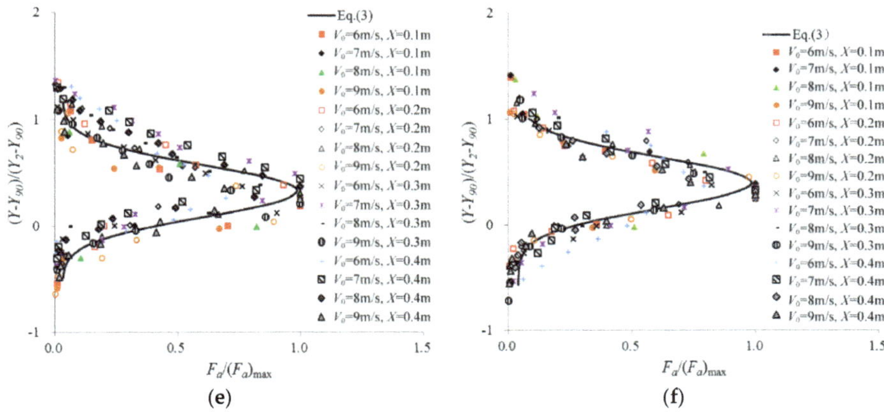

Figure 5. Air bubble frequency distributions compared with the theoretical distribution obtained from Equation (3). Upper nappe (**a**) (h, b) = (4.5 cm, 4.5 cm), θ = 25%. (**b**) (h, b) = (6.5 cm, 6.5 cm), θ = 10%; lower nappe (**c**) (h, b) = (4.5 cm, 4.5 cm), θ = 25%. (**d**) (h, b) = (6.5 cm, 6.5 cm), θ = 10%; side nappe (**e**) (h, b) = (4.5 cm, 4.5 cm), θ = 25%. (**f**) (h, b) = (6.5 cm, 6.5 cm), θ = 10%.

For all experimental results, the air bubble frequency distributions correspond well with the data reported by Toombes and Chanson [19], as illustrated in Figure 5. Figure 6 presents a plot of experimental data, $F_a/(F_a)_{max}$, with calculated non-dimensional air bubble frequency using Equation (3). The calculated results at upper, lower, and lateral nappes were in qualitative agreement with experimental data, although there was some scatter. It is noted that the characteristic values of μ_0 and σ_0 differed among the upper, lower, and side nappes (Table 2). For the upper nappe, the air bubbles do not easily diffuse and instead transport to the inside of the water due to the buoyancy. This result in the air bubbles being drawn closer to the free surface ($\mu_0 = 0.210$). For the side nappe, the air bubbles can easily diffuse and transport to the interior of the water ($\mu_0 = 0.283$) because of the effect of large transverse turbulent diffusion [15]. For the lower nappe, the air bubbles are dragged by the buoyant force to the interior of the water ($\mu_0 = 0.326$).

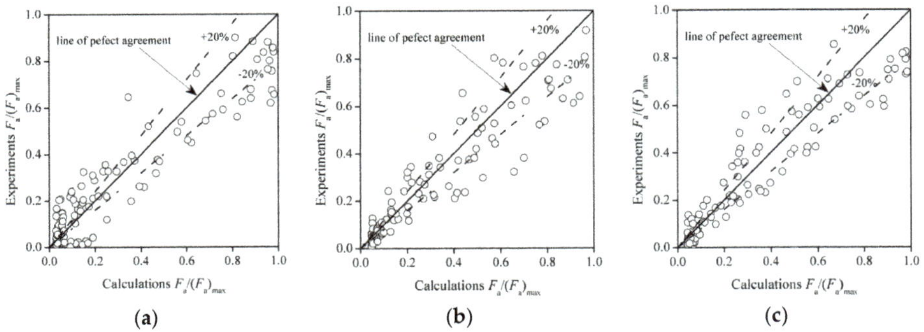

Figure 6. Comparison of Equation (3) with the experimental data. (**a**) upper nappe; (**b**) lower nappe; (**c**) side nappe.

Table 2. Values of μ_0 and σ_0.

Nappe	μ_0	σ_0
lower	0.326	0.419
upper	0.210	0.413
side	0.283	0.423

Air concentration distribution is just the external manifestation of air–water flows, whereas the internal factors include the count rate and size of the air bubbles. So, it is crucial to understand the features of the air bubble frequency and the air bubble chord length.

The air bubble frequency distributions at various positions along the jet obtained in this study and those reported by Chanson [12] and Toombes and Chanson [19] are shown in Figure 7. It is evident that the air bubble frequency initially increases and then decreases with the increase of air concentrations in the upper, lower, and side nappes. At each cross-section, the air bubble frequency profiles reach to an apex which corresponds to air concentrations of approximately 50%. The profiles tend to zero at extremely low and extremely high air concentrations. Overall, the distributions of the dimensionless air bubble frequency can be well fitted by the parabolic function:

$$\frac{F_a}{(F_a)_{max}} = 4C(1-C) \tag{4}$$

Qualitatively, the calculated results of Equation (4) correspond with the results of Chanson [12] and Toombes and Chanson [19] (Figure 7). It is worth noting that the data [12] slightly deviate from the fully developed supercritical flows, suggesting that Equation (4) has broad applicability.

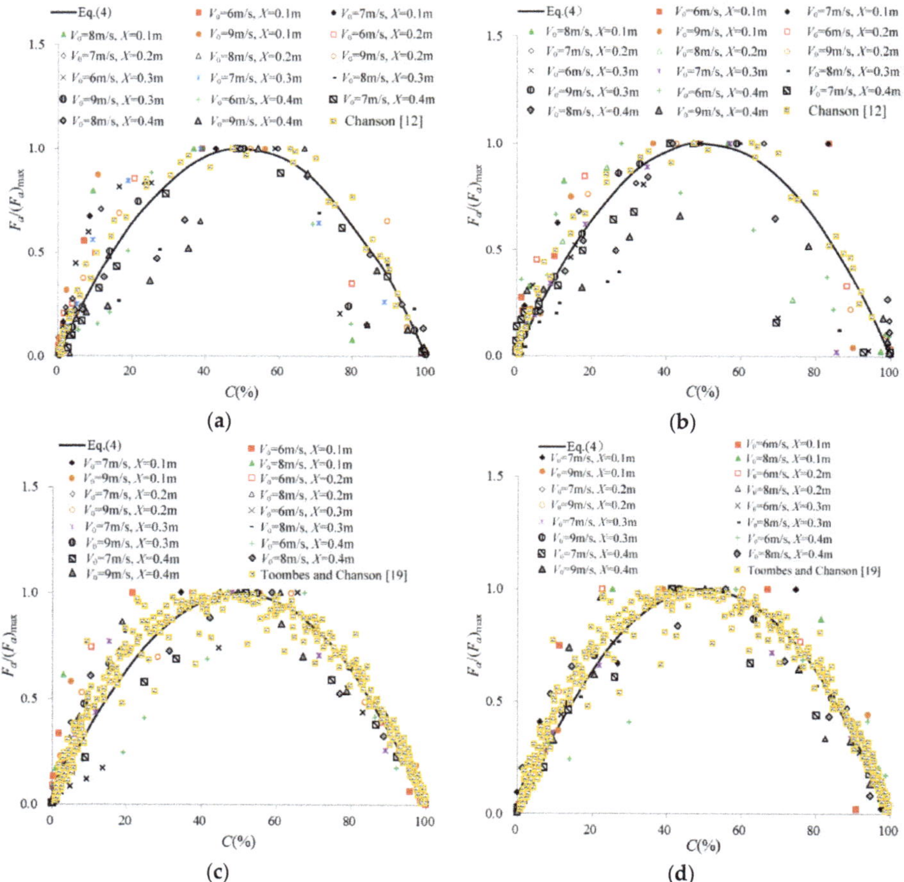

Figure 7. *Cont.*

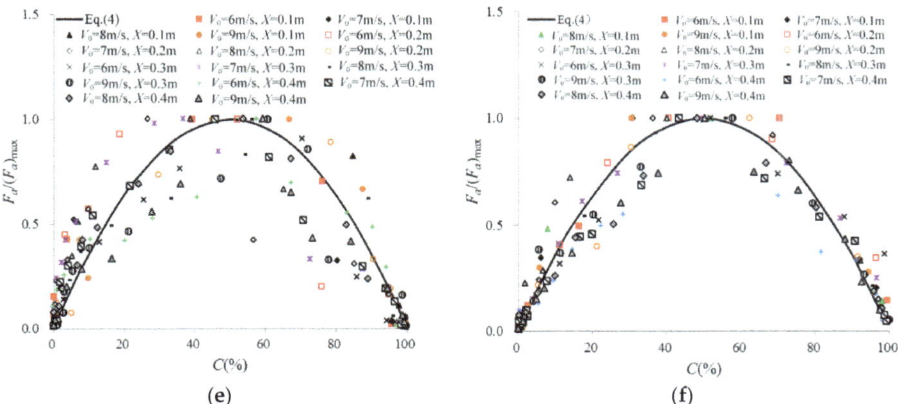

Figure 7. Dimensionless air bubble frequency as a function of the air concentration fitted with Equation (4). Upper nappe (**a**) $(h, b) = (4.5$ cm, 4.5 cm$)$, $\theta = 25\%$. (**b**) $(h, b) = (6.5$ cm, 6.5 cm$)$, $\theta = 10\%$; lower nappe (**c**) $(h, b) = (4.5$ cm, 4.5 cm$)$, $\theta = 25\%$. (**d**) $(h, b) = (6.5$ cm, 6.5 cm$)$, $\theta = 10\%$; side nappe (**e**) $(h, b) = (4.5$ cm, 4.5 cm$)$, $\theta = 25\%$. (**f**) $(h, b) = (6.5$ cm, 6.5 cm$)$, $\theta = 10\%$.

The majority of the data fall within the $\pm20\%$ error lines as shown in Figure 8. The calculated non-dimensional air bubble frequency using Equation (4) are in reasonably good agreement with the experimental data at upper, lower, and lateral nappes.

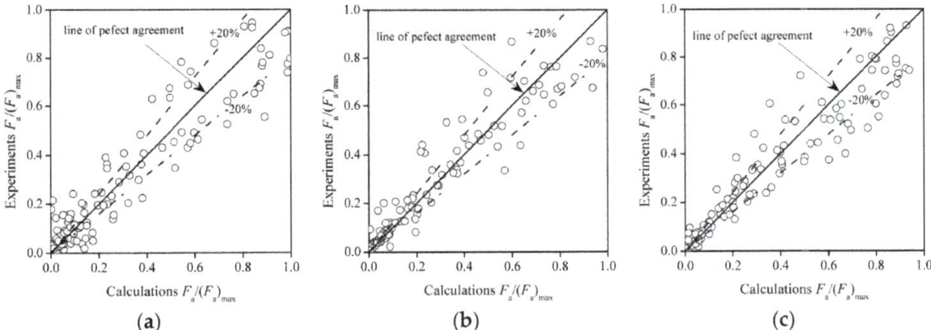

Figure 8. Comparison of experimental data with calculated values of dimensionless air bubble frequency. (**a**) upper nappe; (**b**) lower nappe; (**c**) side nappe.

The relationships between air concentration and the relative distance, which corresponds to the maximum air bubble frequency are presented in Figure 9 where a large fluctuation of air concentration ($C = 30\% \sim 60\%$) in the initial regime of aeration can be observed. The fluctuation may be due to the aeration layer located close to the pressure outlet, which may be thin and unstable. The air–water layer becomes stable with the development of the aeration further downstream. The air concentration corresponding to the maximum air bubble frequency gradually moves toward to 50% line. The result is consistent with the findings of Figure 7. Actually, the maximum air bubble frequency does not always coincide with $C = 50\%$ accurately. Factors such as the average size and length scales of discrete air and water elements may affect the local air concentration and flow conditions in air–water flows [13,31].

The relative location corresponding to the maximum air bubble frequency is shown in Figure 10. It can be seen that the relative location of the maximum air bubble frequency is 0.210, 0.326 and 0.283 times the thicknesses of the air–water layers in the upper, lower and side nappes, respectively. The results are consistent with the results of Figure 5. For the upper nappe, the air can easily be drawn into the water because the air–water external interface directly interacts with the atmosphere. However, the air bubbles cannot easily diffuse and transport to the inside of the jet because of the effect of buoyancy. For the lower nappe, the air bubbles can easily diffuse and move into the interior of the jet due to the local air rotation and eddy currents, which are driven by the air–water interfacial turbulence coupled with the positive effect of buoyancy.

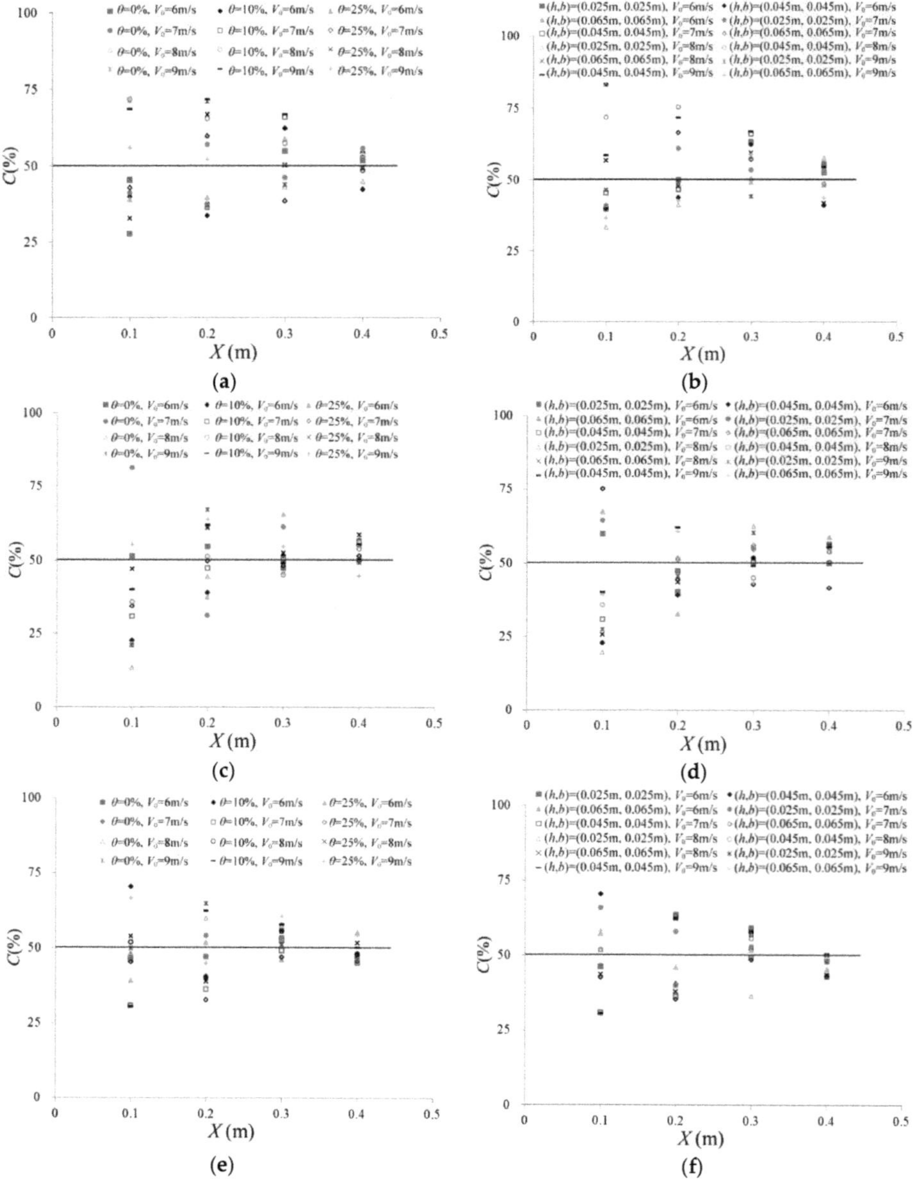

Figure 9. Air concentration corresponding to the maximum air bubble frequency. Upper nappe (**a**) $(h, b) = (4.5$ cm, 4.5 cm$)$. (**b**) $\theta = 10\%$; lower nappe (**c**) $(h, b) = (4.5$ cm, 4.5 cm$)$. (**d**) $\theta = 10\%$; side nappe (**e**) $(h, b) = (4.5$ cm, 4.5 cm$)$. (**f**) $\theta = 10\%$.

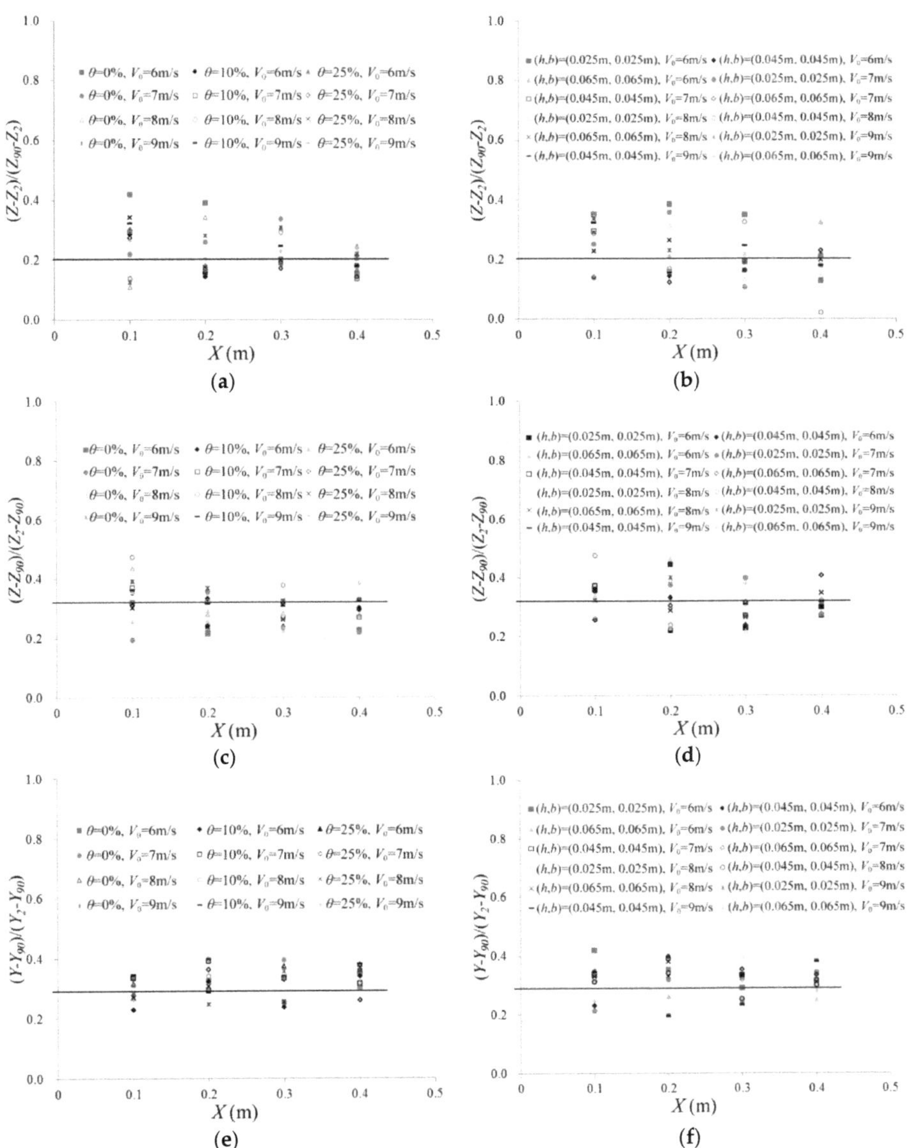

Figure 10. Relative location corresponding to the maximum dimensionless air bubble frequency. Upper nappe (**a**) (h, b) = (4.5 cm, 4.5 cm). (**b**) θ = 10%; lower nappe (**c**) (h, b) = (4.5 cm, 4.5 cm). (**d**) θ = 10%; side nappe (**e**) (h, b) = (4.5 cm, 4.5 cm). (**f**) θ = 10%.

3.2. Air Bubble Chord Length Distributions

Air bubble size is another parameter reflecting the characteristics of air–water flows. It is difficult to measure the size of air bubbles since their shapes vary greatly, are complex and highly changeable. This value can only be indirectly determined from the bubble chord length which is obtained using the double-tip conductivity probe. Here, the chord length $(ch_{ab})_{mean}$ is adopted to assess the size of the air bubbles, which is defined as follows:

$$(ch_{ab})_{\text{mean}} = \frac{\sum\limits_{j=1}^{N} (ch)_j n_j}{\sum\limits_{j=1}^{N} n_j} \tag{5}$$

where n_j is the count of air bubbles in the bubble chord length interval $\Delta(ch_{ab})_j$ ($\Delta(ch_{ab})_j = 0.1$ mm in the experiments); and $(ch_{ab})_j$ is the average value of the chord length interval (e.g., the probability of chord length from 1.0 to 1.1 mm is represented by the label 1.05 mm).

The air bubble chord lengths $(ch_{ab})_{\text{mean}}$ are presented in Figure 11 at various positions in the upper, lower and side nappes. It is note that the air bubble chord length distributions obtained at the upper, lower, and side nappes have similar shapes. The air bubble chord length in the vicinity of the air–water interface is largest and then gradually decreases toward the inside of the water. A broad spectrum of air bubble chord lengths, i.e., from less than 0.1 mm to greater than 100 mm, is observed at each location and cross-section. At high air concentration (i.e., $C \geq 90\%$), chord length of many air bubbles can reach up to 100 mm or even more. These large values may be large air packets and air volumes surrounding the water structure (e.g., droplets). The reason is that the regime with high air concentration is always close to the air–water interfaces, which become uneven under the influence of turbulent forces. Air in the vicinity of the free surface that are trapped by the generated waves were treated as air bubbles by the conductivity probe. In addition, the shape of the air bubble is not completely spherical, in most cases, it is oval or banding, causing it to be measured to be much larger than it actually is. Indeed, it is nearly impossible to distinguish between air-bubbles and an un-enclosed bubble structure. The results indicate that the large air bubbles are constantly sheared and tore as they move towards the interior of the water. This results in air bubbles located deeper inside the body of water to be much smaller in size. Figure 9 also suggests that for $C < 90\%$, bubble chord lengths will most likely be no more than 20 mm and that bubbles close to the pressure outlet will have small sizes.

A positive correlation is observed between the air concentration and air bubble chord length among the upper, lower and side nappes (Figure 12). The distribution can be defined according to the function:

$$(ch_{ab})_{\text{mean}} = \frac{1}{a_1 + a_2 * C^{a_3}} \tag{6}$$

where a_1, a_2 and a_3 are empirical coefficients that may be related to the interaction. For the present experimental data, $a_1 = 0.212$, $a_2 = -0.006$ and $a_3 = 0.767$ were used. The coefficients of determination were set to be 0.85. 0.93 and 0.91 for the upper, lower and side nappes respectively.

Figure 13 presents a plot of experimental data, $(ch_{ab})_{\text{mean}}$, with calculated results using Equation (6). Since most of the values of $(ch_{ab})_{\text{mean}}$ calculated from Equation (6) fall around the line of perfect agreement in an area within $\pm 20\%$ error lines, although some discrepancy exists. The results exhibit agreement with the theoretical distribution curve of Equation (6) for all flow conditions at various positions.

As the relationships between the air concentration and air bubble frequency, as well as the air bubble chord length have been discussed above, the relationship between the air bubble count rate and bubble size is also considered (Figure 14). It is evident that the correlations are clearly non-symmetric, unimodal and positively skewed and may therefore be represented by a probability density function defined by the Gamma distribution. According to mathematical statistics, the gamma distribution is a continuous probability function that can be written as follows:

$$Ga(x) = \frac{1}{\Gamma(\alpha)} \beta^{-\alpha} x^{\alpha-1} e^{-\frac{x}{\beta}}, x > 0 \tag{7}$$

where α is the shape parameter, β is the scale parameter, and $\Gamma(x) = \int_0^{\infty} t^{x-1} e^{-t} dt$ is the gamma Function, and $\Gamma(x+1) = x\Gamma(x)$.

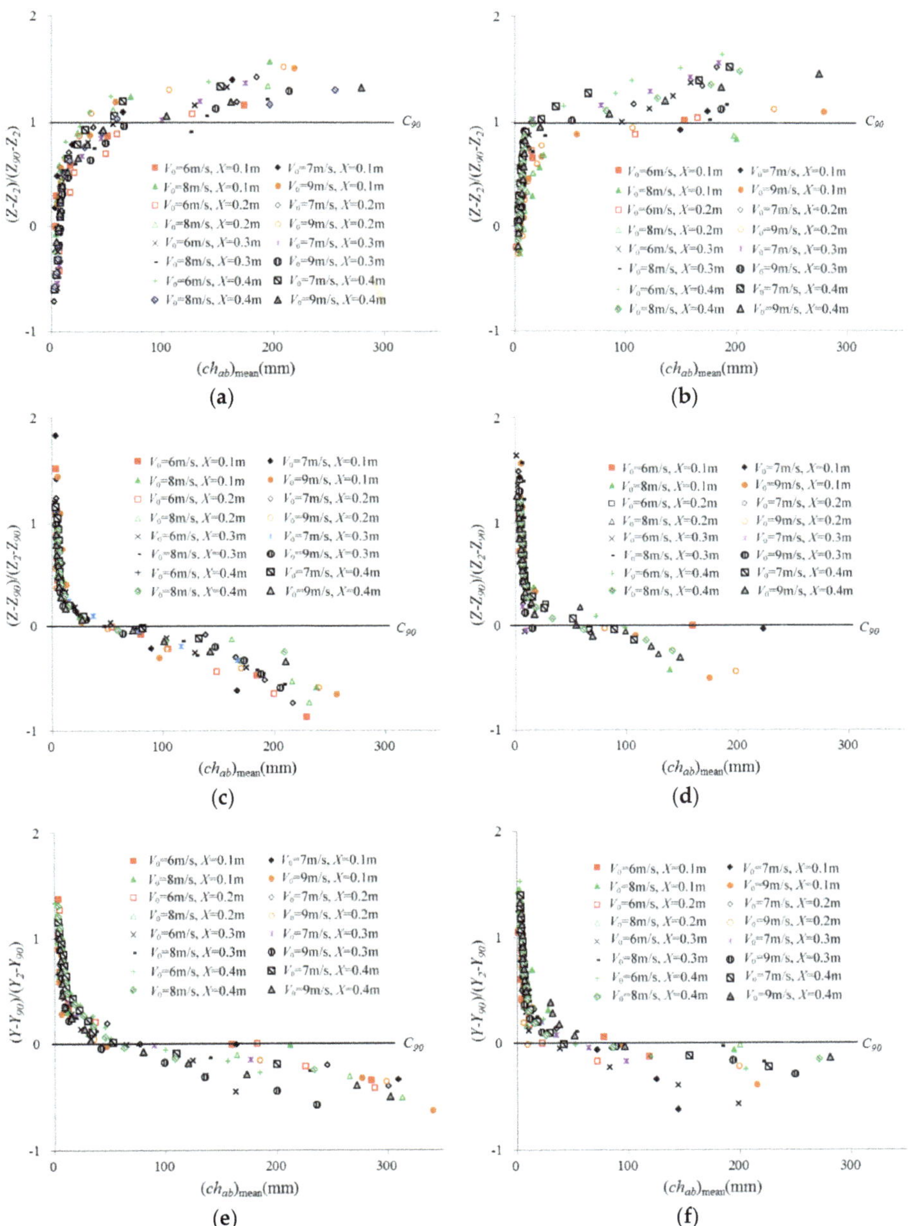

Figure 11. Air bubble chord length distributions. Upper nappe (**a**) $(h, b) = (4.5 \text{ cm}, 4.5 \text{ cm})$, $\theta = 0\%$. (**b**) $(h, b) = (2.5 \text{ cm}, 2.5 \text{ cm})$, $\theta = 10\%$; lower nappe (**c**) $(h, b) = (4.5 \text{ cm}, 4.5 \text{ cm})$, $\theta = 0\%$. (**d**) $(h, b) = (2.5 \text{ cm}, 2.5 \text{ cm})$, $\theta = 10\%$; side nappe (**e**) $(h, b) = (4.5 \text{ cm}, 4.5 \text{ cm})$, $\theta = 0\%$. (**f**) $(h, b) = (2.5 \text{ cm}, 2.5 \text{ cm})$, $\theta = 10\%$.

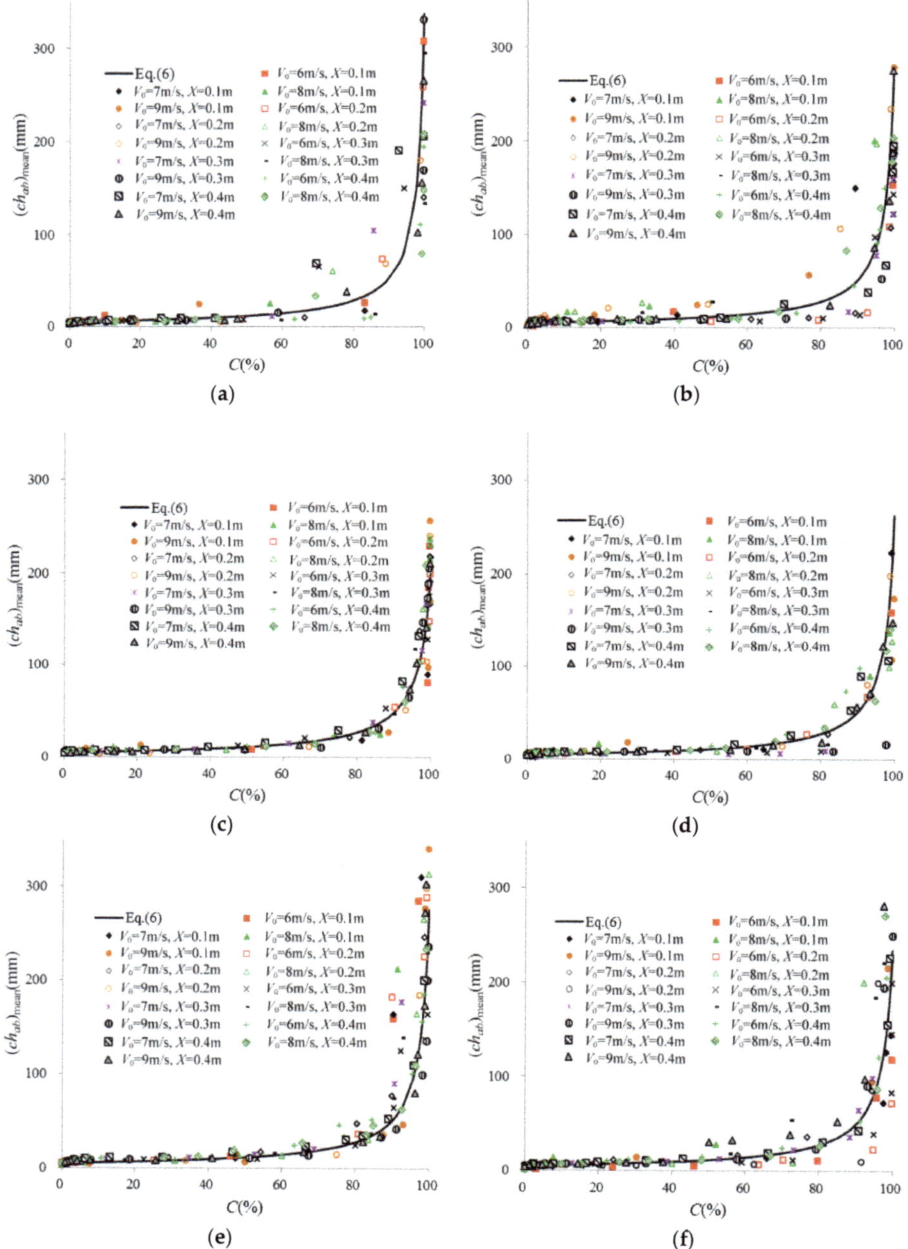

Figure 12. Measured air bubble chord length as a function of the air concentration. The solid lines are determined from Equation (6). Upper nappe (**a**) $(h, b) = (4.5$ cm, 4.5 cm$)$, $\theta = 0\%$. (**b**) $(h, b) = (2.5$ cm, 2.5 cm$)$, $\theta = 10\%$; lower nappe (**c**) $(h, b) = (4.5$ cm, 4.5 cm$)$, $\theta = 0\%$. (**d**) $(h, b) = (2.5$ cm, 2.5 cm$)$, $\theta = 10\%$; side nappe (**e**) $(h, b) = (4.5$ cm, 4.5 cm$)$, $\theta = 0\%$. (**f**) $(h, b) = (2.5$ cm, 2.5 cm$)$, $\theta = 10\%$.

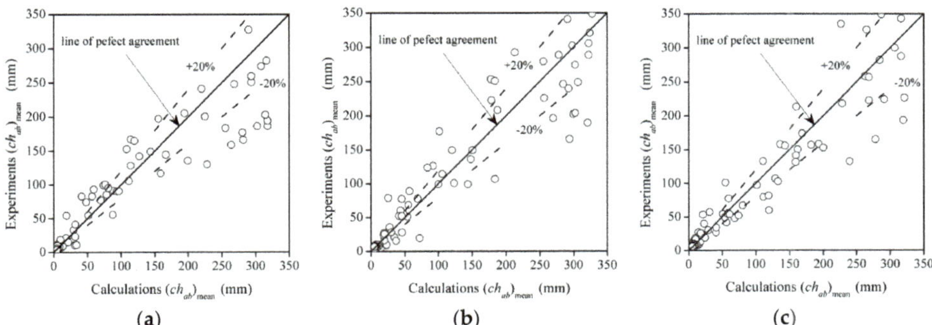

Figure 13. Comparison of experimental data with calculated values of air bubble chord length. (**a**) upper nappe; (**b**) lower nappe; (**c**) side nappe.

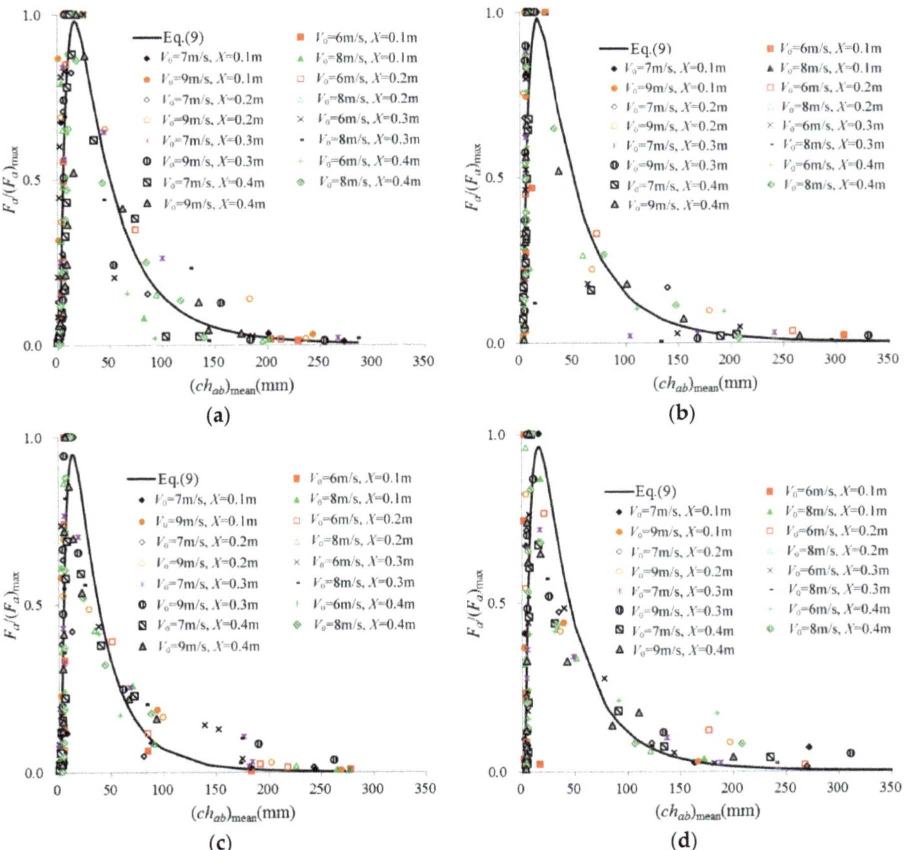

Figure 14. *Cont.*

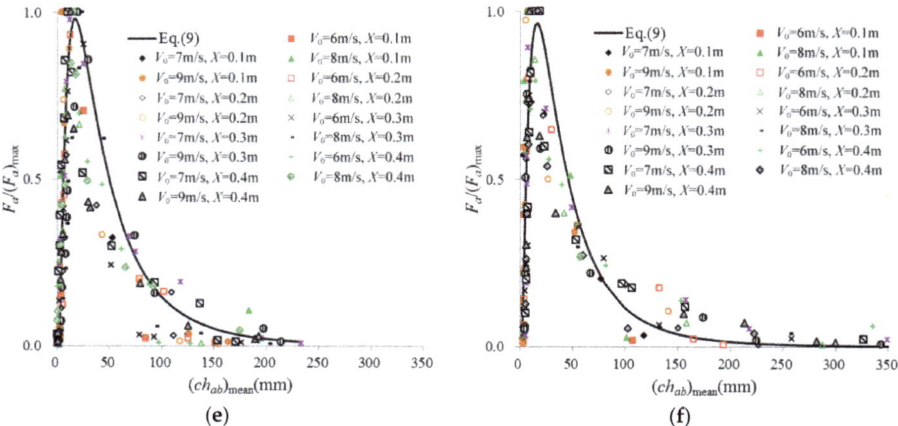

Figure 14. Dimensionless bubble frequency as a function of the bubble chord length compared with the theoretical curve derived from Equation (9). Upper nappe (**a**) (h, b) = (4.5 cm, 4.5 cm), θ = 10%. (**b**) (h, b) = (6.5 cm, 6.5 cm), θ = 10%; lower nappe (**c**) (h, b) = (4.5 cm, 4.5 cm), θ = 10%. (**d**) (h, b) = (6.5 cm, 6.5 cm), θ = 10%; side nappe (**e**) (h, b) = (4.5 cm, 4.5 cm), θ = 10%. (**f**) (h, b) = (6.5 cm, 6.5 cm), θ = 10%.

Equation (7) is a classic two-parameter gamma distribution. Considering the complexity of the air bubble frequency and chord length in aeration flows, the five-parameters form of the generalized gamma distribution (also known as the generalized gamma function with four parameters after translation) is used in this study. The probability density function is written as follows:

$$Ga(x) = \frac{1}{\Gamma(a_1)} a_2^{\eta_1 a_1} (x - a_3)^{\eta_2 a_1} e^{-a_4 * (x - a_3)^{a_5}} \qquad x > 0 \tag{8}$$

where η_1 and η_2 are the coefficients; a_1, a_2, a_3, a_4, and a_5 are the five parameters, of which a_1 and a_5 denote the shape parameters, a_2 and a_4 denote the scale parameters, and a_3 is the threshold parameter; and $\Gamma(\cdot)$ is the gamma function.

Based on the experimental data and the theoretical distribution curve of Equation (8), the relationship between the air bubble frequency and air bubble chord length can be fitted as follows:

$$\frac{F_a}{(F_a)_{max}} = 115[(ch_{ab})_{mean} - 1.5]^{3.8} * e^{-7.65[(ch_{ab})_{mean} - 1.5]^{0.25}} \tag{9}$$

Examples of a "modified" five-parameters gamma distribution that were used to fit the air bubble chord length distributions (Figure 14). The data is reasonably well represented by a "modified" gamma function distribution, although there is significant scatter.

Additionally, Figure 15 presents a plot of experimental data, $F_a/(F_a)_{max}$, with calculated non-dimensional air bubble frequency using Equation (9). Since most of the values of $F_a/(F_a)_{max}$ calculated from Equation (9) fall around the line of perfect agreement in an area within $\pm20\%$ error lines, suggesting that Equation (9) can be considered adequate in calculating the air bubbles chord length distribution. The "modified" gamma function is skewed to the left with a peak that lies in the small chord size values. The mode was found to be between 0.5 and 30 mm.

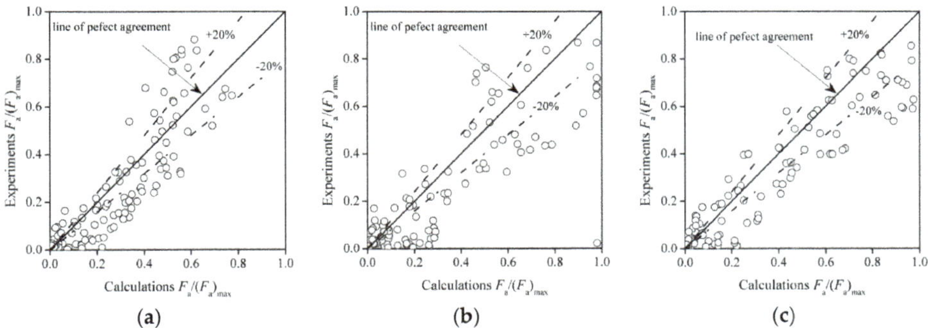

Figure 15. Comparison of experimental data with calculated values of dimensionless bubble frequency.
(**a**) upper nappe; (**b**) lower nappe; (**c**) side nappe.

4. Discussion

Air concentration is a macro parameter that is used to describe air–water flows. However, it is still insufficient in providing insight that would lead to a deeper understanding of the said phenomena. The process of aeration defines the general developing process of air moving into a body of water, i.e., the air bubbles. As such, the micro characteristics of air bubbles such as its count rate and size are crucial parameters in the study of aeration flows.

A double-tip conductivity probe used in present study to detect the air–water interfaces at different fixed points downstream of a three-dimensional (vertical drop and lateral enlargement) aerator. Some simple expressions were developed to simulate the distributions of air bubble frequency and bubble chord sizes. The relationships between the air concentration, the air bubble frequency and the air bubble chord length were also established. The obtained air concentration distributions satisfy the analytical model of Chanson [12], Equation (4) for air bubble frequency distributions as observed in supercritical open channel flows and backward-facing step flows (Toombes and Chanson [19]). A "modified" gamma probability density function was found to provide a good fit to the air bubble chord length distributions measured from three-dimensional aerator flows. This is different from the log-normal probability density functions that Chanson (1997) used for supercritical open channel flows. The distributions of the air bubble chord length showed a positive correlation with the void fraction.

Physical modelling may provide some information on the flow motion if a suitable dynamic similarity is selected. Air–water flows depend on Froude, Reynolds, and Weber numbers. Strictly speaking, dynamic similarity of air concentration and air bubble dissipation for free-surface aeration on aerator flows on a reduced-scale model is not possible (Chanson [32]) because it is impossible to satisfy simultaneously Froude and Reynolds similarities. Hence, the experimental results cannot be directly extrapolated to prototypes unless working at full scale. Nevertheless, with the assumption of Froude similarity, the Reynolds number (Re) should be larger than 1×10^5 to minimize the scale effects (Chanson [32]; Heller [33]). This limit is considered in the present study (Re ~0.89–1.34 × 10^6). Actually, results from scale experiments using Froude similarity are on the safe side for engineers with respect to air concentrations due to aeration tends to be underestimated in the smaller experiments.

Due to the complexity of bubble distribution in the air–water interfaces, the present results only serve as a preliminary investigation of the air bubble properties in the air–water mixtures downstream of a three-dimensional aerator. Hence, more detailed research is needed to study the air bubbles in aeration flows.

5. Conclusions

A number of experiments were performed under various configurations and approach flow velocities for a 3D aerator. The air concentration, air bubble frequency and bubble chord length were recorded in the developing shear layer of the upper, lower and side nappes downstream. The results are summarized as follows:

(1) The air bubble frequency distributions (yielding to the Gauss function) exhibit a unitary self-similarity along the air–water layer, presenting a trend which initially increases and then decreases from the air–water interface to the inside of the water. A quasi-parabolic relationship between the air bubble frequency and the air concentration is obtained for the upper, lower and side nappes.

(2) The air bubble frequency reaches to apex at approximately $C = 50\%$, and then decreases to zero as $C = 0\%$ and $C = 100\%$. The relative location at which the maximum air bubble frequency is observed is at 0.21, 0.326 and 0.283 times of the thickness of the air–water layers for the upper, lower and side nappes, respectively.

(3) The air bubble chord length decreases gradually from the free interface to the interior of the water. When the air bubble chord size is plotted against the air concentration, the resulting distribution follows a power-law function. The relationship between the air bubble frequency and bubble chord size exhibits a "modified" gamma function for the upper, lower and side nappes.

Author Contributions: S.L. and J.Z. proposed the research ideas and methods of the manuscript and writing, X.C. put forward the revise suggestion to the paper. J.C. is responsible for data collection and creating the figures.

Funding: This research was funded by the Chinese Scholarship Council (CSC), grant number 201804910306; the CAS "Light of West China" Program, grant number Y6R2220220; the National Science Fund for Distinguished Young Scholars, grant number 51625901; and the Natural Science Foundation of China, grant number 51579165.

Conflicts of Interest: The authors declare no conflicts of interest.

Notations

The following symbols are used in this paper:

$a_{1,2,3,4,5}$	empirical coefficients
b	the width of sudden fall-expansion aerator
B	width of the pressure outlet
C	air concentration
$(ch_{ab})_{mean}$	the air bubble chord length size
$(ch_{ab})_j$	the average value of air bubble chord length
d_i	air bubble chord length
i	air bubble number
F_a	air bubble frequency
$(F_a)_{max}$	the maximum bubble frequency
F_r	Froude Number
g	gravitational acceleration
h	the height of sudden fall-expansion aerator
H	height of the pressure outlet
q_w	the unit-width discharge
Re	Reynolds number
n_i	the number of detected bubbles
n_j	the count of air bubbles
N_a	the number of air-structures
R_i	a ratio of the bubble measured in the whole measurement time
t	time
V_0	the cross-sectional mean velocity at the pressure outlet
X, Y, Z	the horizontal, transverse and vertical coordinates of the lower nappe profile

θ	the slope with respect to the horizontal downstream bottom plane
Y_2, Z_2	characteristic air–water flow heights where the air concentration $C = 2\%$;
Y_{90}, Z_{90}	characteristic air–water flow heights where the air concentration $C = 90\%$
μ_0	characteristic depth corresponding to the air bubble frequency
σ_0	degree of deviation between the relative depth and its mean value
$Ga(x)$	the probability density function of the gamma function
α	the shape parameter
β	the scale parameter
$\Gamma(\cdot)$	the gamma function

References

1. Cassidy, J.; Elder, R. *Spillways of High Dams, Developments in Hydraulic Engineering—2*; Elsevier: New York, NY, USA, 1984.
2. Kells, J.A.; Smith, C.D. Reduction of cavitation on spillways by induced air entrainment. *Can. J. Civ. Eng.* **1991**, *18*, 358–377. [CrossRef]
3. Hager, W.H.; Pfister, M. Historical advance of chute aerators. In Proceedings of the 33rd IAHR Congress, Vancouver, BC, Canada, 9–14 August 2009; pp. 5827–5834.
4. Pfister, M.; Lucas, J.; Hager, W.H. Chute aerators: Pre aerated approach flow. *J. Hydraul. Eng.* **2011**, *137*, 1452–1461. [CrossRef]
5. Hager, W.H.; Boes, R.M. Hydraulic structures: A positive outlook into the future. *J. Hydraul. Res.* **2014**, *52*, 299–310. [CrossRef]
6. Low, H.S. *Model Studies of Clyde Dam Spillway Aerators*; Research Report, No. 86-6; Department of Civil Engineering, University of Canterbury: Christchurch, New Zealand, 1986.
7. Zhang, J.M.; Wang, Y.R.; Yang, Q.; Xu, W.L. Scale effects of incipient cavitation for high-speed flows. *Proc. Inst. Civ. Eng. Water Manag.* **2013**, *166*, 141–145. [CrossRef]
8. Chanson, H. A study of Air Entrainment and Aeration Devices on a Spillway Model. Ph.D. Thesis, Department of Civil Engineering, University of Canterbury, Christchurch, New Zealand, 1988.
9. Chanson, H. *Air Bubble Entrainment in Free-Surface Turbulent Flows: Experimental Investigations*; Research Report No. CH46/95; Department of Civil Engineering, University of Queensland: Brisbane, Australia, 1995.
10. Chanson, H.; Brattberg, T. *Experimental Investigations of Air Bubble Entrainment in Developing Shear Layer*; Research Report, No. CH48/97; Department of Civil Engineering, The University of Queensland: Brisbane, Australia, 1997.
11. Brattberg, T.; Chanson, H.; Toombes, L. Experimental investigations of free-surface aeration in the developing flow of two-dimensional water jets. *J. Fluids Eng.* **1998**, *120*, 738–744. [CrossRef]
12. Chanson, H. Air bubble entrainment in open channels: Flow structure and bubble size distributions. *Int. J. Multiphase Flow* **1997**, *23*, 193–203. [CrossRef]
13. Toombes, L. Experimental Study of Air-Water Flow Properties on Low-Gradient Stepped Cascades. Ph.D. Thesis, Department of Civil Engineering, University of Queensland, Brisbane, Australia, 2002.
14. Li, S.; Zhang, J.M.; Chen, X.Q.; Zhou Gordon, G.D. Air concentration and velocity downstream of an expanding chute aerator. *J. Hydraul. Res.* **2018**, *56*, 412–423. [CrossRef]
15. Li, S.; Zhang, J.M.; Chen, X.Q.; Chen, J.G.; Zhou Gordon, G.D. Cavity length downstream of a sudden fall-expansion aerator in chute. *Water Sci. Technol. Water Supply* **2018**, *18*, 2053–2062. [CrossRef]
16. Pfister, M.; Hager, W.H. Deflector-generated jets. *J. Hydraul. Res.* **2009**, *47*, 466–475. [CrossRef]
17. Kramer, K.; Hager, W.H. Air transport in chute flows. *Int. J. Multiphase Flow* **2005**, *31*, 1181–1197. [CrossRef]
18. Pfister, M.; Hager, W.H. Chute aerators. I: Air transport characteristics. *J. Hydraul. Eng.* **2010**, *136*, 352–359. [CrossRef]
19. Toombes, L.; Chanson, H. Interfacial aeration and bubble count rate distributions in a supercritical flow past a backward-facing step. *Int. J. Multiphase Flow* **2008**, *34*, 427–436. [CrossRef]
20. Zhang, J.M.; Chen, J.G.; Xu, W.L.; Wang, Y.R.; Li, G.J. Three-dimensional numerical simulation of aerated flows downstream sudden fall aerator expansion-in a tunnel. *J. Hydrodyn. Ser. B* **2011**, *23*, 71–80. [CrossRef]
21. Toombes, L.; Chanson, H. Free-surface aeration and momentum exchange at a bottom outlet. *J. Hydraul. Res.* **2007**, *45*, 100–110. [CrossRef]

22. Fuhrhop, H.; Schulz, H.E.; Wittenberg, H. Solution for spillway chute aeration through bottom aerators. *Int. J. CMEM* **2014**, *2*, 298–312. [CrossRef]

23. Bai, Z.L.; Peng, Y.; Zhang, J.M. Three-Dimensional Turbulence Simulation of Flow in a V-Shaped Stepped Spillway. *J. Hydraul. Eng.* **2017**, *143*, 06017011. [CrossRef]

24. Duarte, R. Influence of Air Entrainment on Rock Scour Development and Block Stability in Plunge Pools. Ph.D. Thesis, EPFL-LCH, Lausanne, Switzerland, 2014.

25. Peng, Y.; Mao, Y.F.; Wang, B.; Xie, B. Study on C-S and P-R EOS in pseudo-potential lattice Boltzmann model for two-phase flows. *Int. J. Mod. Phys. C* **2017**, *28*. [CrossRef]

26. Peng, Y.; Wang, B.; Mao, Y.F. Study on force schemes in pseudopotential lattice Boltzmann model for two-phase flows. *Math. Probl. Eng.* **2018**, *2018*, 6496379. [CrossRef]

27. Peng, Y.; Zhang, J.M.; Meng, J.P. Second order force scheme for lattice Boltzmann model of shallow water flows. *J. Hydraul. Res.* **2017**, *55*, 592–597. [CrossRef]

28. Chanson, H. Hydraulics of aerated flows: Qui pro quo? *J. Hydraul. Res.* **2013**, *51*, 223–243. [CrossRef]

29. Chen, X.P.; Shao, D.C. Measuring bubbles sizes in self-aerated flow. *Water Res. Hydropower Eng.* **2006**, *37*, 33–36.

30. Bai, R.; Liu, S.; Tian, Z.; Wang, W.; Zhang, F. Experimental Investigation of Air–Water Flow Properties of Offset Aerators. *J. Hydraul. Eng.* **2017**, *144*, 04017059. [CrossRef]

31. Gonzalez, C.A. An Experimental Study of Free-Surface Aeration on Embankment Stepped Chutes. Ph.D. Thesis, Department of Civil Engineering, The University of Queensland, Brisbane, Australia, 2005.

32. Chanson, H. Turbulent air-water flows in hydraulic structures: Dynamic similarity and scale effects. *Environ. Fluid Mech.* **2009**, *9*, 125–142. [CrossRef]

33. Heller, V. Scale effects in physical hydraulic engineering models. *J. Hydraul. Res.* **2011**, *49*, 293–306. [CrossRef]

Article

Experimental Study on the Air Concentration Distribution of Aerated Jet Flows in a Plunge Pool

Weilin Xu *, Chunqi Chen and Wangru Wei

State Key Laboratory of Hydraulics and Mountain River Engineering, Sichuan University,
Chengdu 610065, China; ccq-cedric@163.com (C.C.); wangru_wei@hotmail.com (W.W.)
* Correspondence: xuwl@scu.edu.cn

Received: 20 October 2018; Accepted: 30 November 2018; Published: 4 December 2018

Abstract: There is a lack of knowledge on the air concentration distribution in plunge pools affected by aerated jets. A set of physical experiments was performed on vertical submerged aerated jet flows impinging a plunge pool. The air concentration distribution in the plunge pool was analyzed under different inflow air concentrations, flow velocities, and discharge rate conditions. The experimental results show that the air concentration distribution follows a power-law along the jet axis, and it is independent of the initial flow conditions. A new hypothetical analysis model was proposed for air diffusion in the plunge pool, that is, the air concentration distribution in the plunge pool is superposed by the lateral diffusion of three stages of the aerated jet motion. A set of formulas was proposed to predict the air concentration distribution in the plunge pool, the results of which showed good agreement with the experimental data.

Keywords: submerged jets; aerated flow; air concentration; plunge pool

---

1. Introduction

The impingement of a jet flow on a floor is a common issue that is important for many engineering applications, such as the flood discharging of hydraulic structures, and heat and mass transfer in industrial operations. Due to the presence of the floor being located at a distance from the nozzle exit, the diffusion of jets in a plunge pool is different from that of submerged free jets. Based on how the jet impacts the floor, the flow field can be divided into the free jet region, the impingement region, and the free shear region. The jet flow velocity conforms to the distribution law of submerged free jets in the free jet region. In the impingement region, the jet flow velocity decreases rapidly and reaches zero at the stagnation point [1,2].

Aerated and non-aerated impinging jet flows have been studied by many associated investigators [3–5]. Ervine summarized the characteristics of free turbulent jets, and compared the jet diffusion in the plunge pool with submerged and impinging jets [6]. Ervine collected data for both mean and fluctuating components of the pressure field on a plunge pool floor subjected to jet impingement, and compared the data for circular jets with those for wide rectangular nappes and rectangular slot jets [7]. Castillo analyzed the degree of break-up of a rectangular jet before entering the basin, and its relationship to the pressure fluctuation on the plunge pool floor [8]. Chanson discussed the characteristics and similitude of the air–water flow caused by impinging jets [9]. Wei analyzed the influence of aeration and initial water thickness on the axial velocity attenuation of jet flows using experiments and numerical calculations [10,11]. The research concludes that increasing the air concentration of jet flows and decreasing the initial jet thickness are effective ways to improve the axial velocity attenuation of jet flows. Some researchers focused on the pressure on the floor of the plunge pool by an aerated and non-aerated jet flow [12–15]. The results showed that the decreasing effect of air concentration on the time-averaged impact pressure is minimal, if any. However, the fluctuating pressure rises as the air concentration increases. Duarte compared the jet impingement on the center and side of a block embedded on the floor using plunging and submerged jets [16,17]. Others have inquired into the scouring of the plunge pool floor due to jets [18–20]. In addition to the velocity and pressure fields in the plunge pool, the process of air bubble entrainment at the impinging point of the water surface has also been studied extensively [21,22]. From the interaction of the jet flow and the plunge pool surface, an air sheet forms around the impinging point before air bubbles are entrained into the water surface [23]. Several correlative and predictive models on the air entrainment ratio have been reported, due to plunging jets under different working conditions [24–26]. Some researchers used numerical simulations to study aerated jet flows. Brouillio performed a series of two-dimensional numerical experiments on a translating impacting jet, and studied the dynamics of air entrainment by the jets impacting on the surface of the water [27]. Samanta studied the effect of the permeable wall on secondary cross-stream flow by performing direct numerical simulations of the fully developed turbulent flow through a porous square duct [28]. In model tests, the jet velocity profiles and pressure fluctuations on the plunge floor have been studied extensively, with detailed results available. Numerical simulations can essentially reflect characteristics such as the flow field and air entrainment process on the water surface. However, for mass and heat transfer characteristics at the air–water interface under complex conditions, a large number of systematic tests are still needed.

Few researchers have studied the air concentration distribution in the entire plunge pool, including the free shear region. Dong described the concentration diffusion of an aerated jet in the plunge pool with different plunging angles [29]. The air concentration along the jet axis gradually decreased, following a hyperbolic curve. Chanson conducted separate experiments with a vertical supported jet and a horizontal hydraulic jump [30]. The results showed that the air concentration in both cases exhibits a Gaussian distribution, and the longitudinal maximum air content declines exponentially. Khaled used the Particle Image Velocimetry (PIV) technique to characterize the flow field of both aerated and non-aerated jet flows beneath the water surface [31]. However, analyses and predictions of air concentration distribution in the entire plunge pool, especially where affected by the swirling jet, are still missing.

In the present paper, measurements and analyses of the air concentration distribution are presented for a plunge pool. A new analysis model is proposed to describe the air diffusion, with jet swirling in the entire plunge pool. A set of formulas has been built up to predict the air concentration distribution, and the results are believed to be useful for understanding and predicting the air concentration distribution in both the plunge pool and downstream.

2. Experimental Methods

Experiments were carried out at the State Key Laboratory of Hydraulics and Mountain River Engineering (Sichuan University). An experimental facility was set up, as shown in Figure 1. The jet was aerated before being plunged vertically into the still water in a rectangular glass flume of 4.00 m × 0.25 m × 0.80 m. The airflow was provided using an air compressor with a power rating of 7.5 kW at a pressure of 0.8 MPa. The inflow jet was controlled using valves and measured with a TDS-100H ultrasonic flowmeter, of which the measurement range was 0.8–128.0 L/s, and the relative error of measurement was ±1%. The air was controlled using valves, and measured with a SLDLB-Y150D vortex flowmeter, of which the measurement range was 2.0–60.0 L/s, and the relative error of measurement was ±1%. To form a stable and well-mixed air-water flow, the air was forced through micro-perforated plates with a diameter of 1 mm, and mixed into the water flow in the air entrainment region. The water in the flume flowed out at both ends, where the water level was maintained at $H = 0.55$ m. The incident angle of the jet flow was 90 degrees. The rectangular nozzle, 0.20 m in width and 0.02 m–0.05 m at the opening, was installed slightly below the water surface to ensure a precise, stable initial air concentration.

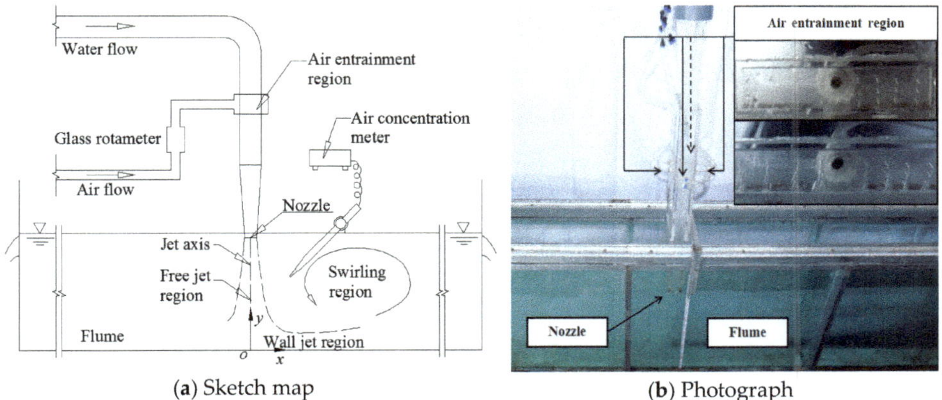

|(a) Sketch map|(b) Photograph|

Figure 1. Experimental facility.

The coordinate origin of an x-0-y coordinate system was located on the flume floor, the horizontal x-axis faced the downstream direction of the plunge pool, and the vertical y-axis coincided with the jet axis. In total, there were 264 measurable points for the air concentration in the plunge. A total of 22 rows were arranged along the x-axis direction. First, six rows were set at an interval of 5 cm (0 cm $\leq x \leq$ 25 cm), and the following 16 rows were set at an interval of 10 cm (30 cm $\leq x \leq$ 180 cm). In total, 12 rows of measuring points were arranged along the y-axis direction, with the first row at y = 1 cm and the last row at y = 54 cm (1 cm from the flume floor). The remaining 10 rows of measuring points along the y-axis were set between y = 5 cm and y = 50 cm at intervals of 5 cm. The air concentration was tested using a CQ6-2004 resistance-type air concentration meter (China Institute of Water Resources and Hydropower Research) with a resolution of 1%. The measurement principle of the resistance-type air concentration meter was to determine the air concentration by detecting the clear water resistance and the aerated water resistance between the two electrodes, and averaging the air concentration values in the selected integration time. Accurate flow measurements can be an important factor in obtaining adequate conclusions [32]. Different integration times for the air concentration measurements were tested beforehand, as shown in Figure 2. It can be seen that the mean air concentration data fall in the measurement accuracy zone if the integration time of the air concentration meter exceeds 60 seconds. Therefore, the sampling period for each measurement point was set to 60 seconds to minimize random error.

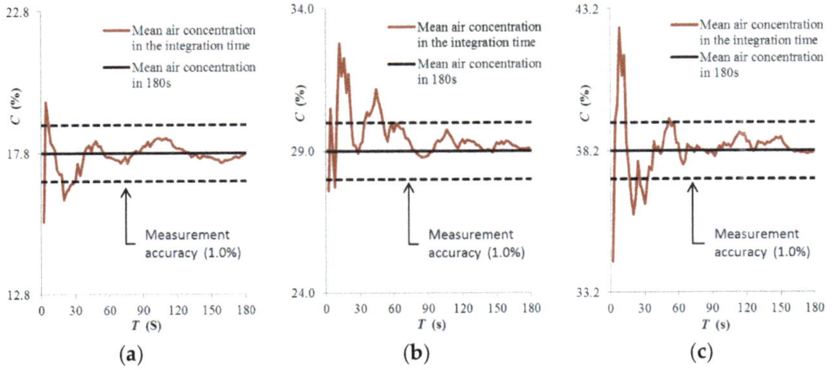

Figure 2. The uncertainty of air concentration measurements in the plunge pool.

The experiments were performed under different conditions, including the jet flow thickness d_0 (2 cm, 3 cm, 4 cm, 5 cm), the initial air concentration C_0 (10%, 15%, 30%, 40%, 50%, 60%), the clear water flow rate Q_w (14.0 L/s–35.2 L/s), and the air flow rate Q_a (2.1 L/s–42.0 L/s), as listed in Table 1.

Table 1. Experimental conditions.

No.	d_0 (cm)	C_0 (%)	Q_w (L/s)	No.	d_0 (cm)	C_0 (%)	Q_w (L/s)
1	2	10	18.6	16	3	30	21.1
2	2	15	16.1	17	3	40	21.1
3	2	30	14.0	18	3	40	23.9
4	2	30	16.8	19	3	50	23.9
5	2	30	19.6	20	4	15	27.8
6	2	30	22.4	21	4	30	21.1
7	2	40	16.7	22	4	30	28.2
8	2	50	14.7	23	4	40	28.2
9	2	60	14.5	24	4	50	21.1
10	2.3	10	21.1	25	4	50	23.9
11	2.3	15	23.9	26	4	50	27.8
12	2.5	15	21.1	27	5	30	35.2
13	2.5	30	21.1	28	5	30	21.1
14	3	10	21.3	29	5	40	35.2
15	3	15	21.1	30	5	50	23.9

In the experiment, the aerated jet was vertically injected and outflowed simultaneously at both ends of the flume, which was slightly wider than the nozzle exit. Jets could develop on the bottom plate of the flume under all experimental conditions. Thus, the development of the aerated jet in the plunge pool could be simplified as a two-dimensional flow, which is symmetrical to the jet axis (y-axis). For typical working conditions ($d_0 = 2$ cm, $C_0 = 30\%$, $Q_w = 16.8$ L/s), an experimental photograph and a contour map of the air concentration distribution on one side of the jet axis are shown in Figure 3. The majority of the air bubbles flow with the jet flow in the free jet region, before being blocked by the flume floor and deflected by 90 degrees into the wall jet region on both sides. Afterwards, the air bubbles flow gradually towards the water surface, with some eventually escaping into the surrounding air, whereas the rest are carried into the swirling region.

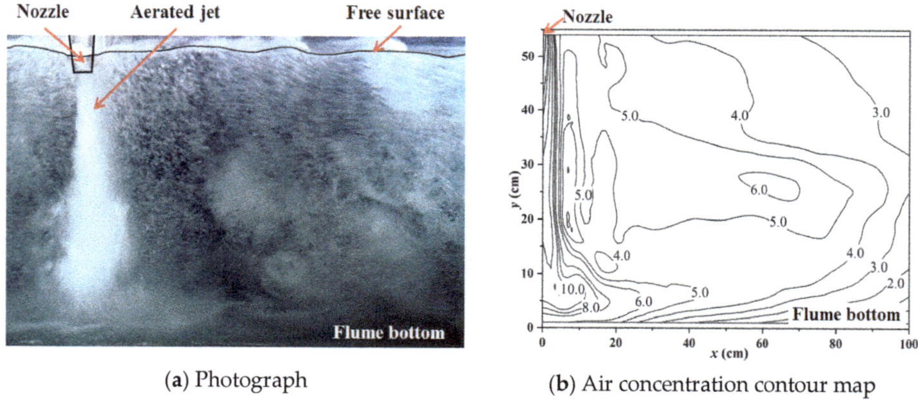

(a) Photograph　　　　　　　　(b) Air concentration contour map

Figure 3. Air concentration distribution for $d_0 = 2$ cm, $C_0 = 30\%$, $Q_w = 16.8$ L/s.

3. Results and Discussion

3.1. Air Concentration Distribution in the Free Jet Region

Along the jet axis, the air concentration distribution should be closely related to the jet velocity distribution. The jet velocity decreases sharply as the aerated jet approaches the plunge pool floor. Due to the pressure gradient, the jet carrying air bubbles drifts toward both sides of the plunge pool, resulting in a steep decrease in air concentration in the impingement region. The region higher than one-fifth of the water depth is considered to be the free jet region, where the attenuation of air concentration was studied in the test.

Considering the influencing factors, the initial water velocity V_w, the initial air concentration C_0, and the clear water flow rate Q_w were adopted, and a series of experiments were performed to test each effect on the air concentration attenuation, as shown in Figure 4. Under a different initial water velocity (3 m/s < V_w < 8 m/s) and a clear water flow rate (21.1 L/s < Q_w < 35.2 L/s), the attenuation of C_m/C_0 varies in a similar pattern along the jet axis. The reason for the results is that the velocity attenuation along the jet axis changes little with different flow conditions, including the initial jet velocity and the clear water flow rate [33].

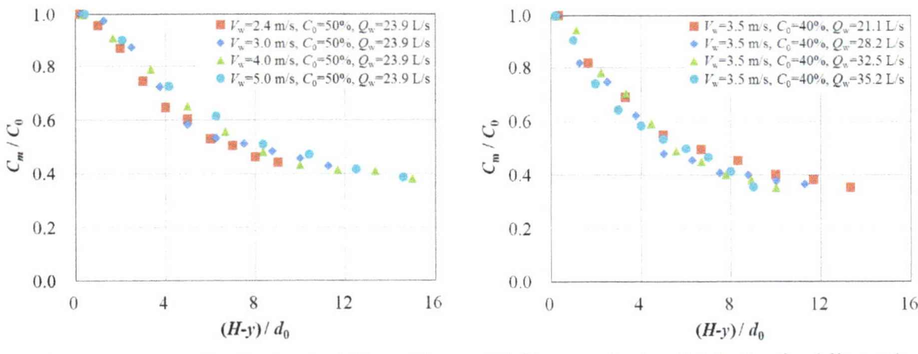

(**a**) Air concentration distribution for different V_w (**b**) Air concentration distribution for different Q_w

Figure 4. Influence of V_w and Q_w on axial air concentration distribution.

By theoretically analyzing the concentration flux of the aerated jet based on the self-similarity of the velocity and air concentration distribution, the power-law relation was found between the axial air concentration and the flow distance in the free jet region [29]. The relationship between the axial air concentration C_m and the distance from the nozzle $x' = H - y$, normalized by the initial air concentration C_0 and the jet flow thickness d_0 in the plunge pool, respectively, is written as:

$$\frac{C_m}{C_0} = k_0 \times \left(\frac{x'}{d_0}\right)^{-0.5} = k_0 \times \sqrt{\frac{d_0}{H-y}} \tag{1}$$

where k_0 is a coefficient, suggested as approximate 1.2 in this study.

The calculated results of the axial air concentration distribution in the free jet region using Equation (1) was compared with the experimental data, as shown in Figure 5a. Independent of the initial jet velocity and the clear water flow rate, the axial air concentration sharply decreased to about 50% of C_0 in a short distance ($\frac{H-y}{d_0} \approx 5$) from the nozzle exit. Due to the influence of the pressure gradient near the plunge floor, further from the nozzle exit, the air concentration distribution was more dispersed in the free jet region. It can be derived from the experimental results that the air concentration distribution along lateral cross-sections in the free jet region conformed to the Gaussian distribution law, as shown in Figure 5b. In the free jet region, no significant influence of different initial flow conditions was observed on the cross-sectional air concentration distribution, which tended to be self-similar. Downstream of the free jet region, the cross-sectional air concentration distribution was affected by the swirling jet in the plunge pool.

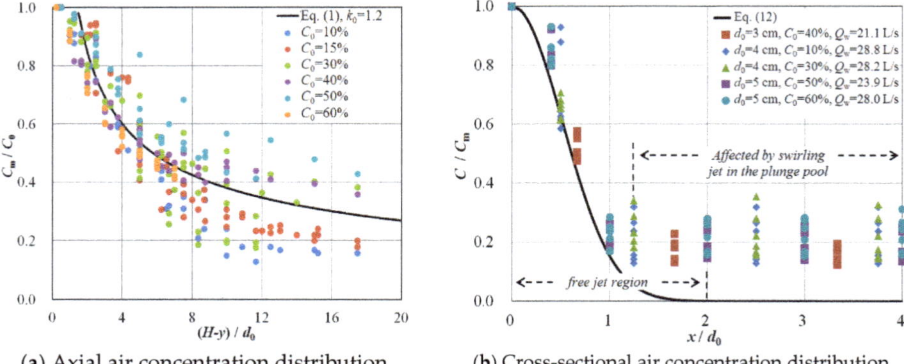

(a) Axial air concentration distribution

(b) Cross-sectional air concentration distribution

Figure 5. Air concentration distribution in the free jet region.

For a two-dimensional constant jet along the x' direction in an x'–0–y' coordinate system, the continuity equation of the air concentration diffusion is:

$$\frac{\partial(CU)}{\partial x'} + \frac{\partial(CV)}{\partial y'} = \frac{\partial}{\partial y'}\left(D_y\frac{\partial C}{\partial y'}\right) \tag{2}$$

where C is the air concentration, U is the velocity in the x' direction, V is the velocity in the y' direction, and D_y is the turbulent diffusion coefficient.

The longitudinal velocity U can be approximated using the average velocity $\bar{u}$. The transverse velocity can be negligible, i.e., $V \approx 0$.

According to Prandtl's hypothesis of free turbulence theory [34], D_y can be approximated as:

$$D_y = k\bar{u}b \tag{3}$$

where k is the coefficient, and b is the thickness of the jet, which is assumed to expand linearly and can be written as:

$$b = mx' = \tan\theta \times x' \tag{4}$$

where θ is the diffusion angle.

Previous research suggests that with respect to the conventional diffusion angle (about eight degrees), the diffusion angle of aerated water increases due to the influence of bubbles [10]. Twelve degrees is taken as the diffusion angel in this study.

Due to hardly any influence of initial flow conditions on both the axial attenuation and the lateral distribution of air concentration in the free jet region, it is assumed that the air concentration distribution in the free jet region is self-similar, resembling that in the velocity distribution, which can be written as:

$$\frac{C}{C_m} = f_1(\eta), \quad \eta = \frac{y'}{b} \tag{5}$$

Combining Equations (2), (3) and (5), $\frac{C}{C_m}$ can be expressed as:

$$f_1 = \frac{C}{C_m} = A \times \exp\left[-\frac{m}{2k} \times \eta^2\right] \tag{6}$$

where A is the integral constant and determined by boundary conditions. When $\eta = \frac{y'}{b} = 0, C = C_m$, $A = 1$, Equation (6) can be rewritten as:

$$\frac{C}{C_m} = \exp\left[-\frac{m}{2k} \times \left(\frac{y'}{b}\right)^2\right] \tag{7}$$

The concentration half-width is defined as:

$$b_{1/2} = y'\big|_{c=0.5C_m} \tag{8}$$

It is assumed that the concentration half-width extends linearly along the path, thus:

$$b_{1/2} = \frac{1}{2}d_0 + \tan\theta \times x' \tag{9}$$

Combining Equations (7) and (8):

$$\ln\left(\frac{1}{2}\right) = -\frac{m}{2k} \times \left(\frac{b_{\frac{1}{2}}}{b}\right)^2 = -0.693 \tag{10}$$

Considering:

$$-\frac{m}{2k} \times \left(\frac{y'}{b}\right)^2 = -\frac{m}{2k} \times \left(\frac{b_{\frac{1}{2}}}{b}\right)^2 \times \left(\frac{y'}{b_{\frac{1}{2}}}\right)^2 = -0.693 \times \left(\frac{y'}{b_{1/2}}\right)^2 \tag{11}$$

Combining Equations (7) and (11), the air concentration distribution on the flow cross-section can be calculated by:

$$\frac{C}{C_m} = \exp\left[-0.693 \times \left(\frac{y'}{b_{1/2}}\right)^2\right] \tag{12}$$

where C is the transverse air concentration diffusion, C_m is the corresponding air concentration along the jet axis, and y' is the transverse distance from the jet axis.

The calculated result of Equation (12) was compared with the experimental results, as shown in Figure 5b. In the free jet region, the calculated results of the lateral diffusion were essentially consistent with the experimental data. Further away from the axis, due to the influence of the swirling jet, the experimental value of the air concentration was greater than the calculated result obtained using Equation (13). Therefore, the next section will focus on the establishment of an analysis model for air concentration diffusion in the plunge pool.

In the x–0–y coordinate system in this study, the air concentration distribution in the free jet region can be calculated by:

$$\begin{aligned}
C &= C_0 \times 1.2\sqrt{\tfrac{d_0}{x'}} \times \exp\left[-0.693\left(\tfrac{y'}{b_{\frac{1}{2}}}\right)^2\right] \\
&= C_0 \times 1.2\sqrt{\tfrac{d_0}{H-y}} \times \exp\left[-0.693\left(\tfrac{x}{b_{1/2}}\right)^2\right]
\end{aligned} \tag{13}$$

3.2. Air Concentration Distribution in the Swirling Region of the Plunge Pool

For one side of the plunge pool, the experimental photographs of the submerged jet motion and air concentration distribution are shown in Figure 6. After being plunged into the still water, the aerated jet moves along the vertical axis, and the aerated wall jet forms after a 90-degree deflection of the jet flow near the flume floor. While the wall jet moves downstream, it is gradually developed toward the free surface under the combined effects of turbulence and buoyancy. Due to the existence of the downstream board, part of the aerated flow moves upstream along the water surface and forms a swirling region in the plunge pool. At the downstream end of the swirling region, the direction of the aerated flow movement is almost vertically upward. While moving with the water flow, the air bubbles expand continuously and laterally into the swirling region under the effects of turbulence.

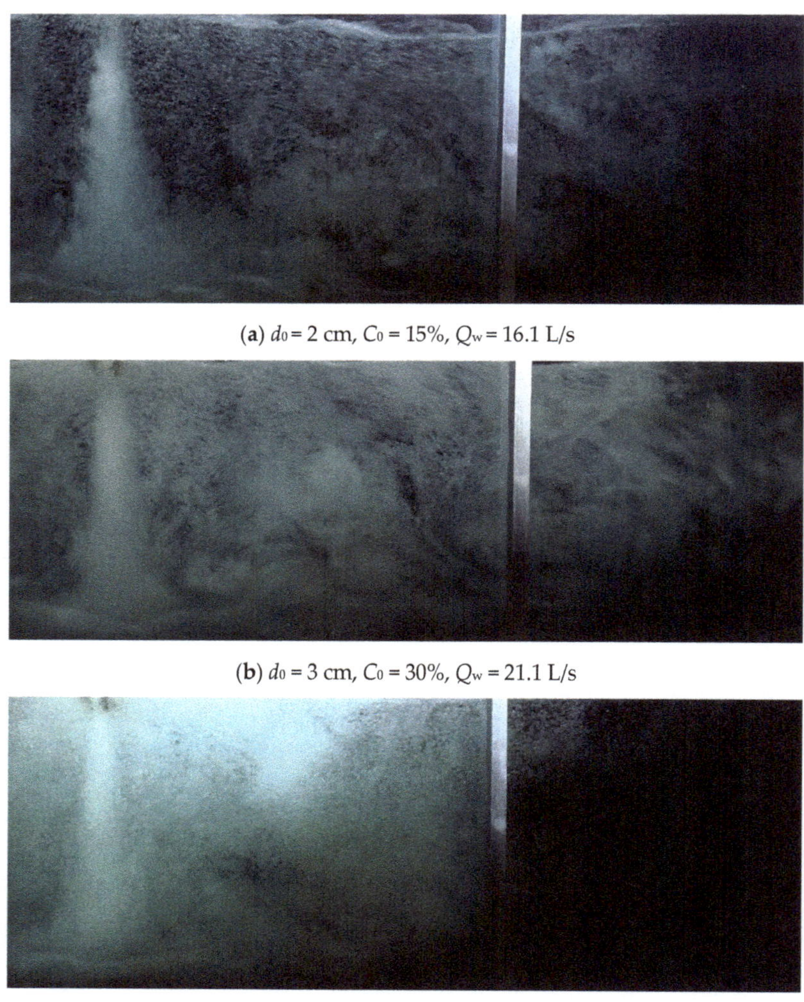

(**a**) $d_0 = 2$ cm, $C_0 = 15\%$, $Q_w = 16.1$ L/s

(**b**) $d_0 = 3$ cm, $C_0 = 30\%$, $Q_w = 21.1$ L/s

(**c**) $d_0 = 2$ cm, $C_0 = 60\%$, $Q_w = 14.5$ L/s

Figure 6. Air concentration distribution in the plunge pool.

According to Melo [13], the flow patterns in the plunge pool impinged by vertical jets are composed of three distinct regions, as shown in Figure 7a. In the free jet region, the vertical jet develops by shearing with the surrounding water body without the influence of the plunge floor. The impact of the jet against the floor builds up the pressure gradient and deflects the jet parallel to the plunge floor into the wall jet region. Due to the limited area of the plunge pool, the horizontal wall jet develops upwards and flows into the swirling region.

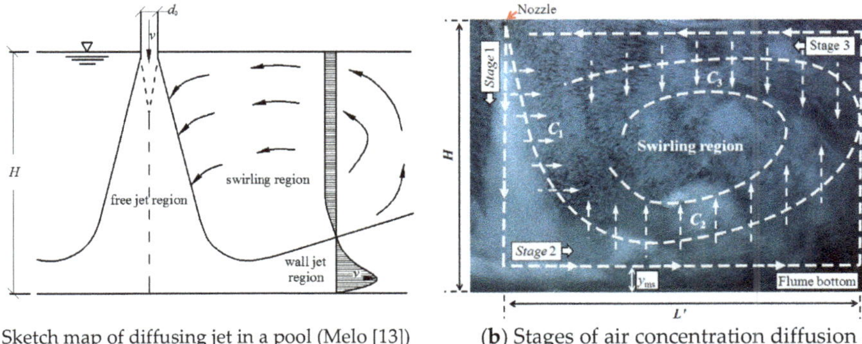

(a) Sketch map of diffusing jet in a pool (Melo [13])   (b) Stages of air concentration diffusion

Figure 7. Generalized calculation of air concentration in the swirling region.

When calculating the concentration field in the swirling region, the motion and diffusion of the air bubbles in the plunge pool are generalized into the following three stages, as shown in Figure 7b:

Stage 1: The aerated jet moves along the jet axis until it reaches the floor of the plunge pool, categorized as the free jet region in Figure 7a. The attenuation and lateral diffusion of the air concentration occurs along the path, which is consistent with the distribution law in the free jet region mentioned above. At this stage, the air concentration that diffuses laterally into the swirling region is taken as C_1;

Stage 2: The aerated flow is deflected by 90 degrees at the bottom of the plunge pool, and moves along the axis near the bottom plate, categorized as the wall jet region in Figure 7a. Regardless of the local and instantaneous air concentration loss, the initial air concentration at this stage is assumed to be the same as the air concentration at the end of the axis in the previous stage. The air concentration of the aerated jet continues to decrease along the path, until the jet reaches the end section of the swirling region. At this stage, the air concentration that diffuses vertically into the swirling region is taken as C_2;

Stage 3: The aerated flow moves upstream along the free surface until it reaches the original jet axis at stage 1, categorized as the swirling region in Figure 7a. The initial air concentration value in this stage is obtained by approximating the axis of the aerated jet in the swirling region to a rectangle with a length of L' and a width of H. The air concentration that diffuses vertically into the swirling region is taken as C_3.

In the plunge pool, excluding the jet axis, the air concentration at a certain point can be considered as the superposition of C_1, C_2, and C_3. When calculating the axial air concentration attenuation and the lateral diffusion in the three stages mentioned using Equation (13), separately, the jet flow distance x' and the transverse distance from the axis y' can be expressed as:

$$\text{stage 1}: \quad x_1' = H - y, \quad y_1' = x \tag{14}$$

$$\text{stage 2}: \quad x_2' = H + x, \quad y_2' = y \tag{15}$$

$$\text{stage 3}: \quad x_3' = 2H + 2L' - x, \quad y_3' = H - y \tag{16}$$

In order to locate the end section in stage 2 of the air concentration diffusion, the swirling region length L' is defined as the distance between the jet axis and the vertical cross-section, where the mean air concentration equals 5% of C_0.

$$L' = x\big|_{c_{mean}=0.05c_0} \tag{17}$$

where C_{mean} is the mean air concentration.

Under current experimental conditions, the relationship between the swirling region length and the initial water flow velocity is shown in Figure 8, which can be correlated with the linear function as:

$$\frac{L\prime}{d_0} = 8 \times \frac{V_w^2}{gH} + 11, \quad 0.6 < \frac{V_w^2}{gH} < 6 \tag{18}$$

where V_w is the water velocity, and g is the acceleration of gravity. The correlation coefficient is 0.80. It should be noted that the length of the swirling region was deliberately defined using the statistical mean air concentration of the vertical cross-section. The swirling region was actually determined by the flow field. In the downstream end of the plunge pool, the decrease in flow velocity led to changes in the following behaviors of the air bubbles. This resulted in the deviation of measurements in the experiment.

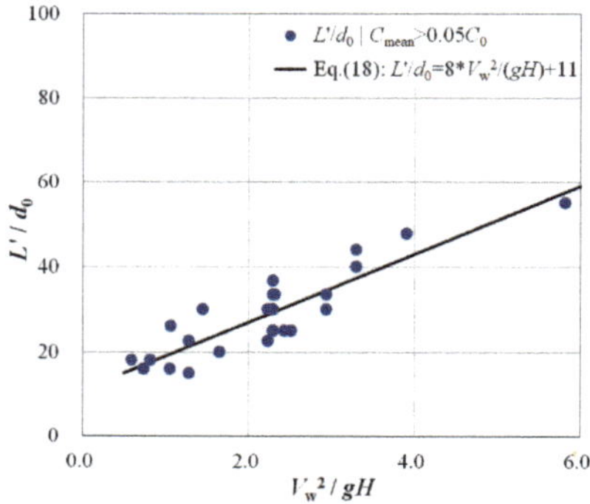

Figure 8. Distribution of swirling region length L'.

In stage 2 of the air concentration diffusion, the position of the maximum air concentration values along the vertical cross-sections in the wall jet can be estimated as:

$$y_{ms} = \frac{1}{3} \times b \tag{19}$$

where b is the thickness of the jet.

According to Beltaos [35], the relation between the maximum velocity of the wall jet u_{m1} and the initial velocity u_0 is:

$$\frac{u_{m1}}{u_0}\sqrt{\frac{H}{d_0}} = 2.77\left\{1 - \exp\left[-38.5\left(\frac{x}{H}\right)^2\right]\right\}^{\frac{1}{2}} \tag{20}$$

Considering the symmetrical aerated flow on both sides of the plunge pool, and the relation:

$$\frac{1}{2}u_{m1} \times b = \frac{1}{2}u_0 \times d_0 \tag{21}$$

Combining Equations (19)–(21), y_{ms} can be expressed as:

$$y_{ms} = \frac{1}{3} \times \frac{\sqrt{H \times d_0}}{2.77\left\{1 - \exp\left[-38.5(x/H)^2\right]\right\}^{1/2}} \tag{22}$$

The air concentration can be linearly solved on the vertical cross-sections between the axis of the wall jet and the plunge pool floor, where the air concentration value is zero.

According to the above analysis, at any point (x, y) in the swirling region of the plunge pool, the calculation of the lateral air concentration diffusion C_1 from the jet axis can be written as:

$$C_1 = C_0 \times 1.2 \sqrt{\frac{d_0}{H - y}} \times \exp\left[-0.693\left(\frac{x}{b_{1/2}}\right)^2\right] \tag{23}$$

The calculation of the vertical air concentration diffusion C_2 from near the plunge pool floor can be written as:

$$C_2 = C_0 \times 1.2 \sqrt{\frac{d_0}{H + x}} \times \exp\left[-0.693\left(\frac{y - y_{ms}}{b_{1/2}}\right)^2\right] \tag{24}$$

The calculation of the vertical air concentration diffusion C_3 from the free surface can be written as:

$$C_3 = C_0 \times 1.2 \sqrt{\frac{d_0}{2H + 2L' - x}} \times \exp\left[-0.693\left(\frac{H - y}{b_{1/2}}\right)^2\right] \tag{25}$$

Thus, the air concentration value at the point (x, y) can be obtained by:

$$C\big|_{(x,y)} = C_1 + C_2 + C_3 \tag{26}$$

with 12 degrees of diffusion angles used in calculating C_1 and C_2, and zero degrees in calculating C_3.

Combing Equations (23)–(26), it is possible to calculate the air concentration at any point in the swirling region of the plunge pool.

3.3. Comparison of Calculated and Experimental Results

The calculated and experimental results of the air concentration distribution along the vertical cross-sections in the plunge pool are compared in Figure 9. When the initial air concentration was small, such as $C_0 < 40\%$, the calculated results agreed well with the experimental measurements. When the initial air concentration was large, such as $C_0 > 40\%$, the calculated results could essentially reflect the variation of the air concentration distribution in the region close to the jet axis, such as $x/d_0 < 15$–20. In the downstream region, on the other hand, such as $x/d_0 > 15$–20, the calculated results were somewhat different from the experimental measurements.

The reason for the deviation is that the bubble motion and diffusion process in the hypothetical analysis model is different from that in the actual jet flow. The influence of the bubble floating process by buoyancy on the actual jet flow was not considered, leading to a decrease in the adaptability of the idealized hypothetical analysis in this paper. However, the absolute value of the air concentration in the downstream region of the plunge pool was relatively small. For example, for $d_0 = 2$ cm, $C_0 = 60\%$, $Q_w = 14.5$ L/s, the absolute value of the air concentration was essentially less than 0.05 in the downstream region, such as $x/d_0 > 10$. Consequently, the present analysis model can essentially predict the air concentration distribution of aerated jet flows in the swirling region of the plunge pool.

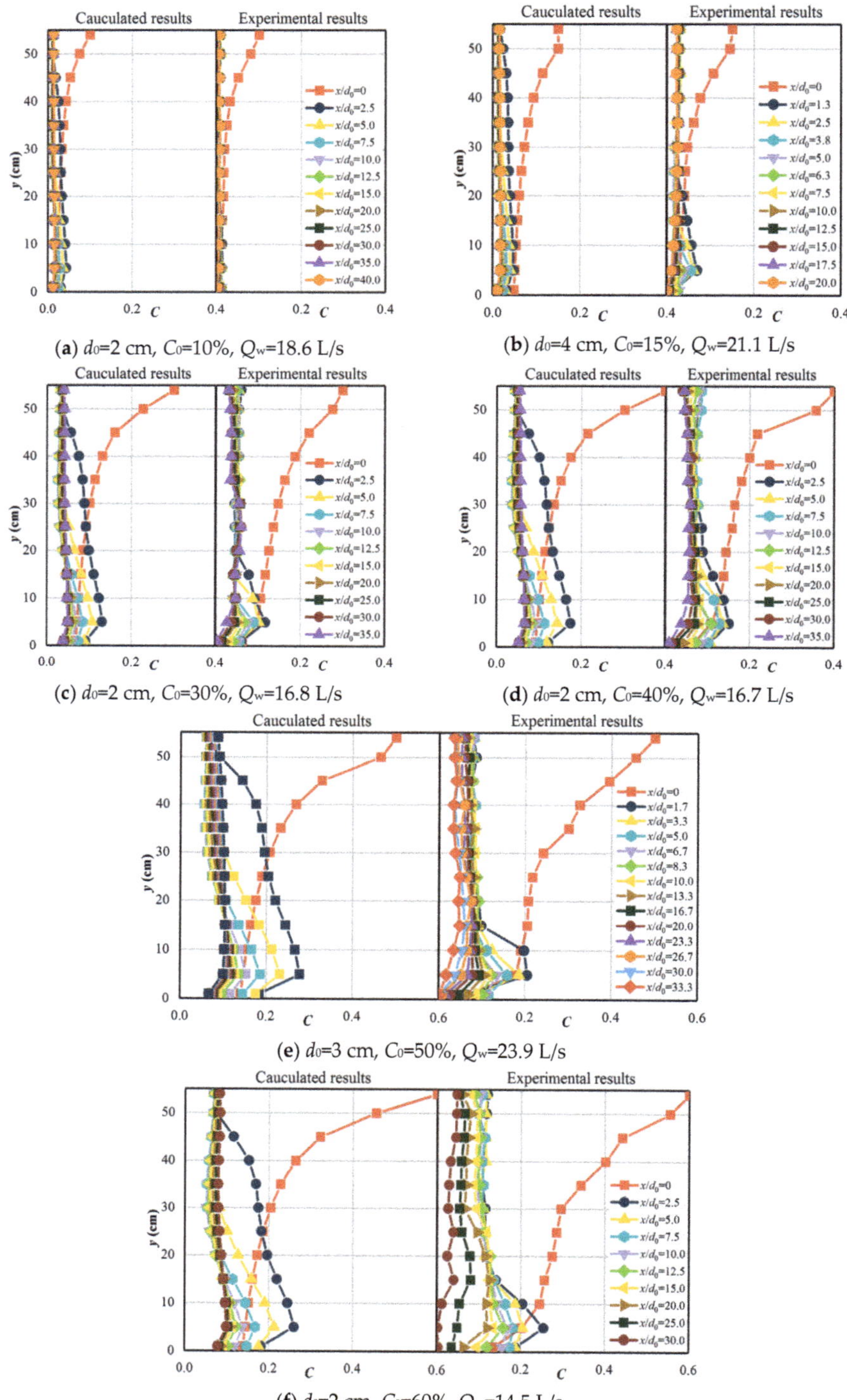

Figure 9. Comparison of calculated and experimental results.

4. Conclusions

In this paper, a series of experimental studies on the air concentration distribution of vertical submerged aerated jet flows in a plunge pool were carried out. Based on the distribution law of the air concentration attenuation along the axial and lateral directions, an analysis model of the air concentration distribution in the plunge pool was proposed. The conclusions were as follows.

In the free jet region, the axial air concentration attenuation followed a power-law distribution, and the air concentration distribution along the lateral cross-sections conformed to the Gaussian distribution law. Both the axial attenuation and the lateral distribution of air concentration in the free jet region were almost unchanged for different initial water velocities, air concentrations, and flow rate conditions. Based on the observation and analysis of the flow patterns in the swirling region in the plunge pool, the air concentration was assumed to be a superposition of lateral diffusion in three stages of the jet flow. The position of the maximum air concentration values along the vertical cross-sections in the wall jet and the length of the swirling region, as well as their relationship with the initial flow conditions, were introduced. A set of formulas was proposed for the prediction of the air concentration distribution in the plunge pool. The present analysis model, the results of which have been compared with experimental data, can essentially predict the air concentration distribution of aerated jet flows in the swirling region of a plunge pool.

In water conservancy projects, the jet flows are aerated, and they form an air concentration distribution inside the plunge pool, which has a significant effect on the downstream energy dissipation and the total dissolved gas. Meanwhile, the increase in aeration results in changes to the mass and heat transfer process at the air–water interface. This research improves our understanding of the air dissolution process downstream practical applications, and provides a new model to predict the air concentration distribution in plunge pools caused by aerated jets. The results of this study reveal the complexity of the air concentration distribution in a plunge pool caused by aerated jets. In order to further improve the prediction accuracy for the air concentration distribution in plunge pools, coupling analyses of bubble floating by buoyancy and jet motion in the air diffusion process should be the focus of future studies.

Author Contributions: The paper is the product of the joint efforts of the authors who worked together on experimental model tests. W.X. conceived the experiments, having a scientific background in applied hydraulics. C.C. and W.W. conducted the experimental investigations and analyzed the data under the supervision of W.X. C.C. wrote the manuscript with the co-authors.

Funding: Resources to cover the Article Processing Charge were provided by the National Natural Science Foundation of China (Grant No. 51609162).

Conflicts of Interest: The authors declare no conflict of interest.

References

1. Beltaos, S.; Rajaratnam, N. Impinging Circular Turbulent Jets. *J. Hydraul. Div.* **1974**, *100*, 1313–1328.
2. Beltaos, S.; Rajaratnam, N. Impingement of Axisymmetric Developing Jets. *J. Hydraul. Res.* **2010**, *15*, 311–326. [CrossRef]
3. Cooper, D.; Jackson, D.C.; Launder, B.E.; Liao, G.X. Impinging jet studies for turbulence model assessment—I. Flow-field experiments. *Int. J. Heat Mass Transfer* **1993**, *36*, 2675–2684. [CrossRef]
4. Baydar, E.; Ozmen, Y. An experimental investigation on flow structures of confined and unconfined impinging air jets. *J. Heat Mass Transfer* **2006**, *42*, 338–346. [CrossRef]
5. Schmocker, L.; Pfister, M.; Hager, W.H.; Minor, H.E. Aeration characteristics of ski jump jets. *J. Hydral. Eng.* **2008**, *98*, 90–97. [CrossRef]
6. Ervine, D.A.; Falvey, H.T. Behaviour of turbulent water jets in the atmosphere and in plunge pools. *Proc. Inst. Civil Eng.* **1987**, *83*, 295–314. [CrossRef]
7. Ervine, D.A.; Falvey, H.T.; Withers, W. Pressure fluctuations on plunge pool floors. *J. Hydraul. Res.* **1997**, *35*, 257–279. [CrossRef]

8. Castillo, L.G.; Carrillo, J.M.; Blázquez, A. Plunge pool dynamic pressures: A temporal analysis in the nappe flow case. *J. Hydraul. Res.* **2015**, *53*, 101–118. [CrossRef]

9. Chanson, H.; Aoki, S.; Hoque, A. Physical modelling and similitude of air bubble entrainment at vertical circular plunging jets. *Chem. Eng. Sci.* **2004**, *59*, 747–758. [CrossRef]

10. Wei, W.R.; Deng, J.; Liu, B. Influence of aeration and initial water thickness on axial velocity attenuation of jet flows. *J. Zhejiang Univ-Sci. A (Appl Phys & Eng)*. **2013**, *14*, 362–370. [CrossRef]

11. Wei, W.R.; Deng, J.; Liu, B.; Zhang, Y.L.; Qian, X.Y.; Liu, X.J. Experimental investigation on the effects of aeration on the velocity attenuation of jet flow. *J. Sichuan Uni.* **2011**, *43*, 29–33. [CrossRef]

12. Luo, M.; Guo, Y.K. Study of diffusion of confined two-dimensional aerated jets under water surface. *J. Hydral. Eng.* **1992**, *7*, 29–34. [CrossRef]

13. Melo, J. Reduction of plunge pool floor dynamic pressure due to jet air entrainment. In Proceedings of the International Workshop on Rock Scour due to Falling High-Velocity Jet, Lausanne, Switzerland, 25–28 September 2002; Schleiss, A.J., Bollaert, E.F., Eds.; Swets and Zei tlinger: Lisse, The Netherlands, 2002; pp. 125–136.

14. Deng, J.; Xu, W.L.; Liu, S.J.; Wang, W. Influence of water jet aeration on pressure in scour pool and plunge pool. *Adv. Water Sci.* **2009**, *20*, 373–378. [CrossRef]

15. Xu, W.; Deng, J.; Qu, J.; Liu, S.; Wang, W. Experimental investigation on influence of aeration on plane jet scour. *J. Hydral. Eng.* **2004**, *130*, 160–164. [CrossRef]

16. Duarte, R.; Schleiss, A.J.; Pinheiro, A. Influence of jet aeration on pressures around a block embedded in a plunge pool bottom. *Environ. Fluid Mech.* **2015**, *15*, 673–693. [CrossRef]

17. Duarte, R.; Pinheiro, A.; Schleiss, A.J. An Enhanced Physically Based Scour Model for Considering Jet Air Entrainment. *Eng.* **2016**, *2*, 294–301. [CrossRef]

18. Sarkar, A.; Dey, S. Review on Local Scour Due to Jets. *Int. J. Sediment. Res.* **2004**, *19*, 53–81.

19. Pagliara, S.; Hager, W.; Minor, H.-E. Hydraulics of Plane Plunge Pool Scour. *J. Hydral. Eng.* **2006**, *132*, 450–461. [CrossRef]

20. Bollaert, E.; Schleiss, A. Scour of rock due to the impact of plunging high velocity jets Part II: Experimental results of dynamic pressures at pool bottoms and in one- and two-dimensional closed end rock joints. *J. Hydraul. Res.* **2003**, *41*, 465–480. [CrossRef]

21. Bin, A.K. Gas entrainment by plunging liquid jets. *Chem. Eng. Sci* **1993**, *48*, 136. [CrossRef]

22. Chanson, H. References. In *Air Bubble Entrainment in Free-Surface Turbulent Shear Flows*; Chanson, H., Ed.; Academic Press: San Diego, CA, USA, 1996; pp. 239–261. ISBN 978-0-12-168110-4.

23. Harby, K.; Chiva, S.; Muñoz-Cobo, J.L. An experimental study on bubble entrainment and flow characteristics of vertical plunging water jets. *Exp. Therm. Fluid Sci.* **2014**, *57*, 207–220. [CrossRef]

24. Brattberg, T.; Chanson, H. Air entrapment and air bubble dispersion at two-dimensional plunging water jets. *Chem. Eng. Sci* **1998**, *53*, 4113–4127. [CrossRef]

25. Kiger, K.T.; Duncan, J.H. Air-Entrainment Mechanisms in Plunging Jets and Breaking Waves. *Annu. Rev. Fluid Mech.* **2011**, *44*, 563–596. [CrossRef]

26. Miwa, S.; Moribe, T.; Tsutstumi, K.; Hibiki, T. Experimental Investigation of Air Entrainment by Vertical Plunging Liquid Jet. *Chem. Eng. Sci* **2018**, *181*. [CrossRef]

27. Brouilliot, D.; Lubin, P. Numerical simulations of air entrainment in a plunging jet of liquid. *J. Fluids Struct.* **2013**, *43*, 428–440. [CrossRef]

28. Samanta, A.; Vinuesa, R.; Lashgari, I.; Schlatter, P.; Brandt, L. Enhanced secondary motion of the turbulent flow through a porous square duct. *J. Fluid Mech.* **2015**, *784*, 681–693. [CrossRef]

29. Dong, Z.Y.; Wu, C.G. Characteristics of concentration diffusion of submerged aerated jets. *J. Chengdu Univ. Sci. Technol.* **1994**, *4*, 31–38.

30. Chanson, H. Air entrainment in two-dimensional turbulent shear flows with partially developed inflow conditions. *Int. J. Multiphase Flow* **1995**, *21*, 1107–1121. [CrossRef]

31. Khaled, J.H. Liquid Jet Impingement on a Free Liquid Surface: PIV Study of the Turbulent Bubbly Two-Phase Flow. In Proceedings of the ASME 2010 Joint Us-European Fluids Engineering Summer Meeting Collocated with International Conference on Nanochannels, Microchannels, and Minichannels ASME 2010 Joint Us-European Fluids Engineering Summer Meeting: Volume 1, Symposia—Parts A, B, and C, Montreal, QC, Canada, 1–5 August 2010; pp. 2877–2885.

32. Vinuesa, R.; Schlatter, P.; Nagib, H.M. Role of data uncertainties in identifying the logarithmic region of turbulent boundary layers. *Exp. Fluids* **2014**, *55*, 1751. [CrossRef]
33. Albertson, M.L.; Dai, Y.B.; Jensen, R.A.; Rouse, H. Diffusion of Submerged Jets. *Trans. Amer. Soc. Civil Eng.* **1950**, *115*, 639–664. [CrossRef]
34. Prandtl, L. Bemerkungen zur Theorie der freien Turbulenz. *Z.A.M.M.* **1942**, *22*, 241–243. [CrossRef]
35. Beltaos, S.; Rajaratnam, N. Plane Turbulent Impinging Jet. *J. Hydraul. Res.* **1973**, *11*, 29–59. [CrossRef]

Article

Spatial Distribution Characteristics of Rainfall for Two-Jet Collisions in Air

Hao Yuan, Weilin Xu *, Rui Li, Yanzhang Feng and Yafeng Hao

State Key Laboratory of Hydraulics and Mountain River Engineering, Sichuan University,
Chengdu 610065, China; 18996152721@163.com (H.Y.); li.rui.sai@gmail.com (R.L.); 13835141175@163.com (Y.F.);
13258387216@163.com (Y.H.)
* Correspondence: xuwl@scu.edu.cn

Received: 12 September 2018; Accepted: 4 November 2018; Published: 7 November 2018

Abstract: Many researchers have studied the energy dissipation characteristics of two-jet collisions in air, but few have studied the related spatial rainfall distribution characteristics. In this paper, in combination with a model experiment and theoretical study, the spatial distributions of rainfall intensity of two-jet collisions, with different collision angles and flow ratios, are systematically studied. The experimental results indicated that a larger collision angle corresponds to a larger rainfall intensity distribution. The dimensionless maximum rainfall intensity sharply decreased with the flow ratio, while the maximum rainfall intensity slightly increased when the flow ratio was greater than 1.0. A theoretical equation to compute the location of maximum rainfall intensity is presented. The range of rainfall intensity distribution sharply increased with the flow ratio. When the flow ratio was greater than 1.0, the range of longitudinal distribution slightly increased, whereas the lateral distribution remained unchanged or slowly decreased. A formula to calculate the boundary lines of the x-axis is proposed.

Keywords: rainfall intensity distribution; two water jets; collision in air; Gaussian distribution; trajectory line

---

1. Introduction

At present, the high dam projects that have been built, are under construction, or are planned in China have the common features of large drops, a narrow river valley, and a large flood discharge flow. The flood discharge energy dissipation mode of the dam body has been mostly adopted by two-jet collisions and downstream water reservoirs, such as Ertan (Figure 1), Xiaowan, Xiluodu, Goupitan, and the Jinping grade I project. Two-jet collisions in air change the motion track of each of the two jets, significantly disperse the water flow, effectively alleviate the dynamic pressure impact of the water flow on the bottom plate of the water cushion reservoir, and have a remarkable energy dissipation effect [1–7]. However, compared with the traditional energy dissipation method, two-jet collisions in air appear to be a serious problem with regard to flood discharge atomization in the existing engineering examples, which has caused great damage to the normal operation, traffic safety, surrounding environment, and even the stability of the downstream bank slopes. It is necessary to study the spatial distribution characteristics of rainfall intensity in two-jet collisions in air, in order to improve the prediction accuracy of the discharge atomization rain strength and influence range, as well as make a relevant protection classification and protection design scheme in the design planning [8–10].

Figure 1. Simultaneous discharge of the surface and the deep hole of the Ertan hydropower station in 2010.

Many former researchers who have studied two-jet collisions in air have focused on the energy dissipation effect of collisions. Xiong [11] introduced the main factors that affect energy dissipation during an in-air collision between the surface and the deep hole. Guo [12] theoretically analyzed the effect of the collision angle on energy dissipation. Using the momentum integral equation of fluid mechanics, Liu [13] derived the related formula of energy dissipation for the collision of two water jets in air in detail, and introduced the concept and calculation method for the energy dissipation rate. Diao [14] studied the relationship between the collision angle, the flow ratio, and the collision energy dissipation effect through model testing, and proposed that the main function of jet collisions is to disperse the water flow. Sun [15] analyzed hydraulic characteristics, such as the energy loss of up- and down-collisions, three-dimensional (3-D) diffusion and leakage collision of the surface, and deep orifice slugs. The 3-D collision velocity, collision efficiency, and collision energy loss were derived using the momentum equation and variation law of the air diffusion width of the water tongue, and the optimal hydraulic conditions for collision were obtained. Sun [16] applied the turbulence jet theory and flow momentum equation, in order to introduce the calculation formula of the collision velocity vector and collision energy dissipation efficiency of the combined water tongue when it collides with the left and right sides of the air.

At present, the research on flood discharge atomization is mainly about ski-jump atomization. Liang [17] proposed a calculation model of atomized water flow, and obtained various calculation formulas and methods in the field of atomized water flow influence. Liu [18,19] studies the diffusion law, velocity distribution, and energy loss of a water jet. According to the prototype observation data of the flood discharge atomization and some experimental data of the model, based on the comprehensive analysis of various factors that affect the atomization range, Li [20] discusses the rough estimation method of the range of the energy dissipation atomization precipitation area. Liu [21–24] introduces several key problems with regard to the shape and the numerical simulation of atomized flows, such as the jet calculation of the water tongue, collision between the water tongue and the water surface, and calculation of the fog source quantity. By collecting, inducing, and summarizing the partial flood discharge atomization of engineering prototype observation data, Sun [25] found the longitudinal boundary flood discharge atomization average discharge flow, and a good relationship between the water flow velocity and the water entry angle of the water tongue. Based on the dimensional analysis, a method was established to estimate the rainfall intensity, flood discharge atomization, and longitudinal boundary experience relationship. Liu [26] proposed an atomization forecast model based on artificial neural networks. The Monte Carlo method was applied to add the effect of environmental, wind, and topographic factors to the atomization mathematical model of overpass discharge by Lian [27]. Sun [28] revised the resistance coefficient of particles, using the research results of raindrop movement, and obtained the resistance coefficient of the splashing water droplet movement. Then the effect of the diameter of a splashing water droplet on the splashing length is preliminarily discussed, which deepens the understanding of the splashing movement. Using the generalized model test in 2013 by Wang [29]—under different hydraulic conditions, to select a tongue that fell into the water downstream of an atomized water source area of rainfall intensity—the flood discharge atomization fog source area of the plane distribution features was measured and analyzed, to determine the reasons for the formation of different areas around the atomization source and the plane distribution of rainfall intensity. Lian [30] proposes a method to predict the spray atomization of spillage by combining a physical model and theoretical analysis. Zhang [31] experimentally studied the motion track, distribution range, water point shape, and velocity of single-strand jets under different air dosages. Although their methods are accurate in predicting the atomization of a ski-jump flow, they are not good at predicting the atomization of two-jet collisions, which add a collision zone and change the trajectory of two jets. In rocket propulsion, extensive experiments—mostly based on physical model investigations—have been conducted on two-jet collisions, including the liquid membrane [32–35], droplet size, and velocity distribution [36–41].

Most domestic and foreign scholars have focused on the atomization of flood discharge by single-strand pick flows. However, compared with the single-strand pick flow, two-jet collisions in air increase the area of collision atomization, where both range and intensity of the atomization increase. For example, Ertan applied this method to flood discharge and cause landslide by rainfall. Therefore, in this paper, through theoretical analysis and a physical model experiment, we studied the distribution characteristics of rainfall intensity after the collision of two jets, the trajectory lines of the water tongue, and the range of the x-axis after collision for different flow ratios and collision angles.

2. Model Design and Experiment

The experiment was conducted in the State Key Laboratory of Hydrodynamics and Mountain River Engineering in Sichuan University. Figure 2 shows the schematic diagram of the model test. The water flux was supplied by a large rectangular tank (width of 4.0 m, length of 4.0 m, and height of 5.0 m). When laying the rainfall receiving platform, we considered the symmetry of the surface and deep hole, and only one side was arranged. Water was fed into the iron box through the pump behind the iron box, and the water level of the iron box was maintained through the valve in order to achieve a stable deep and surface flow.

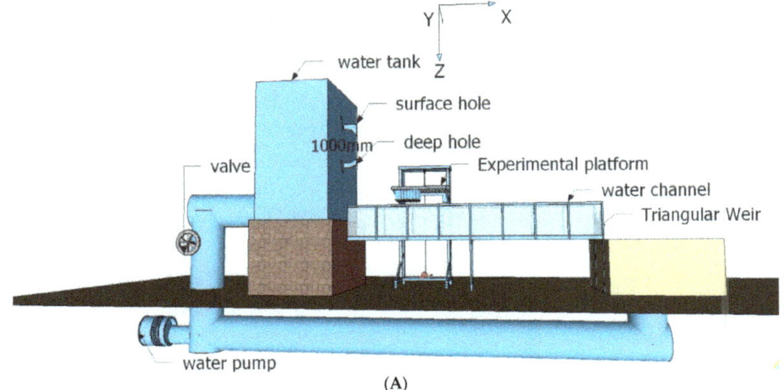

Figure 2. *Cont.*

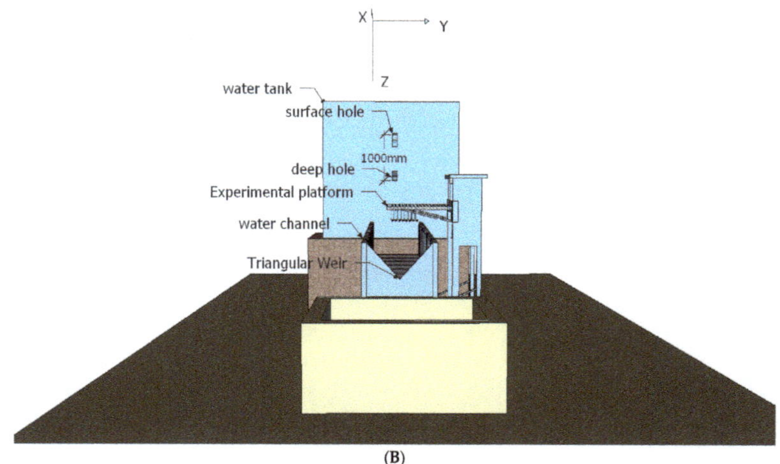

Figure 2. (**A**) Experiment layout: side view; (**B**) experiment layout: front view.

In order to be consistent with the prototype which adopts two-jet collisions in air, this experiment used a constant surface hole angle θ_1 (a pitch-down angle of $-30°$) and adjusted the deep hole pick angle θ_2 ($0°{\sim}45°$) to change the collision angle β. The width of the surface deep hole was identical, the height of the deep hole could be adjusted, and the range was $0{\sim}5$ cm. The maximum single-width flux of the model table hole was 0.1236 m^2/s, the maximum single-width flux of the deep hole was 0.1431 m^2/s, and the flux was measured by a triangular thin-walled weir ($\pm1\%$). Four horizontal planes, under the collision points of 36 cm, 56 cm, 76 cm, and 96 cm, were set to measure the rainfall intensity. The platform was supported by a slide rail system, and the accuracy in the vertical platform position was less than 0.5 cm. Table 1 lists 18 working conditions in the experiment, including almost 30,000 measurement points. To obtain the four different sections, we first used image processing to find the collision point, and then adjusted the height of the measuring platform. Detailed visual observations of the collision points were conducted using a Canon EOS80D DSLR camera (Cannon, Chengdu, China).

Rainwater was collected in the horizontal plane (xy plane in Figure 3) at the different elevations below the collision point and 0.15 m above the faceplate. The rainfall collection platform included the movable base, support, and rainfall collection panel, which was made of an organic glass panel and a PVC pipe. In each plane, up to 121 PVC pipes (each with a diameter of 2.0 cm) were used simultaneously to collect rain within a period varying from 60 s to 1000 s, depending on the rain intensity in that region, which was found to be sufficient to ensure stable results. Zhang and Zhu [31] indicated that the measurement error could be ignored with different pipe diameters. In their test, the bottles (25 mL with a bottle-neck diameter of 1.52 cm) were tested to give the same (3% difference) result of rain intensity distribution as those of much larger plastic bottles (250 mL with a bottle-neck diameter of 3.26 cm). To avoid the impact of water splashing on the back of the faceplate, the faceplate was hollow, except for the holes and supporting frames. The lower end of the short pipe, whose range was 15~3000 mL, used a measuring cylinder to collect rainfall. Rainfall intensity varies from point to point, so different cylinders were used. To obtain the whole region's rainfall intensity, the platform was first placed in one region to collect rainwater, then the entire platform was moved 1.2 m in the longitudinal direction to the next region. To ensure measurement accuracy, two measurements were made in the same working conditions, and the average was taken.

Table 1. Working conditions in experiments.

Run	θ_1 (°)	θ_2 (°)	q_m (m^2/s)	q_c (m^2/s)	f	$Q \times 2/(q_m + q_c)$
T1-1	−30	0	0.0373	0.1236	0.3	103%
T1-2	−30	0	0.0745	0.1236	0.6	95%
T1-3	−30	0	0.0727	0.0800	0.9	102%
T1-4	−30	0	0.1091	0.0800	1.3	104%
T1-5	−30	0	0.1073	0.0436	2.5	98%
T1-6	−30	0	0.1427	0.0436	3.2	104%
T2-1	−30	30	0.0373	0.1236	0.3	95%
T2-2	−30	30	0.0745	0.1236	0.6	110%
T2-3	−30	30	0.0727	0.0800	0.9	102%
T2-4	−30	30	0.1091	0.0800	1.3	105%
T2-5	−30	30	0.1073	0.0436	2.5	103%
T2-6	−30	30	0.1427	0.0436	3.2	92%
T3-1	−30	45	0.0373	0.1236	0.3	111%
T3-2	−30	45	0.0745	0.1236	0.6	94%
T3-3	−30	45	0.0727	0.0800	0.9	103%
T3-4	−30	45	0.1091	0.0800	1.3	106%
T3-5	−30	45	0.1073	0.0436	2.5	108%
T3-6	−30	45	0.1427	0.0436	3.2	111%

q_m is the flow discharge per unit width of deep hole. q_c is the flow discharge per unit width of surface hole. f is the ratio of deep hole flow to surface hole flow, that is $f = q_m/q_c$. Q is the integral of measurement value.

Figure 3. Rainfall collection platform in 2017.

In each test, the total rain flux in the horizontal plane was integrated and compared to the total water flux of the surface and the deep holes. The total water/rain flux Q was calculated from

$$Q = \int_{-\infty}^{+\infty}\int_{-\infty}^{+\infty} I \, dx dy,$$

where I was the measured rain intensity. Comparisons of the integrated values with the water flux at the surface and the deep holes were listed in Table 1. For all tests, the conservation ratio of rain flux on average was 101.4 ± 7.0%. In runs T2-6 and T3-6, the loss of rain flux was slight larger (±9.3%), which may be related to the fact that the rain had a wider range in the test. Hence, the above comparison showed the reliability of the measurements.

3. Experimental Results and Discussion

3.1. Spatial Distribution Characteristics of Rainfall Intensity

Figures 4–6 show the distribution of rainfall intensity for several flow ratios, collision angles, and heights of the rainfall intensity collection platform from the collision point. The rainfall intensity boundary defined 5% of the maximum rainfall intensity under each working condition [31].

In Figure 4, the flow ratios vary, and the height and collision angles from the collision point to the measurement platform are constant. Note that on the section with the same vertical distance from the collision point, the rainfall intensity range showed an increasing trend on the y-axis when the flow ratio increased. On the x-axis, the rainfall intensity range increased first, and subsequently decreased slowly. For $f = 0.3$, the rainfall intensity range was concentrated in the range $0 \text{ cm} \leq x \leq 22 \text{ cm}$; for $f = 0.9$, the rainfall intensity range was concentrated in the range of $10 \text{ cm} \leq x \leq 100 \text{ cm}$; while when $f = 3.2$, the rainfall intensity range was concentrated in the range of $40 \text{ cm} \leq x \leq 115 \text{ cm}$. Under all of the working conditions, the maximum rainfall intensity was on the x-axis.

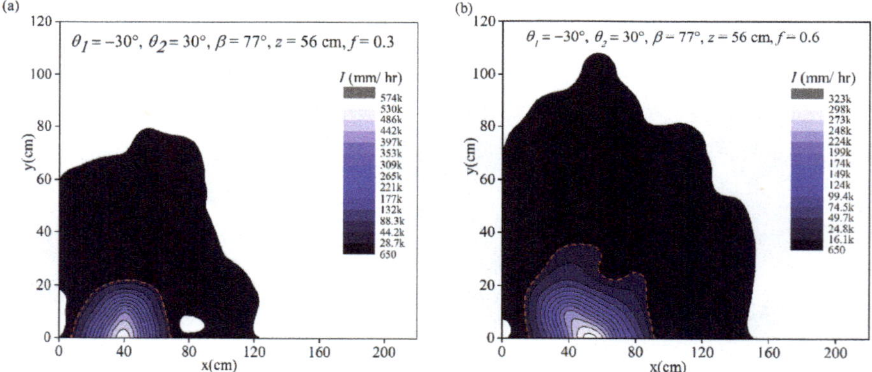

Figure 4. *Cont.*

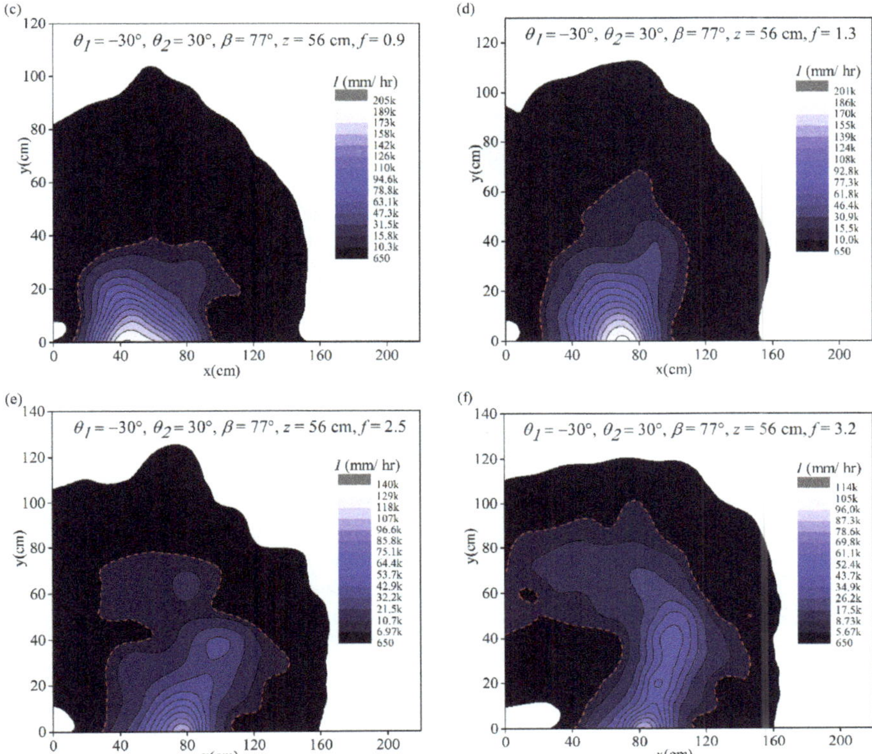

Figure 4. Rainfall intensity distribution at different flow ratios on the *x-o-y* measuring platform. (**a**) θ_1 = $-30°$, θ_2 = $30°$, β = $77°$, z = 56 cm, and f = 0.3; (**b**) θ_1 = $-30°$, θ_2 = $30°$, β = $77°$, z = 56 cm, and f = 0.6; (**c**) θ_1 = $-30°$, θ_2 = $30°$, β = $77°$, z = 56 cm, and f = 0.9; (**d**) θ_1 = $-30°$, θ_2 = $30°$, β = $77°$, z = 56 cm, and f = 1.3; (**e**) θ_1 = $-30°$, θ_2 = $30°$, β = $77°$, z = 56 cm, and f = 2.5; (**f**) θ_1 = $-30°$, θ_2 = $30°$, β = $77°$, z = 56 cm, and f = 3.2.

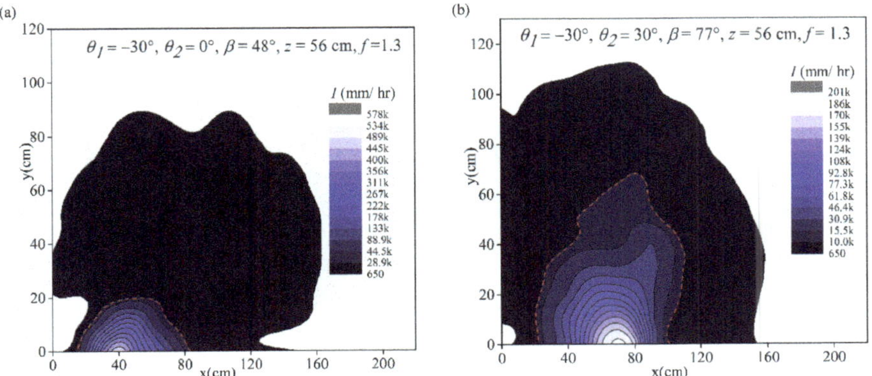

Figure 5. *Cont.*

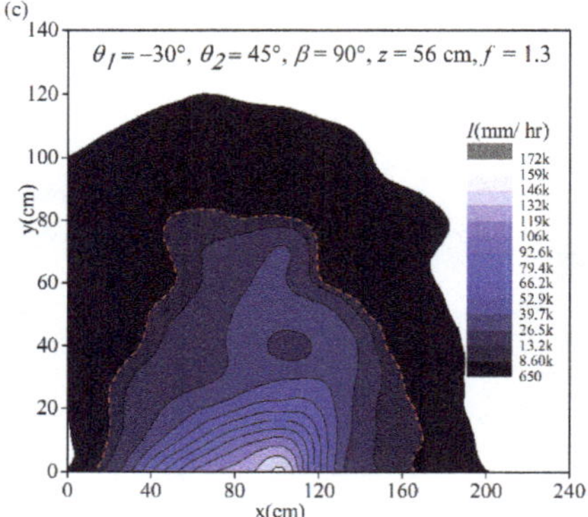

Figure 5. Rainfall intensity distribution at different collision angles on the *x-o-y* measuring platform. (a) $\theta_1 = -30°$, $\theta_2 = 0°$, $\beta = 48°$, $z = 56$ cm, and $f = 1.3$; (b) $\theta_1 = -30°$, $\theta_2 = 30°$, $\beta = 77°$, $z = 56$ cm, and $f = 1.3$; (c) $\theta_1 = -30°$, $\theta_2 = 45°$, $\beta = 90°$, $z = 56$ cm, and $f = 1.3$.

In Figure 5, the flow ratios and height from the collision point to the measurement platform are constant, and the collision angles vary. Note that with the increase in the collision angle, the rainfall intensity range on the *x*- and *y*-axes increased—for example, when $\beta = 48°$, $L_x = 70$ cm, and $L_y = 20$ cm; when $\beta = 77°$, $L_x = 108$ cm, and $L_y = 68$ m; when $\beta = 90°$, $L_x = 163$ cm, and $L_y = 80$ cm. However, the maximum rainfall intensity decreased with the increased collision angle—for example, when $\beta = 48°$, $I_{max} = 5.78 \times 10^5$ mm/h; and for $\beta = 90°$, $I_{max} = 5.78 \times 10^5$ mm/h.

In Figure 6, the flow ratio and collision angle were constant, and the collision point varied from the height of the measurement platform. Note that (1) the range of rainfall intensity on the *x*- and *y*-axes gradually increases along the *z* direction, and (2) the maximum rainfall intensity value continues decreasing. At $z = 56$ cm, the maximum value is 5.74×10^5 mm/h; at $z = 76$ cm, the maximum value is 5.32×10^5 mm/h; and at $z = 96$ cm, the maximum value is 4.22×10^5 mm/h. The maximum value decreases from 5.74×10^5 mm/h to 4.22×10^5 mm/h, which indicates that the maximum rainfall was continuously spreading during the falling process.

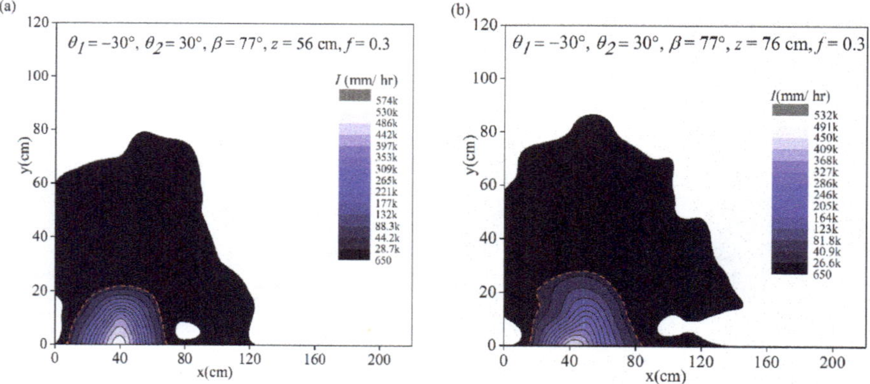

Figure 6. *Cont.*

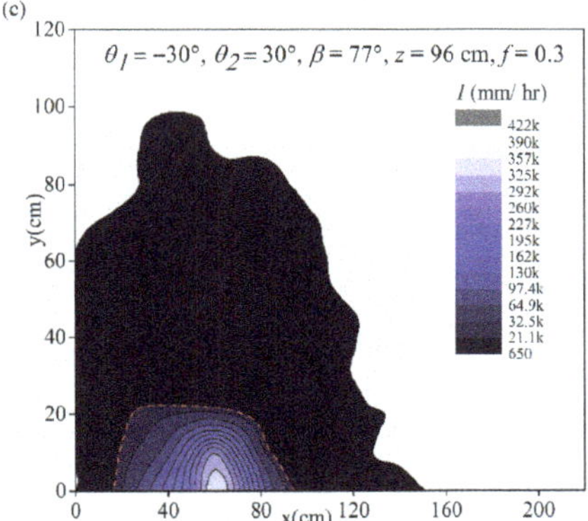

Figure 6. Rainfall intensity distribution at different section heights on the *x-o-y* measuring platform. (**a**) $\theta_1 = -30°$, $\theta_2 = 30°$, $\beta = 77°$, $z = 56$ cm, and $f = 0.3$; (**b**) $\theta_1 = -30°$, $\theta_2 = 30°$, $\beta = 77°$, $z = 76$ cm, and $f = 0.3$; (**c**) $\theta_1 = -30°$, $\theta_2 = 30°$, $\beta = 77°$, $z = 96$ cm, and $f = 0.3$.

The maximum rainfall of each section (q_{max}) is shown in Figure 7. Note that (1) the maximum rainfall intensity decreases with an increase in distance of z, (2) q_{max} sharply decreases for $f < 1.4$ and slightly increases for $f > 1.4$, and (3) the results indicated that q_{max} decreases with an increase of β. In Figure 7a,c, for $f = 1.3$ and $z = 96$ cm, $q_{max}/Q_t \times 10^3 = 1.6$ at $\beta = 48°$, while $q_{max}/Q_t \times 10^3$ decreased to 0.45 at $\beta = 90°$, where Q_t is the sum flow discharge of surface hole and deep hole outlets.

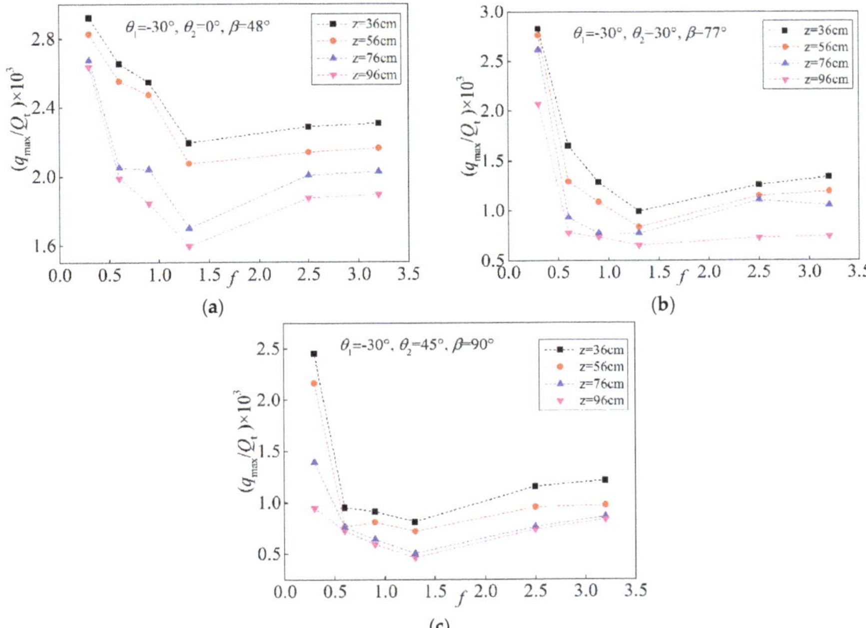

Figure 7. The maximum rainfall varies with the collision angles and flow ratios at different sections. (**a**) $\theta_1 = -30°$, $\theta_2 = 0°$, and $\beta = 48°$; (**b**) $\theta_1 = -30°$, $\theta_2 = 30°$, and $\beta = 77°$; (**c**) $\theta_1 = -30°$, $\theta_2 = 45°$, and $\beta = 90°$.

3.2. Rainfall Intensity Distribution on the x-Axis

The rainfall intensity distributions on the *x*-axis are shown in Figure 8. Note that the rainfall intensity data along the *x*-axis was best correlated with the Gaussian distribution. In addition, the maximum rainfall intensity value gradually decreases with an increase along the *z*-axis. For example, in Figure 8a, the height of the section increases from 36 cm to 96 cm, and the maximum rainfall intensity value decreases from 2.48×10^5 mm/h to 1.43×10^5 mm/h. With an increase in section height, the rainfall intensity distribution along the *x*-axis flattens, and the range on the *x*-axis is bigger.

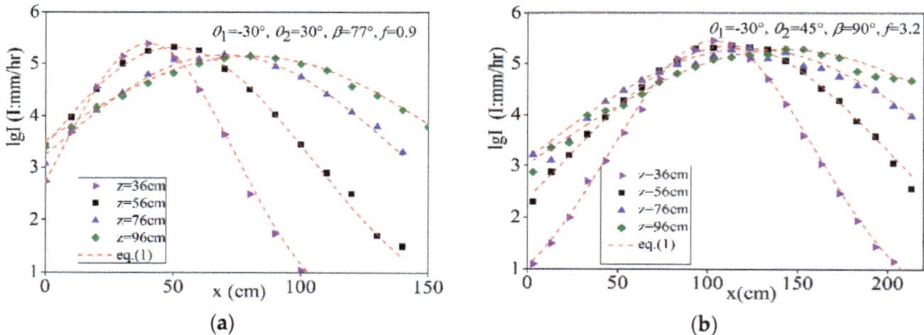

Figure 8. Rainfall intensity distributions along the *x*-axis and comparison with Equations (1) and (2). (a) $\theta_1 = -30°$, $\theta_2 = 30°$, $\beta = 77°$, and $f = 0.9$; (b) $\theta_1 = -30°$, $\theta_2 = 45°$, $\beta = 90°$, and $f = 3.2$.

For two-jet collisions in air on the *x*- and *z*-axis, the rainfall intensity distribution along the *x*-axis after the collision can be expressed by the following formulas (Figure 8):

$$\lg I = \lg I_{\max} \times e^{\left(-a \times \left(\frac{x-x_{\max}}{z}\right)^2\right)} \tag{1}$$

$$\alpha = 1.7e^{(-2.1 \times f)} + 2\cos\beta \tag{2}$$

where I is the rainfall intensity and $I_{\max}$ is the maximum rainfall intensity, α is a coefficient.

3.3. Trajectory Line after the Collision on the x-Axis

As shown in Figure 9, based on previous research on collision flow [9], the surface water tongue and deep water tongue completely collide in air; then, two water tongues combine and shoot in one direction. V_1 is the surface hole water tongue velocity, θ_1 is the angle of depression, and q_1 is the discharge per unit width; in addition, V_1 is the deep hole water tongue velocity, θ_2 is the pick angle, and q_2 is the discharge per unit width. Two water jets meet at the unit of M. At the confluence, the flow velocity of the upper edge of the water tongue at the confluence is V_{1M}; the angle between the upper edge of the water tongue and the horizontal direction is β_1; the flow velocity of the inner edge of the water tongue is V_{2M}; the angle between the inner edge of the water tongue and the horizontal direction is β_2; the flow velocity of the mixed water tongue after the phase confluence is V_M, and the angle between the water tongue and the horizontal direction is β_M.

The spatial position of collision point M can be obtained by calculating the trajectories of the inner edge of the surface-hole jet and outer edge of the deep-hole jet. The trajectories of the deep and surface water tongues were obtained from Equation (3).

$$z(x) = z_0 + \tan(\alpha)x - \frac{gx^2}{2V_0^2\cos^2(\alpha)} \tag{3}$$

where V_0 is the initial speed, g is gravity acceleration.

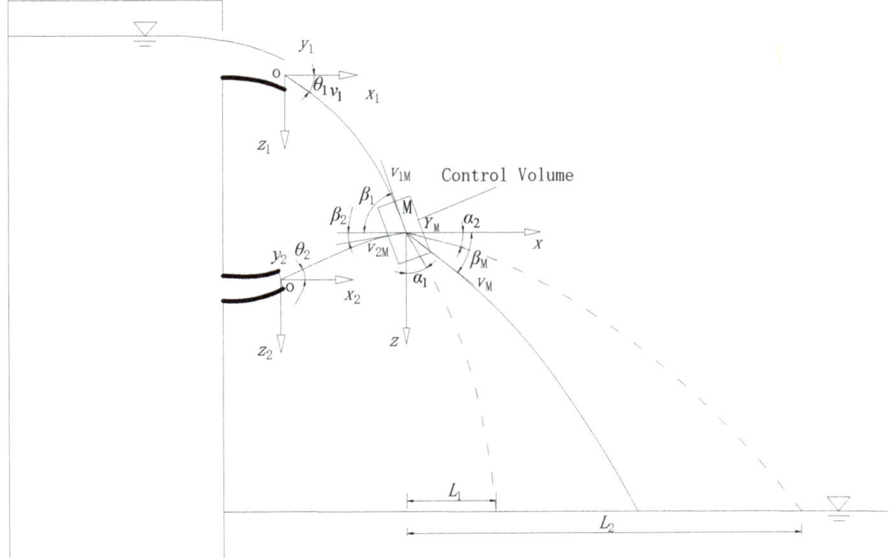

Figure 9. Schematic diagram of the collision between the surface-hole jet and the deep-hole jet in air.

The coordinate systems $x_1 0 z_1$ and $x_2 0 z_2$, in Figure 9 have the following geometric relationship:

$$x_1 = x_2 + \Delta x_0, z_1 = z_2 + \Delta z_0 \tag{4}$$

Using the projectile principle, the trajectory equation of the surface water jet was:

$$z_1 = \frac{g}{2V_1^2 \cos^2 \theta_1} x_1^2 + x_1 tg\theta_1 = A_1 x_1^2 + x_1 tg\theta_1 \tag{5}$$

where V_1 is the surface hole jet initial speed, x_1 is the transverse distance, z_1 is the vertical distance, $A_1 = \frac{g}{2V_1^2 \cos^2 \theta_1}$.

The energy velocities of V_{1M} and V_{2M} before the impact of the surface water jet and deep water jet became:

$$V_{1M} = \varphi_a \sqrt{2g(z_{1M} + \frac{V_1^2}{2g})} \tag{6}$$

$$V_{2M} = \varphi_a \sqrt{2g(z_{2M} + \frac{V_2^2}{2g})} \tag{7}$$

where φ_a is the velocity coefficient of air resistance ($\varphi_a \approx 0.95$), z_{1M} is the vertical distance from the surface hole to control volume, z_{2M} is the vertical distance from the deep hole to control volume. The control volume was taken around point M, as shown in Figure 9, and the momentum integral equation and continuous equation of fluid mechanics were used.

$$\iiint_M \frac{\partial \rho}{\partial t} \vec{v} dV + \oiint_M \rho \vec{v}(\vec{v} \cdot \vec{n}) dA = \Sigma \vec{F} \tag{8}$$

$$\iiint_M \frac{\partial \rho}{\partial t} dV + \oiint_M \rho(\vec{v} \cdot \vec{n}) dA = 0 \tag{9}$$

where ρ is the density of water, and $\vec{F}$ is the external force that acts on the control volume. In the absence of air resistance, β_M and V_M could be obtained with

$$\tan \beta_M = \frac{V_{1M}q_1 \sin \beta_1 - V_{2M}q_2 \sin \beta_2}{V_{1M}q_1 \cos \beta_1 + V_{2M}q_2 \cos \beta_2} \tag{10}$$

$$V_M = \frac{V_{1M}q_1 \cos \beta_1 + V_{2M}q_2 \cos \beta_2}{q_M \cos \beta_M} \tag{11}$$

with the assumption that $V_{1M} = V_{2M}$. Thus, Equations (10) and (11) become

$$\tan \beta_M = \frac{q_1 \sin \beta_1 - q_2 \sin \beta_2}{q_1 \cos \beta_1 + q_2 \cos \beta_2} \tag{12}$$

$$V_M = \frac{V_{1M}(q_1 \cos \beta_1 + q_2 \cos \beta_2)}{q_M \cos \beta_M} \tag{13}$$

The comparisons of the measured locations of I_{max} with Equations (12) and (13) are shown in Figure 10, which indicates good agreement.

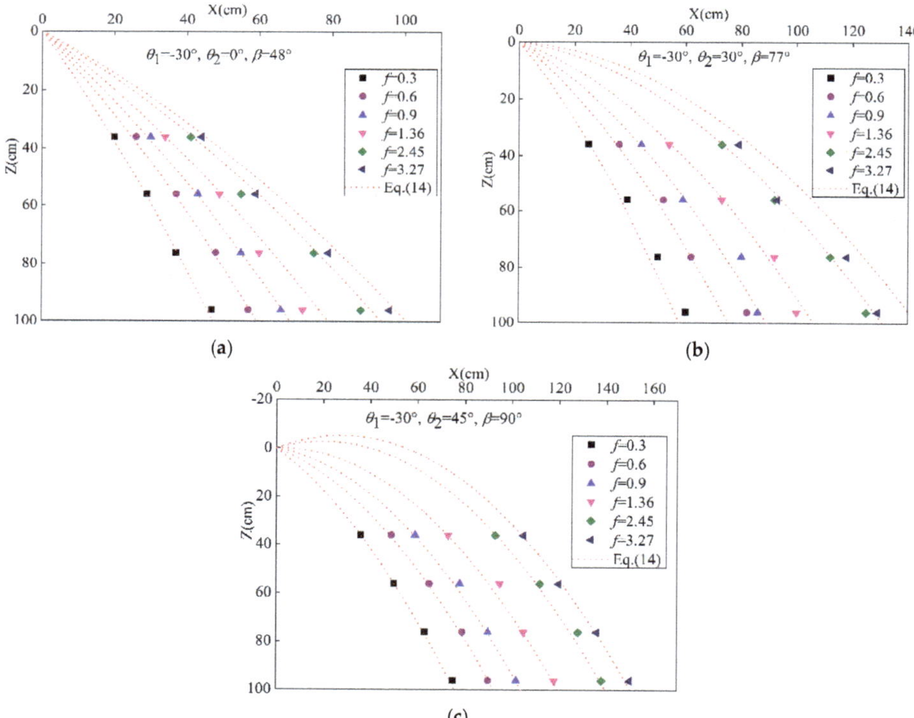

Figure 10. Comparison of experimental and calculated values of the maximum rainfall intensity points. (a) $\theta_1 = -30°$, $\theta_2 = 0°$, and $\beta = 48°$; (b) $\theta_1 = -30°$, $\theta_2 = 30°$, and $\beta = 77°$; (c) $\theta_1 = -30°$, $\theta_2 = 45°$, and $\beta = 90°$.

Based on the above results, we assume that the velocity magnitude was identical in all directions after the collision, but that the directions vary. Therefore, the velocities in all directions were V_M. Downstream of the collision, the trajectory boundaries ($5\% \times I_{max}$) are shown in Figure 11. Hence, using Equation (3), α_1 and α_2 can be obtained—α_1 is the angle between the z-axis and the inner edge line, and α_2 is the angle between the x-axis and the outer edge line.

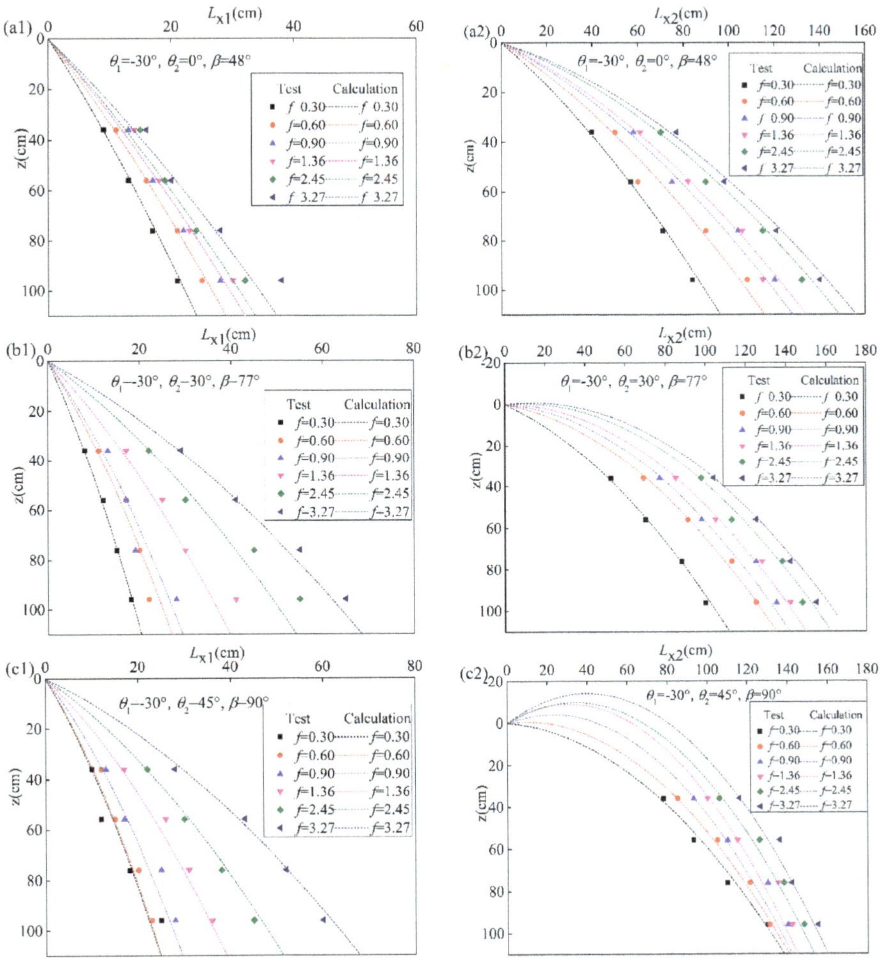

Figure 11. Comparison of the experimental and calculated values of L_{X1} and L_{X2}. L_{X1} is the distance of inner edge water tongue in x axis, L_{X2} is the distance of outer edge water tongue in x axis. (**a1**) $\theta_1 = -30°$, $\theta_2 = 0°$, and $\beta = 48°$; (**a2**) $\theta_1 = -30°$, $\theta_2 = 0°$, and $\beta = 48°$; (**b1**) $\theta_1 = -30°$, $\theta_2 = 30°$, and $\beta = 77°$; (**b2**) $\theta_1 = -30°$ and $\theta_2 = 30°$; (**c1**) $\theta_1 = -30°$, $\theta_2 = 45°$, and $\beta = 90°$; (**c2**) $\theta_1 = -30°$, $\theta_2 = 45°$, and $\beta = 90°$.

The calculated values of α_1 and α_2 are shown in Figure 12. Note that (1) with the identical collision angle β, α_1 increased, and with the increase in flow ratio, α_2 decreased; and (2) with the identical flow ratio, α_1 increased with the increase in collision angle β, and α_2 decreased with the increase in collision angle β. It can also be seen from Figure 12 that α_1 approximately shows exponential growth with the flow ratio, and α_2 shows logarithmic growth with the flow ratio. Data analysis indicates that the values of α_1 and α_2 are obtained as

$$\alpha_1 = 16e^{(f\sin^2\beta)} \tag{14}$$

$$\alpha_2 = \beta \times (-0.182) \times \ln(f) - \beta \times \sin(\beta - 80°) \tag{15}$$

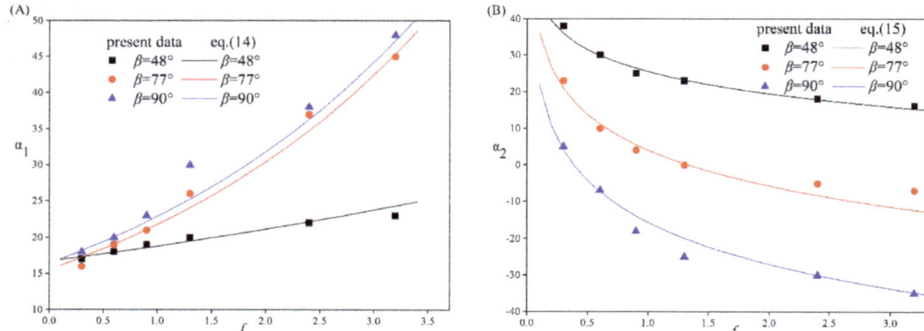

Figure 12. Comparison of the values obtained from Equations (14) and (15) with the present data. (**A**) Inner angle of α_1, and (**B**) outer angle of α_2.

The computed and measured values correlated to $R^2 = 0.97$ (Figure 12A) and 0.96 (Figure 12B). However, up until now, there was a paucity of information on the rainfall characteristics of the two-jet collisions. This present study provides a new method for understanding the rainfall characteristics of the two-jet collision, which will be useful to extend to other models or even prototypes for further research.

4. Conclusions

In this paper, the flow characteristics of two-jet collisions with different collision angles and flow ratios were investigated systematically by a series of model tests. The present investigation illustrated the development of the spatial distribution of rainfall intensity. Observations indicated that the maximum rainfall intensity sharply decreased with an increase in flow ratio, while the maximum rainfall intensity slightly increased, as the flow ratio was greater than 1.0.

On the x-axis, the distribution of rainfall intensity followed a Gaussian distribution. The range of rainfall intensity tended to reach the maximum for the flow ratio = 1.0. A theoretical equation to compute the locations of maximum rainfall intensity was presented, which is in good agreement with the model tests. The formulas to calculate the boundary lines of the x-axis were proposed. The present analysis was based upon experiments performed in a model with $48° < \beta < 90°$, $0.3 < f < 3.2$, and 36 cm $< z <$ 96 cm. The rainfall intensity distributions on the y-axis will be further studied.

Author Contributions: Conceptualization, H.Y. and W.X.; Methodology, W.X.; Software, Validation, H.Y.; W.X. and R.L.; Formal Analysis, H.Y.; Experiment, H.Y., Y.F., Y.H.; Writing-Original Draft Preparation, H.Y.; Writing-Review & Editing, H.Y.; Supervision, W.X.; Project Administration, W.X.; Funding Acquisition, W.X.

Funding: This work was supported by the National Key Research and Development Program of China (Grant No. 2016YFC0401707) and National Natural Science Foundation of China (Grant 51609162 and Grant 51409181).

References

1. Li, N.-W. *Study on the Non-Collision Way of Flood Discharge and Energy Dissipation in High Arch Dams with Contracting Spillways and Middle Level Outlets*; Sichuan University: Chengdu, China, 2008.

2. Li, N.-W.; Xu, W.-L.; Zhou, M.-L.; Tian, Z. Experimental study on energy dissipation of flood discharge in high arch dams without impact of jets in air. *J. Hydraul. Eng.* **2008**, *39*, 927–933.

3. Shuangke, S. Summary of research on flood discharge and energy dissipation of high dams in China. *J. China Inst. Water Resour. Hydropower Res.* **2009**, *7*, 89–95.

4. Bai, R.; Zhang, F.; Liu, S.; Wang, W. Air concentration and bubble characteristics downstream of a chute aerator. *Int. J. Multiph. Flow* **2016**, *87*, 156–166. [CrossRef]

5. Bai, R.; Liu, S.; Tian, Z.; Wang, W.; Zhang, F. Experimental investigations on air-water flow properties of offse-aerator. *J. Hydraul. Eng.* **2017**, *144*, 04017059. [CrossRef]

6. Li, S.; Zhang, J.; Xu, W.; Chen, J.; Peng, Y. Evolution of Pressure and Cavitation on Side Walls Affected by Lateral Divergence Angle and Opening of Radial Gate. *J. Hydraul. Eng.* **2016**, *142*, 05016003. [CrossRef]

7. Bai, Z.; Peng, Y.; Zhang, J. Three-dimensional turbulence simulation of flow in a V-shaped stepped spillway. *J. Hydraul. Eng.* **2017**, *143*, 06017011. [CrossRef]

8. Huang, G.; Wu, S.; Chen, H. Model test studies of atomized flow for high dams. *Hydro-Sci. Eng.* **2008**, *2008*, 91–94.

9. Xue, L.-F. Influence of atomization on environment during flood discharge from dam and reducing measures. *Sichuan Water Power* **2005**, *24*, 104–107.

10. Bai, R.; Liu, S.; Tian, Z.; Wang, W.; Zhang, F. Experimental investigation of the dissipation rate in a chute aerator flow. *Exp. Therm. Fluid Sci.* **2019**, *101*, 201–208. [CrossRef]

11. Xiong, X.-L.; Ge, G. The collision energy dissipation between surface hole's water tongue and middle hole's water tongue of Ertan hydropower station. *J. Hydropower Eng. Res.* **1991**, 1–10. (In Chinese)

12. Chi-gong, G.Y.W. An optimal research on the energy dissipation by collision of two free jets in ertan hydroelectric plant. *J. Chengdu Univ. Sci. Technol.* **1992**, *18*, 17–24.

13. Liu, P.; Dong, J.; Liu, Y. Computation of energy dissipation for impact of two jets in air. *J. Hydraul. Eng.* **1995**, *7*, 38–44.

14. Diao, M.J.; Yang, Y.Q. Experimental study on two jets impact in air for energy dissipation. *J. Sichuan Univ. (Eng. Sci. Ed.)* **2002**, *34*, 13–15.

15. Sun, J.; Li, Y.-Z. Hydraulic characteristics of water jets by air-collision in lateral direction and estimation on limiting riverbed scour. *Chin. J. Appl. Mech.* **2004**, *21*, 134–137.

16. Sun, J.; Li, Y.-Z.; Yu, C.-Z. Limiting scour depth by the interaction of water jets from surface outlets with jets from mid-level outlets in high arch dams. *J. Tsinghua Univ. (Sci. Technol.)* **2002**, *42*, 564–568.

17. Liang, Z.-C. A Computation Model for Atomization Flow. *J. Hydrodyn.* **1992**, *7*, 247–255.

18. Liu, X.-L.; Zhang, W. The investigation of dynamic characteristics of jet-flow in open air. *J. Hydroelectr. Eng.* **1988**, *2*, 46–54. (In Chinese)

19. Liu, X.-L.; Liu, J. Experiment study on the diffusion and aeration of three-dimensional jet. *J. Hydraul. Eng.* **1989**, *11*, 9–17. (In Chinese)

20. Li, W.-X.; Wang, W.; Xu, W.-L.; Diao, M.-J. A rough method to calculate the range of atomized flow caused by jet impact. *J. Sichuan Union Univ. (Eng. Sci. Ed.)* **1999**, *3*, 17–23.

21. Liu, S.-H.; Liang, Z.-C. On atomizing problems by jumping jet in hydropower station at narrow and deep valley. *J. Wuhan Univ. Hydraul. Electr. Eng.* **1997**, *30*, 6–9.

22. Liu, S.-H.; Qu, B. Investigation on splash length of aerated jet. *J. Wuhan Univ. Hydraul. Electr. Eng.* **2003**, *36*, 5–8.

23. Liu, S.-H. Study of the atomized flow in hydraulic engineering. *J. Hydrodyn.* **1999**, *2*, 77–83.

24. Liu, S.-H.; Yin, S.-R.; Luo, Q.-S. Numerical simulation of atomized flow diffusion in deep and narrow goeges. *Conf. Glob. Chin. Scholar Hydrodyn.* **2006**, *18*, 515–518.

25. Sun, S.-K.; Liu, Z.-P. Longitudinal range of atomized flow forming by discharge of spillways and outlet works in hydropower stations. *J. Hydraul. Eng.* **2003**, *34*, 53–58.

26. Liu, H.-T.; Sun, S.-K.; Liu, Z.-P.; Wang, X.-S. Atomization prediction based on artificial neural networks for flood releasing of high dams. *J. Hydraul. Eng.* **2005**, *36*, 1241–1245.

27. Lian, J.-J.; Liu, F.; Huang, C.-Y. Numerical study on effects of environmental wind and terrain on spray caused by jet from flip bucket. *J. Hydraul. Eng.* **2005**, *36*, 1147–1152.

28. Sun, X.-F.; Liu, S.-H. Investigation on the motion of splash droplets in atomized flow. *J. Hydrodyn.* **2008**, *23*, 61–66.

29. Wang, S.-Y.; Chen, D.; Hou, D.-M. Experimental Research on the Rainfall intensity in the Source Area of Flood Discharge Atomization. *J. Yangtze River Sci. Res. Inst.* **2013**, *30*, 70–74.

30. Lian, J.-J.; Li, C.-Y.; Liu, F.; Wu, S.-Q. A prediction method of flood discharge atomization for high dams. *J. Hydraul. Res.* **2014**, *52*, 274–282. [CrossRef]

31. Zhang, W.; Zhu, D. Far-field properties of aerated water jets in air. *Int. J. Multiph. Flow* **2015**, *76*, 158–167. [CrossRef]

32. Taylor, G. Formation of thin flat sheets of water. *Proc. R. Soc.* **1960**, *259*, 1–17. [CrossRef]

33. Dombrowski, N.; Hooper, P.C. A study of the sprays formed by impinging jets in laminar and turbulent flow. *J. Fluid Mech.* **1963**, *18*, 392–440. [CrossRef]

34. Anderson, W.E.; Ryan, H.M.; Pal, S.; Santoro, R.J. Fundamental studies of impinging liquid jets. In Proceedings of the 30th AIAA, Aerospace Sciences Meeting and Exhibit, Reno, NV, USA, 6–9 January 1992; Volume 1, pp. 6–9.

35. Ibrahim, E.A.; Przekwas, A.J. Impinging jets atomization. *Phys. Fluids Fluid Dyn.* **1991**, *3*, 2981–2987. [CrossRef]

36. Choo, Y.J.; Kang, B.S. Parametric study on impinging-jet liquid sheet thickness distribution using an interferometric method. *Exp. Fluids* **2001**, *31*, 56–62. [CrossRef]

37. Choo, Y.J.; Kang, B.S. The velocity distribution of the liquid sheet formed by two low-speed impinging jets. *Phys. Fluids* **2002**, *14*, 622–627. [CrossRef]

38. Couto, H.S.; Bastos-Netto, D. Modeling droplet size distribution from impinging jets. *J. Propul. Power* **1989**, *7*, 654–656. [CrossRef]

39. Choo, Y.J.; Kang, B.S. A study on the velocity characteristics of the liquid elements produced by two impinging jets. *Exp. Fluids* **2003**, *34*, 655–661. [CrossRef]

40. Lee, C.H. An experimental study on the distribution of the drop size and velocity in asymmetric impinging jet sprays. *J. Mech. Sci. Technol.* **2008**, *22*, 608. [CrossRef]

41. Fore, L.B.; Dukler, A.E. The distribution of drop size and velocity in gas-liquid annular flow. *Int. J. Multiph. Flow* **1995**, *21*, 137–149. [CrossRef]

Article

Development of Bubble Characteristics on Chute Spillway Bottom

Ruidi Bai [1], Chang Liu [1], Bingyang Feng [1], Shanjun Liu [1] and Faxing Zhang [1,*]

State Key Laboratory of Hydraulics and Mountain River Engineering, Sichuan University, Chengdu 610065, China; bairuidiscu@163.com (R.B.); sculiuchangvip@163.com (C.L.); 18625801588@163.com (B.F.); drliushanjun@vip.sina.com (S.L.)
* Correspondence: zhfx@scu.edu.cn; Tel.: +86-159-2803-8306

Received: 30 July 2018; Accepted: 23 August 2018; Published: 24 August 2018

Abstract: Chute aerators introduce a large air discharge through air supply ducts to prevent cavitation erosion on spillways. There is not much information on the microcosmic air bubble characteristics near the chute bottom. This study was focused on examining the bottom air-water flow properties by performing a series of model tests that eliminated the upper aeration and illustrated the potential for bubble variation processes on the chute bottom. In comparison with the strong air detrainment in the impact zone, the bottom air bubble frequency decreased slightly. Observations showed that range of probability of the bubble chord length tended to decrease sharply in the impact zone and by a lesser extent in the equilibrium zone. A distinct mechanism to control the bubble size distribution, depending on bubble diameter, was proposed. For bubbles larger than about 1–2 mm, the bubble size distribution followed a—5/3 power-law scaling with diameter. Using the relationship between the local dissipation rate and bubble size, the bottom dissipation rate was found to increase along the chute bottom, and the corresponding Hinze scale showed a good agreement with the observations.

Keywords: chute aerator; spillway bottom; air concentration; air bubble frequency; air bubble chord length

---

1. Introduction

Chute aerators, regarded as a cost-effective method, have been widely used owing to their proven success as a countermeasure in the Grand Coulee dam since 1960 [1]. Previous studies have provided a significant amount of information on the macroscopic air-water flow properties [2–13], such as the air concentration, air entrainment and detrainment. Rutschmann and Hager [2], combining both model and prototype data, proposed two approaches to calculate the air entrainment coefficient $\beta = q_a/q$, where q_a = air discharge and q = water discharge. The air entrainment coefficient was also investigated by, among others, Shi [14], Pinto [15,16], and Low [17]. Chanson [3] investigated the characteristics of chute aerator flow on the distributions of various bubble parameters (void fraction, velocity, air discharge, turbulent diffusivity, etc.) for different inflow Froude numbers. He also provided a better understanding of the air entrainment processes and found a strong detrainment process occurred in the impact zone. Kramer et al. [5] divided the chute aerator into five zones: inflow region; air detrainment region; minimum air concentration region; air entrainment region; and uniform mixture flow region, and systematically investigated the air detrainment gradients in the far-field of chute aerators. Pfister and Hager [6–8] measured the air concentration and divided the chute aerator into three zones: jet zone, reattachment and spray zone, and far–field zone. The authors also discussed the air entrainment and proposed two comprehensive methods to calculate the average and bottom air concentration development using deflectors and offsets along the chute. Avoiding the upper aeration effect, Bai et al. [9] divided the chute aerator into three zones and found that an intensive air detrainment occurred in the impact zone and the detrained air escaped into the cavity zone. The previous experimental investigations are summarized in Table 1.

Table 1. Experimental investigations of Chute Spillways.

Reference	V_0 (m/s)	h_0 (m)	F_0	Remarks
Chanson [3]	9.2 6.2–11.3 5.3	0.023 0.035 0.081	19.5 10.5–19.5 6.0	Model ($W = 0.25$ m)
Prifser [6–8]	4.7–7.5	0.041–0.066	7.4–9.3	Model ($W = 0.30$ m)
Bai [9,10]	4.0–9.0	0.15	3.3–7.4	Model ($W = 0.25$ m)
Shi [14]	14.0	0.058	18.6	Model ($W = 0.20$ m)
Pinto [15,16]	19.5–27.7 19.9–32.7 19.9–36.2	0.38–1.06 0.38–1.43 0.38–1.29	7.1–10.3 8.7–11.1 10.3–11.4	Prototype (Foz do Areia)
Low [17]	4.2–9.5	0.05	6–13.5	Model ($W = 0.25$ m)

Many studies concerning the minimum or critical air concentration to avoid cavitation erosion have been conducted [18,19]. Peterka [18] found that when the average air concentration was between 0.01 and 0.06, no cavitation erosion was observed. Up to now, there is no reliable design guideline for the distance required between two aerators to produce sufficient bottom air concentration [5]. Based on prototype observations on Russian dams, Semenkovand Lantyaev [20] provided an average bottom air concentration decay of 0.40% to 0.80% per chute meter.

However, air concentration governs the comprehensive performance of the bubbles, including their chord length and frequency form. In particular, Robinson [21], Xu et al. [22], and Brujan [23] adopted advanced techniques to investigate the mechanism of erosion damage and found that the interaction between the air bubbles and cavitation bubbles was the mechanism that assisted in the prevention of erosion damage. These studies indicate that the bubble size and frequency are significant parameters for the mechanism of erosion damage. However, negligible research has been conducted on the spillway bottom air microcosmic characteristics such as the bottom bubble frequency and chord size. Such investigations are difficult because of the complex nature of the flow and strong impaction of the bottom. In the present study, the development of the bottom bubble characteristics was measured and analyzed.

2. Hydraulics Model

The experiments were conducted in a rectangular chute that was 0.25-m wide with a variable bottom slope $5.71° \leq \alpha \leq 18.2°$ (Figure 1). The measured section was 5-m long and consisted of a smooth convergent nozzle with a width of 0.25 m and height of 0.15 m. The approach flow depth was $h_0 = 0.15$ m and the emergence angle was $0° \leq \theta_0 \leq 14.1°$. The water discharge was supplied by a large tank (3.1-m wide, 3.2-m long, and 9.2-m high) and was measured using a rectangular sharp-crested weir with an accuracy of approximately 1%. In order to gain sufficient velocity, the nozzle was located in the bottom of the tank. A range of approach flow velocities was set from 4.0 m/s to 9.0 m/s. Observations indicated that the approach of the flow was stable. Hence, the Froude number F_0 ($V_0/(gh_0)^{0.5}$) was within the range of 3.3 to 7.4 (Table 2), where g was the gravity. x was the streamwise coordinate along the chute bottom, h_s was the offset height, L was the cavity length, and L_m and L_D were the specific lengths [9].

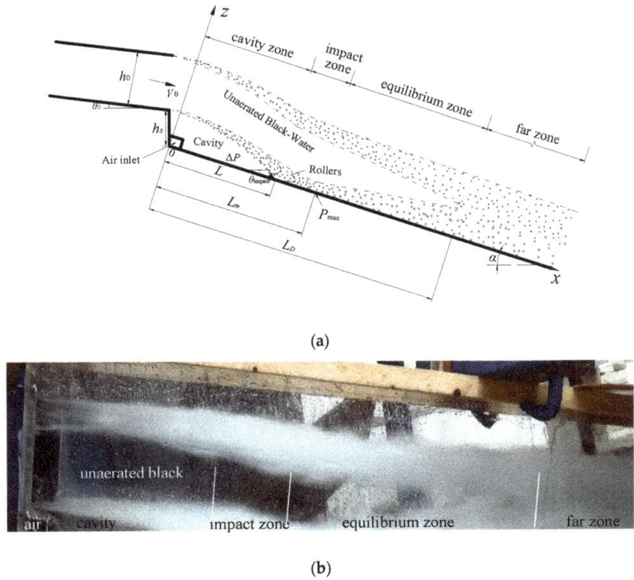

Figure 1. Definition sketch with relevant parameters (**a**) and side view of chute aerator flow with high speed camera (**b**).

Table 2. Experimental conditions of chute aerator flow.

Run	V_0 (m/s)	h_s (m)	θ_0 (°)	α (°)	F_0
S1	5.0–7.0	0.03	5.71	12.5	3.8–5.8
S2	4.0–6.0	0.05	12.5	18.2	3.3–4.9
S3	5.0–7.0	0.045	0	5.7	3.8–5.8
S4	4.0–6.0	0.045	5.7	14.1	3.3–4.9
S5	6.0–9.0	0.045	14.1	14.1	4.9–7.4

All measurements were performed on the channel centre line. The air-water flow properties were measured with a phase-detection needle probe. The working principle of a conductivity probe was based on the difference in the voltage indices at the platinum tip between the air and water phases. The response time of this sensor was less than 10 μs. The signals from the conductivity probe were recorded at a scan rate of 100 kHz per for a 40 s scan period. The accuracy of the air concentration was $\Delta C/C = 3\%$, and the bubble frequency was $\Delta f/f = 1\%$, where C was the air concentration and f was the bubble frequency. The length of the impact zone was very short. A few studies in the impact zone have been performed with a space of 0.03 m along the chute bottom.

3. Basic Chute Aerator Flow Properties

Air Concentration and Bubble Frequency Profiles

A typical photo of the offset-aerator flow is shown in Figure 2, where four zones were introduced. The typical distributions of the air concentration and bubble frequency in the different zones are depicted in Figure 2 as a function of the dimensionless distance perpendicular to the bottom, i.e., z/h_0. Owing to a sufficient depth of the approach flow, the unaerated water prevented the upper aeration effect.

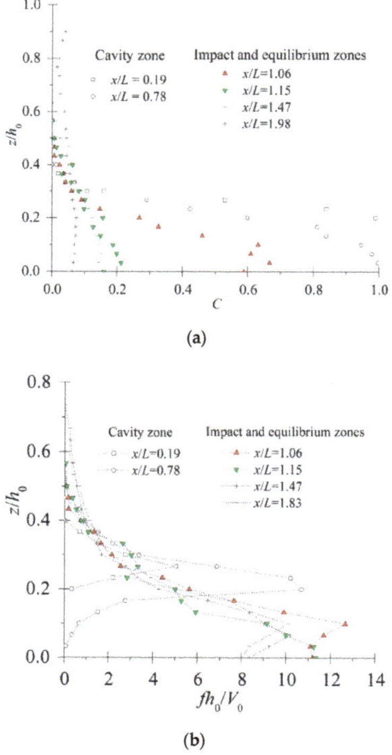

Figure 2. Distributions of air concentration (**a**) and bubble frequency (**b**) along the chute (S5: $V_0 = 9$ m/s, $F_0 = 7.4$).

In the cavity zone ($0 < x/L < 1$), air entrainment occurs in the form of air bubbles entrapped into the lower flow and the air concentration decreased from the lower surface to the vertical direction. At $x/L = 1$, the lower jet impacted on the chute bottom. Further downstream ($x/L > 1$), the distributions of air concentration were different from the cavity zone. Figure 2a shows that the air concentration distributions downstream of the cavity zone, exhibiting an air concentration peak in the turbulent shear region, whereas the air concentration increases from the chute bottom to its maximum value, and then decreases in the vertical direction.

Note that: (1) for $x > L$, the maximum value air concentration C_m decreased sharply in the impact zone, while in the equilibrium zone, the peak air concentration C_m decreased slightly. (2) Experimental results indicated the existence of a maximum value f_m in bubble frequency (Figure 2b), but its location differed from the locus of the maximum void fraction as noticed by Bai et al. [10]. In the cavity, the location of f_m approached $C = 0.50$ with the development of the low jet aeration. The location of f_m was approximately the same as that of the section maximum air concentration C_m at a section in the impact zone, but lower than that in the equilibrium zone.

For more than 25 experiments conducted in this model, only one set of experiments is shown herein to illustrate the basic chute aerator flow property [9]. However, equations presented hereafter are based on all experiments.

4. Bottom Air Bubble Characteristics

The entrained air from the chute aerator aims to prevent erosion damage. Recent investigations have indicated that the bubble size and frequency are the significant factors preventing erosion damage [22,23]. The air bubbles generated in the cavity zone travelled through the impact zone to continue downstream. For the chute aerator flows, a large amount of air entrained in the cavity zone was only partially transported to the downstream flow. Under the effect of rollers, large bubbles broke into small bubbles before being transported downstream zone simultaneously in the impact zone.

4.1. Bottom Air Concentration and Bubble Frequency

Bottom air concentration C_b was defined experimentally using the values measured closest to the model chute bottom (z = 1–1.5 mm). The experimental data of bottom air concentration C_b is presented in Figure 3. Note the following: (1) the decay rate of C_b was different in the impact ($L < x < L_m$) and equilibrium zones ($L_m \leq x < L_D$); C_b decreased sharply in the impact zone and decreased slightly in the equilibrium zone. In the impact zone, bottom air concentration C_b = 0.64 at x/L = 1.06 and it decreased to 0.18 at x/L = 1.16. In contrast, the bottom air concentration decreased slightly in the equilibrium zone, with concentration C_b = 0.145 at x/L = 1.31 decreasing to 0.064 at x/L = 1.91. (2) With the increase in F_0, the values of C_b increase in the present experiments, in agreement with the results of Pfister and Hager [7]. (3) The comparison of the measured values of C_b with the values from the formula proposed by Pfister and Hager [7] is shown in Figure 3, which displays a similar trend downstream of the chute aerator. However, the present values are generally larger than those of Pfister and Hager [7]. At a sufficient water depth, the detrained air bubbles cannot escape into the atmosphere through the upper surface; it can only escape into the cavity zone, which may cause the level of C_b to be larger than that of previous studies. In addition, for a shorter length of the impact zone, the fewer measurement sections may cause an insufficient sharp decrease. (4) It can be concluded that the air transportation regularity is different in the impact and equilibrium zones.

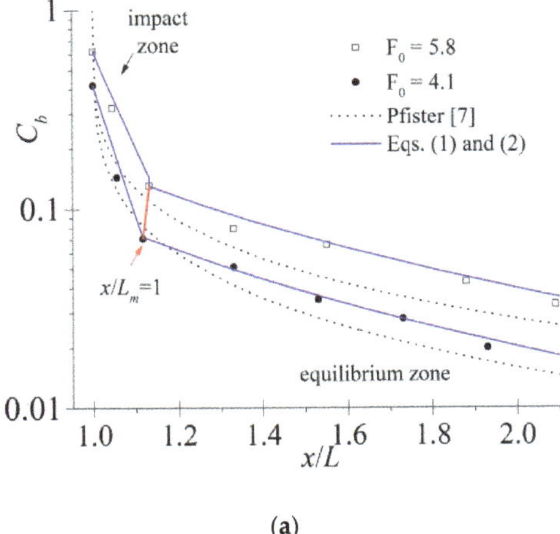

(a)

Figure 3. *Cont.*

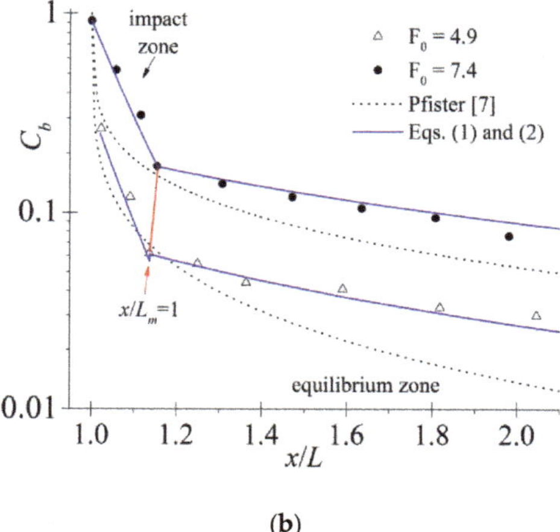

(b)

Figure 3. Bottom air concentration along the chute and compared with Pfister [7] (**a**): S2; (**b**): S5.

From the experimental data analysis, bottom air concentration C_b can be expressed by the following equation (Figure 3):

$$\frac{C_b}{(C_b)_L} = \left(\frac{x}{L}\right)^{0.985F_0 - 17.5} \text{ for } L < x < L_m,\ R^2 = 0.89 \tag{1}$$

$$\frac{C_b}{(C_b)_{Lm}} = \left(\frac{x}{L_m}\right)^{0.071F_0 - 1.73} \text{ for } x > L_m,\ R^2 = 0.97 \tag{2}$$

where R is the correlation coefficient, $(C_b)_L$ is the bottom air concentration at $x = L$, $(C_b)_{Lm}$ is the bottom air concentration at $x = L_m$.

The experimental data of bottom bubble frequency $f_b h_0 / V_0$ along the chute are displayed in Figure 4. There are a few reports on the chute bottom bubble frequency of the lower jet downstream of the chute aerator. Note the following: (1) $f_b h_0 / V_0$ decreased in both the impact and equilibrium zones, whereas decay rate of $f_b h_0 / V_0$ differed slightly. For $F_0 = 7.4$, $f_b h_0 / V_0 = 11.2$ at $x/L = 1.06$, and it decreased to 10.88 at $x/L = 1.17$ in the impact zone (Figure 4d). In contrast, the bottom bubble frequency decreased slightly more in the equilibrium zone, with $f_b h_0 / V_0 = 9.44$ at $x/L = 1.31$ decreasing to 6.22 at $x/L = 1.91$. (2) In contrast, $f_b h_0 / V_0$ decreased slightly in the impact zone, and there was a large variation in air bubble concentration C_b. (3) Furthermore, $f_b h_0 / V_0$ increased with increasing F_0.

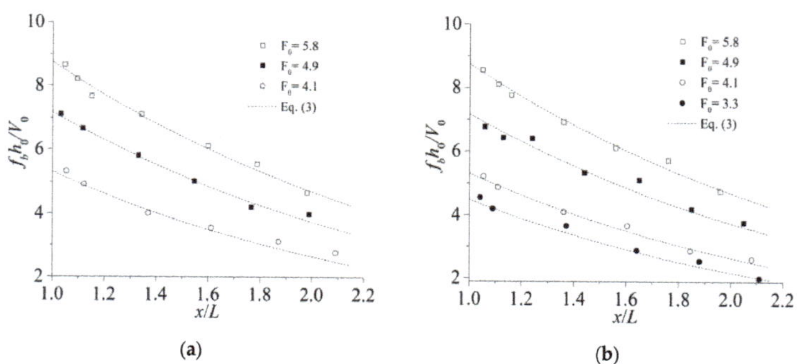

(a)   **(b)**

Figure 4. *Cont.*

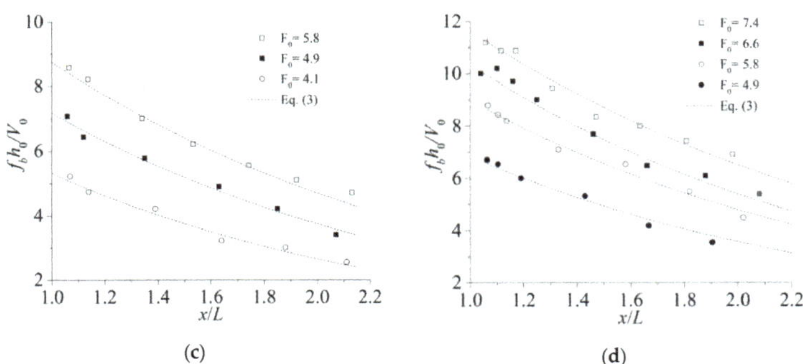

Figure 4. Bottom air bubble frequency along the chute (**a**): S2; (**b**): S3; (**c**): S4; (**d**): S5.

Overall, the present data could be described by

$$\frac{f_b h_0}{V_0} = 2.81 F_0 \times \exp\left(-\frac{x}{L} F_0^{-0.27}\right) \text{ for } x > L, \ R^2 = 0.94 \tag{3}$$

where R was the correlation coefficient.

4.2. Bottom Bubble Chord Length Distributions

The chord length distribution of the air bubbles shows microscopic variations passing a measurement point in the stream-wise direction. For all the flow conditions, the experimental results exhibited a broad range of the chord length, varying from less than 0.1 mm to more than 30.0 mm. In this study, the chord length interval was 0.1 mm. The chord lengths are not the bubble diameters, but are the characteristic stream-wise bubble sizes.

Garrett et al. [24] and Deane et al. [25] indicated that the bubble chord length distribution N_a depended on the dissipation rate ε and bubble diameter d_{ab}, and the scaling law followed as

$$N_a \propto A \varepsilon^{1/3} d_{ab}^{-5/3} \tag{4}$$

where A was a constant, $N_a = N_t/t$, N_t was the number of air bubbles detected in the scan period t (t = 40 s).

The distributions of the bottom bubble chord length in the impact and equilibrium zones are depicted in Figure 5. In the impact zone, the air concentration was within the range of $0.10 < C < 0.90$, such that the intermediate region neither corresponded to the high-air concentration region nor to the bubbly flow region. In the equilibrium zone, the maximum air concentration was less than 0.3, so that the zone corresponded to the bubbly flow region. Note the following: (1) the distributions of bottom bubble chord length were skewed with a large proportion of small bubbles. Compared with the impact region, the range of bubble chord length distributions decreased. In the impact zone, the largest probability of bubble chord lengths was concentrated in the range $0.1 \leq d_{ab} \leq 30$ mm, while in the equilibrium zone, the highest probability of bubble chord lengths was concentrated in the range $0.1 \leq d_{ab} \leq 10$ mm. (2) The curve of distribution of the bottom bubble chord sizes was also sharper and narrower along the chute. The power-law scaling of bubble density on diameter is about $\beta = -5/3$ for larger than about 1–2 mm bubbles in the impact and equilibrium zones. (3) In the impact zone, β showed a slight increase from -1.2 at $x/L = 1.06$ to $-5/3$ at $x/L = 1.15$ due to the air detrainment. In the equilibrium zone, β showed a slightly increase from $-5/3$ at $x/L = 1.21$ to -2.4 at $x/L = 1.64$ due to the buoyancy effect.

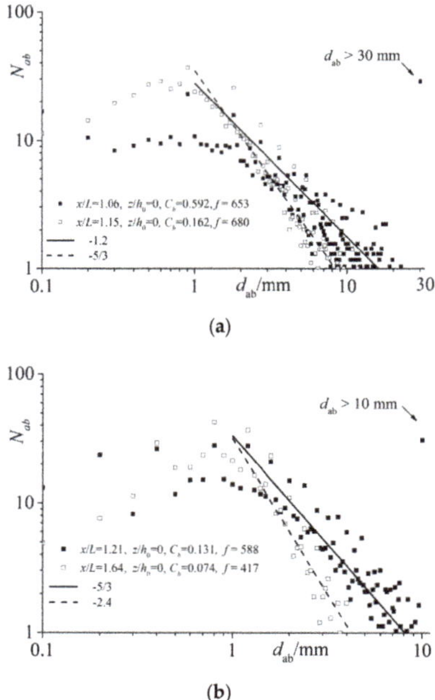

Figure 5. Bottom bubble chord length distributions in the impact zone (**a**) and equilibrium zone (**b**) (S5, $V_0 = 8$ m/s, $F_0 = 6.6$).

4.3. Bottom Turbulent Dissipation Rate

Xu et al. [22,23] and Brujan et al. proved that the size of air bubbles near the bottom was a significant parameter for the collapse process of cavitation bubbles. There was little information to investigate the air bubble variations on the chute bottom. The equilibrium zone belonged to the bubbly flow region and the dynamic behavior of the flow was considered to be a steady state air-water two-phase flow.

Hinze [26] indicated that any fluid gas bubbles or droplets were likely to fragment if the pressure forced on its surface exceeds the restoring forces associated with surface tension.

$$\rho \frac{u^2}{\sigma} d_c = W_c \tag{5}$$

where W_c = the critical Weber numbers, u = the turbulent velocity field on the scale of the bubble. If the fluctuating velocity field was assumed to be described by Kolmogorov's inertial subrange, so that $u^2 = 2\varepsilon^{2/3} d_{ab}^{2/3}$, then only bubbles with diameters larger than the Hinze scale would fragment [17–19].

$$d_c = \left(\frac{W_c \sigma}{2\rho}\right)^{3/5} \varepsilon^{-2/5} \tag{6}$$

where d_c = the Hinze scale, ε = turbulent dissipation rate, which was estimated rather than measured directly. Recent experiments suggest that W_c lied in the range 0.585–4.7, and we have used the value $W_c = 1.0$ (Table 3). For this to exceed a critical value, W_c required that $d > d_c$.

Table 3. Critical Weber numbers for the splitting of air bubbles in water flows.

Reference	W_c	Fluid	Flow Situation	Comments
Hinze [26]	0.585		Two co-axial cylinders, the inner one rotating	Dimensional analysis.
Killen [27]	1.017	Air bubbles in water	Turbulent boundary layer	Experimental data. V in the range 3.66 to 18.3 m/s.
Lewis and Davidson [28]	2.35	Air and Helium bubbles in water and Fluorisol	Circular jet	Experimental data. V in the range 0.9 to 2.2 m/s.
Evans et al. [29]	0.60	Air bubbles in water	Confined plunging water jet	Experimental data. V in the range 7.8 to 15 m/s.
Chanson [3]	1.00	Air bubbles in water	Self-aerated flow	Experimental data. V in the range 3.2 to 5.3m/s.
Martínez-Bazán et al. [30,31]	1.00	Air bubbles in water	Submerged water jet	Experimental data. $V = 17.0$ m/s.
Deane and Stokes [25]	4.7	Air bubbles in water	Breaking waves	Experimental data.
Bai et al. [9]	1.00	Air bubbles in water	Chute aerator flow	Experimental data. V in the range 4.0 to 9.0 m/s.

If the Hinze scale was known, the turbulent energy dissipation rate ε on the chute bottom could be estimated from Equation (7). Hinze [26] used the definition that 95% of the air is contained in bubbles with a diameter less than d_c and the calculated turbulent energy dissipation rate ε for Deane's [18] data was much smaller. Garrett et al. [24] used the results of Martínez-Bazán et al. [30,31] and established a one-to-one relationship between the local dissipation rate and bubble size.

$$\varepsilon = (1/2)^{5/2}\left(\frac{\sigma}{\rho}\right)^{3/2}\int_0^\infty \left(\frac{d_{ab}}{2}\right)^{1/2} N_a \mathrm{d}d_{ab} \bigg/ \int_0^\infty \left(\frac{d_{ab}}{2}\right)^3 N_a \mathrm{d}d_{ab} \tag{7}$$

The bottom dissipation rate ε calculated from Equation (7) is shown in Figure 6 and tends to increase in the measured distance. From Equation (6), a change in dissipation rate ε from 0.10 to 0.75 m^2 s^{-3} corresponded to a change in Hinze scale from 5.5 to 2.4 mm. Thus, the estimate of 2.4 mm for the Hinze scale was in good agreement with the observed change in power-law scaling of the bubble distribution at around 1–2 mm (Figure 5b). Because bubbles smaller than the Hinze scale are stabilized by surface tension forces, a reasonable estimate for this scale is the diameter of the smallest bubbles observed to fragment. In the present study, the bubble chord lengths were smaller than the bubble diameters, due to the measurement technique utilizing conductivity probes. A challenge remains to observe the bubble diameter in a model facility, or even a prototype facility.

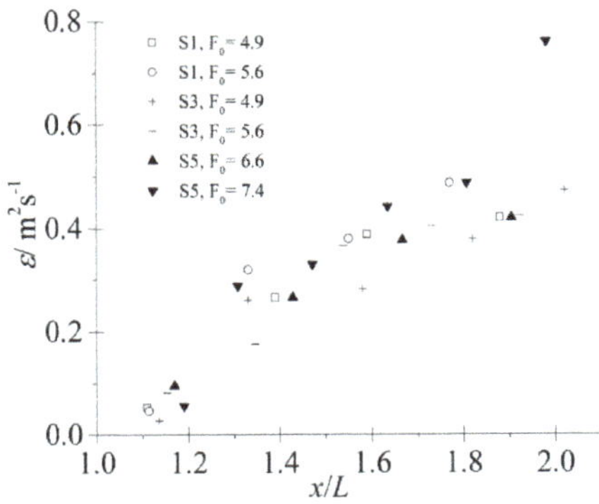

Figure 6. Calculated values of turbulent energy dissipation rate ε on the chute bottom.

5. Conclusions

This study presented new information on the bottom air bubble characteristics in the downstream of chute offset-aerators, which were measured with an accurate measurement technique utilizing conductivity probes. For eliminating the upper aeration, the bottom concentration, bubble frequency, bubble chord length, and bottom dissipation rate were analyzed along the chute longitudinal direction. The present study complemented the existing studies on the bubble properties of chute aerator flows.

The bottom air concentration was found to decrease sharply in the impact zone and to decrease slightly in the equilibrium zone. The bottom air bubble frequency was found to decrease slightly along the chute bottom. A formula to predict the bottom bubble frequency in the impact and equilibrium zones was proposed.

Observations showed that the range of probability of bottom bubble chord lengths tended to decrease sharply in the impact zone and to a lesser extent in the equilibrium zone. The bubbles larger than the Hinze scale were subject to fragmentation and showed $-5/3$ power-law scaling with diameter. Using the relationship between the local dissipation rate and bubble size [24], the bottom dissipation rate was found to increase along the chute bottom. The Hinze scale estimated from Equation (6) showed a good agreement with the observed change in power-law scaling of the bubble distribution.

Author Contributions: Conceptualization, R.B., F.Z. and S.L.; Methodology, R.B. and F.Z.; Software, C.L.; Validation, R.B.; Formal Analysis, R.B. and F.Z.; Investigation, R.B.; Resources, S.L.; D.C., B.F.; Writing-Original Draft Preparation, R.B.; Writing-Review & Editing, R.B.; Visualization, B.F.; Supervision, F.Z.; Project Administration, R.B., F.Z. and S.L.; Funding Acquisition, R.B., F.Z. and S.L.

Funding: This work was supported by the National Natural Science Foundation of China (Grant No. 51709293 and Grant No. 51679157), the Fundamental Research Funds for the Central Universities (Grant No. 20826041A4305), and the National Key Research and Development Program (Grant No. 2016YFC0401707).

Conflicts of Interest: The authors declare no conflict of interest.

References

1. Pan, S.; Shao, Y. Scale effects in modeling air demand by a ramp slot. *Scale Eff. Model. Hydraul. Struct.* **1984**, *4*, 1–5.
2. Rutschmann, P.; Hager, H.W. Air entrainment by spillway aerators. *J. Hydraul. Eng.* **1990**, *116*, 765–782. [CrossRef]
3. Chanson, H. Air bubble entrainment in turbulent water jets discharging into the atmosphere. *Aust. Civ. Eng. Trans.* **1996**, *39*, 39–48.
4. Kramer, K.; Hager, W.H. Air transport in chute flows. *Int. J. Multiph. Flow* **2005**, *31*, 1181–1197. [CrossRef]
5. Kramer, K.; Hager, W.H.; Minor, H.-E. Development of air concentration on chute spillways. *J. Hydraul. Eng.* **2006**, *132*, 908–915. [CrossRef]
6. Pfister, M.; Hager, W.H. Chute aerators I: Air transport characteristics. *J. Hydraul. Eng.* **2010**, *136*, 352–359. [CrossRef]
7. Pfister, M.; Hager, W.H. Chute aerators II: Hydraulic design. *J. Hydraul. Eng.* **2010**, *136*, 360–367. [CrossRef]
8. Pfister, M.; Hager, W.H. Chute aerators: pre aerated approach flow. *J. Hydraul. Eng.* **2011**, *137*, 1452–1461. [CrossRef]
9. Bai, R.; Zhang, F.; Liu, S.; Wang, W. Air concentration and bubble characteristics sownstream of a chute aerator. *Int. J. Multiph. Flow* **2016**, *87*, 156–166. [CrossRef]
10. Bai, R.; Liu, S.; Tian, Z.; Wang, W.; Zhang, F. Experimental investigations on air-water flow properties of offse-aerator. *J. Hydraul. Eng.* **2018**, *144*, 04017059. [CrossRef]
11. Li, S.; Zhang, J.M.; Xu, W.L.; Chen, J.G.; Peng, Y. Evolution of pressure and cavitation on side walls affected by lateral divergence angle and opening of radial gate. *J. Hydraul. Eng.* **2016**, *142*, 05016003. [CrossRef]
12. Bai, Z.L.; Peng, Y.; Zhang, J.M. Three-dimensional turbulence simulation of flow in a V-shaped stepped spillway. *J. Hydraul. Eng.* **2017**, *143*, 06017011. [CrossRef]
13. Zhang, J.M.; Chen, J.G.; Wang, Y.R. Experimental study on time-averaged pressures in stepped spillway. *J. Hydraul. Res.* **2012**, *50*, 236–240. [CrossRef]

14. Shi, Q.; Pan, S.; Shao, Y.; Yuan, X. Experimental investigation of flow aeration to prevent cavitation erosion by a deflector. *J. Hydraul. Eng.* **1983**, *3*, 1–13.
15. Pinto, N.L.S.; Neidert, S.H.; Ota, J.J. Aeration at high velocity flows—Part one. *Int. Water Power Dam Constr.* **1982**, *34*, 34–38.
16. Pinto, N.L.S.; Neidert, S.H.; Ota, J.J. Aeration at high velocity flows—Part two. *Int. Water Power Dam Constr.* **1982**, *34*, 42–44.
17. Low, H.S. Model Studies of Clyde Dam Spillway Aerators. Ph.D. Thesis, University of Canterbury, Christchurch, New Zealand, 1986.
18. Peterka, A.J. The effect of entrained air on cavitation pitting. In *Proceedings of the Minnesota International Hydraulic Convention*; ASCE: Reston, VA, USA, 1953; pp. 507–518.
19. Russell, S.O.; Sheenan, G.J. Effect of entrained air on cavitation damage. *Can. J. Civ. Eng.* **1974**, *1*, 97–107. [CrossRef]
20. Semenkov, V.M.; Lentyaev, L.D. Spillway with nappe aeration. *Hydrotech. Constr.* **1973**, *7*, 437–441. [CrossRef]
21. Robinson, P.B.; Blake, J.R.; Kodama, T.; Shima, A.; Tomita, Y. Interaction of cavitation bubbles with a free surface. *J. Appl. Phys.* **2001**, *89*, 8225–8237. [CrossRef]
22. Xu, W.; Bai, L.; Zhang, F. Interaction of a cavitation bubble and an air bubble with a rigid boundary. *J. Hydrodyn.* **2010**, *22*, 503–512. [CrossRef]
23. Brujan, E.A.; Matsumoto, Y. Collapse of micrometer–sized cavitation bubbles near a rigid boundary. *Microfluid. Nanofluid.* **2012**, *13*, 957–966. [CrossRef]
24. Garrett, C.; Li, M.; Farmer, D. The connection between bubble size spectra and energy dissipation rates in the upper ocean. *J. Phys. Oceanogr.* **2000**, *30*, 2163–2171. [CrossRef]
25. Deane, G.B.; Stokes, M.D. Scale dependence of bubble creation mechanisms in breaking waves. *Nature* **2002**, *418*, 839–844. [CrossRef] [PubMed]
26. Hinze, J.O. Fundamentals of the hydrodynamic mechanism of splitting in dispersion processes. *AIChE J.* **1955**, *1*, 289–295. [CrossRef]
27. Killen, J.M. Maximum stable bubble size and associated noise spectra in a turbulent boundary layer. In *Proceedings of the Cavitation and Polyphase Flow Forum*; ASME: New York, NY, USA, 1982; pp. 1–3.
28. Lewis, D.A.; Davidson, J.F. Bubble Splitting in Shear Flow. *Trans. I. Chem. Eng.* **1982**, *60*, 283–291.
29. Evans, G.M.; Jameson, G.J.; Atkinson, B.W. Prediction of the bubble size generated by a plunging liquid jet bubble column. *Chem. Eng. Sci.* **1992**, *47*, 3265–3272. [CrossRef]
30. Martínez-Bazán, C.; Montañés, J.L.; Lasheras, J.C. On the break up of an air bubble injected into a fully developed turbulent flow. Part 1. Breakup frequency. *J. Fluid Mech.* **1999**, *401*, 157–182. [CrossRef]
31. Martínez-Bazán, C.; Montañés, J.L.; Lasheras, J.C. On the break up of an air bubble injected into a fully developed turbulent flow. Part 2. Size PDF of the resulting daughter bubbles. *J. Fluid Mech.* **1999**, *401*, 183–207. [CrossRef]

Article

Experimental Study on the Impact Characteristics of Cavitation Bubble Collapse on a Wall

Jing Luo, Weilin Xu *, Jun Deng, Yanwei Zhai and Qi Zhang

State Key Laboratory of Hydraulics and Mountain River Engineering, Sichuan University, Chengdu 610065, China; luojing@scu.edu.cn (J.L.); djhao2002@scu.edu.cn (J.D.); zaiyanwei0422@163.com (Y.Z.); scu_zq@163.com (Q.Z.)
* Correspondence: xuwl@scu.edu.cn; Tel.: +86-028-8540-1301

Received: 20 August 2018; Accepted: 14 September 2018; Published: 15 September 2018

Abstract: As a hydrodynamic phenomenon, cavitation is a main concern in many industries such as water conservancy, the chemical industry and medical care. There are many studies on the generation, development and collapse of cavitation bubbles, but there are few studies on the variation of the cyclic impact strength on walls from the collapse of cavitation bubbles. In this paper, a high-speed dynamic acquisition and analysis system and a pressure measuring system are combined to study the impact of a cavitation bubble generated near a wall for various distances between the cavitation bubble and the wall. The results show that (1) with the discriminating criteria of the impact pressure borne by the wall, the critical conditions for the generation of a micro-jet in the collapse process of the cavitation bubbles are obtained, and therefore collapses of cavitation bubbles near the wall are mainly divided into primary impact area collapses, secondary impact area collapses and slow release area collapses; (2) it can be seen from the impact strength of the cavitation bubble collapse on the wall surface that the impact of cavitation bubbles on the wall surface during the first collapse decreases as γ (the dimensionless distance between the cavitation bubble and the wall) increases, but the impact of the second collapse on the wall surface increases first and then decreases sharply. When γ is less than 1.33, the impact on the wall surface is mainly from the first collapse. When γ is between 1.33 and 2.37, the impact on the wall surface is mainly from the second collapse. These conclusions have potential theoretical value for the utilization or prevention and control technologies for cavitation erosion.

Keywords: cavitation bubble; high-speed photography; impact pressure; micro-jet

---

1. Introduction

When the collapse of cavitation bubbles occurs near a wall, the wall will bear the cyclic impact from the collapse of the cavitation bubble. The characteristic of the interaction between the cavitation bubble and the wall has become the focus of many technical fields, such as water conservancy, shipbuilding, the chemical industry and many other industries.

Theoretical study about cavitation bubbles can be tracked back to the last century [1–3]. The study of the cavitation field mainly focuses on the damage and destruction of a solid wall surface upon cavitation bubble collapse [4,5].

Shaw et al. [6] used a laser to generate a cavitation bubble and combined this with Schlieren photography to conduct a study on the impact performance of the impact wave and micro-jet generated during the first collapse of a cavitation bubble on a wall. Adopting high-speed photographic equipment with a camera speed of up to 100 million frames per second to study the evolution properties of cavitation bubbles around a wall, Lindau and Lauterborn [7] discovered the annular form during the collapse of a cavitation bubble, the formation of a contrajet and the shock wave. Liu et al. [8] used the method of laser-inducing a cavitation bubble to conduct a study on the impacts of a micro-jet caused by laser-induced cavitation bubble collapse on a metallic copper surface, and found that the liquid-jet was the main damage mechanism in cavitation erosion. Xu et al. [9] studied the cavitation bubble dynamic characteristics on an aluminum surface by using the technology of a sensitive fiber-optic sensor based on the optical beam deflection principle. Ren et al. [10] used a hydrophone to conduct a study on the strength mechanism of impact from cavitation collapse within the range of the relative bubble–wall dimensionless distance γ ($\gamma = h/R_{max}$, R_{max} is the maximal radius of the cavitation bubble, h is the distance between the center of the cavitation bubble and the wall) which is between 0.5 and 2.5. The above studies on the collapse of cavitation bubbles either focus on the first collapse or adopt the indirect method of hydrophone measurement. In addition, from the research on the impact of the micro-jet or high-velocity droplet on a wall with the use of high-speed photography, only qualitative rules can be obtained. However, measuring the microscopical area of the wall under the impact of the micro-jet or high-velocity droplet with the pressure test system provides a new quantitative method for solving millimeter-sized impact-related issues [11].

In recent years, some new research directions on cavitation have been proposed, such as extracorporeal shock-wave lithotripsy [12] and ultrasonic cleaning [13,14]. In addition, studies on the interaction between cavitation bubbles and cavitation bubbles, cavitation bubbles and air bubbles, cavitation bubbles and particles, and cavitation bubbles and ice blocks have also been reported. Fong et al. [15] adopted the underwater discharge technique to induce multiple cavitation bubbles so as to study the coupling effect between two cavitation bubbles generated at different instants. Pain et al. [16] studied the jet flow in an air bubble induced by a nearby cavitation bubble and discovered that the speed of such a jet flow can be up to 250 m/s. Luo et al. [17,18] used the method of high-voltage discharge technology to generate a cavitation bubble to study the interaction between a cavitation bubble and an air bubble, and the interaction between two cavitation bubbles and a rigid wall. Goh et al. [19] studied the interactions between cavitation bubbles horizontally placed below the underwater slab and air bubbles attached to it and discovered that the ratio between the oscillation time of the air bubbles and the oscillation time of the cavitation bubbles is the important parameter impacting the micro-jet caused by cavitation bubble collapse. The interaction between a cavitation bubble and a particle was researched by adopting a high-voltage discharge technology to induce a cavitation bubble [20] and low-voltage underwater discharge technology to induce a cavitation bubble [21]. The interaction between ice and cavitation bubbles was researched by Cui et al. [22], and the direction of the micro-jet and the propagation of shock waves were captured.

Due to the very small volume and short life cycle of cavitation bubbles during the evolution, great difficulties have been imposed on related studies. Therefore, numerical simulation studies have been performed by various researchers. Coupled boundary element and finite element methods have been performed on bubble–structure interactions [23–25]. The boundary element method is used for research on the toroidal configuration and jet impact characteristics in the later stage of the collapse of cavitation bubbles in the inviscid and irrotational field [26–28] numerically simulated the late rebound phenomena of a cavitation bubble by the method of vortex lines. Instead of using a vortex surface, Klaseboer et al. [29] simulated the ring stage of a cavitation bubble with a vortex ring, and profoundly understood the dynamic characteristics of a cavitation bubble and the interaction with walls for different distances between the bubble and the wall, which was in agreement with the experimental results.

For the past few decades, many researchers have tried to find out the impulses associated with cavitation bubble collapse. In most existing studies, these results are only qualitative due to the non-uniform spatial sensitivity of the transducer. Therefore, in this paper, cavitation bubbles are induced by adopting a low voltage spark-discharge system, the cavitation bubble evolution and corresponding wall pressure change processes are measured simultaneously by the piezoresistive pressure sensor. The impact strength of the first collapse and rebound–regeneration collapse of the cavitation bubble on the wall surface is mainly analyzed with the high-speed photographic image and pressure testing data of the wall surface, and the influence of the dimensionless distance of the bubble wall on the impact strength of the wall surface is revealed.

2. Experimental Setups

2.1. Cavitation Bubble Induction Device

In order to get the impact characteristics of the cavitation bubble collapse on the wall, the experimental setup requires a cavitation bubble induction device, a high-speed dynamic collection and analysis system and a transient pressure testing system, as shown in Figure 1.

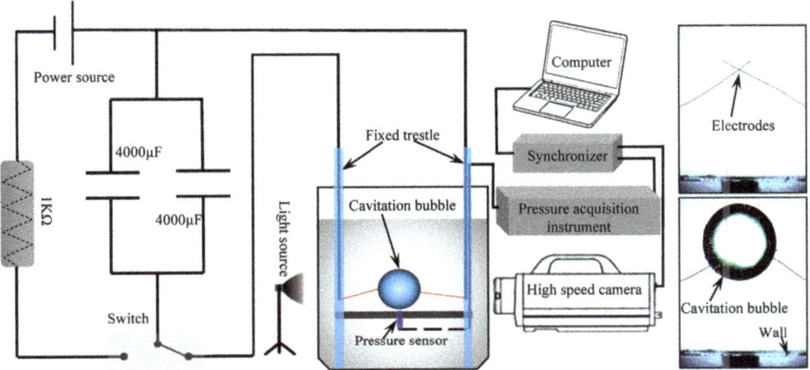

Figure 1. Schematic diagram of experimental facilities.

There are three ways to induce cavitation bubbles in quiescent water: by focusing a laser, by a high-frequency ultrasound wave or by an underwater discharge. The laser technique to induce a cavitation bubble has many advantages, such as its higher accuracy in controlling the focal spot, but the focusing method of the laser is rather complex and costly [30]. In the current work, the method of low-voltage underwater discharge was adopted. The method can accurately control the position and size of the cavitation bubble. To minimize possible interference, the diameter of the electrodes used is 0.15 mm, and the bubbles created have a maximum radius of 30 times the diameter of the electrodes. The copper electrodes (length 50 mm) are fixed in the trestles. The trestles are placed in a water tank (40 cm × 30 cm × 40 cm), which is filled with twice deionized water (at a constant temperature of 22 °C). To create the bubble, two capacitors are fully charged to the specified voltage (the adjustable range is 0 V~100 V) through a 1 KΩ resistor, then switched to the discharge circuit, and the capacitor is discharged through the pair of crossed electrodes.

The life cycle of cavitation bubbles is very short; in order to study the impact process of the collapse of the cavitation bubble on the wall surface, a high-speed dynamic collection and analysis system and a transient pressure testing system must be adopted for synchronized recording. The high-speed dynamic collection and analysis system is composed of a high-speed camera, lens assembly, lighting equipment, etc., and the transient pressure testing system is composed of the resistance-type pressure sensor and the data acquisition instrument. These two systems are respectively connected to a computer through a synchronous trigger device. The Fastcam SA-Z high-speed camera (maximum acquisition rate: 1,000,000 fps, Photron Inc., Tokyo, Japan,) is used. The radius is of millimeter-size when the cavitation bubble generated after the discharge between electrodes in the water body evolves to its maximum volume. Thus, in order to obtain the clear outline of the cavitation bubbles, a micro-lens and LED lamp (150 W) must be adopted in the acquisition system and synchronized with the high-speed photography system. A framing rate of 150,000 frames per second is used to capture the bubble dynamics in this paper, with a shutter speed of 5.06 μs. The width of the frame is 256 pixels.

2.2. Measurement of Cavitation Bubbles During the Evolution Process and the Impact Process

Considering the size of the cavitation bubbles (the radius of cavitation bubbles produced from the system in the experiment ranges from 5.00 mm to 12.00 mm) and the time for the impact of the collapse of the bubble on the wall surface being very short, the size of the sensor in the transient pressure testing system must be small enough, and the sampling frequency and accuracy must be high enough; in addition, by observing the high-speed photographic images, the time for the impact of the micro-jet on the wall due to the collapse of the cavitation bubbles induced by the experimental system lasted about 100 μs. The minimum rise time was at least 10 μs. The pressure sensor used in the experimental system has a high-frequency bandwidth (200 kHz). Therefore, with the existing sensor technology, the piezoresistive pressure sensor used in this paper can meet the requirements for the measurement of the impact of the collapse of cavitation bubbles on the wall surface. The radius of the sensing area of the sensor adopted in the test is 1.50 mm (the maximum measuring range is 40 MPa). The sampling frequency is 2 MHz, and the accuracy is 0.5% of the maximum measuring range.

The method of image gray processing is used to obtain the maximum radius of the cavitation bubble. In this paper, a circle radius with the same number of the pixels is regarded as the equivalent radius R_{eq} of the cavitation bubble (R_{max} is R_{eq} when the cavitation bubble radius reaches maximum), and the centroid of the cavitation bubble in the high-speed photographic image is regarded as the center of the cavitation bubble. The dimensionless distance γ is the ratio of the distance h between the centers of the cavitation bubble and pressure sensor and the maximum value of the maximal equivalent radius R_{max} of the cavitation bubble. Figure 2 shows the data processing process: therein, the scatter diagram shows the relationship between the first cycle of the cavitation bubble time and the corresponding R_{max}, and it can be seen that this data processing method provides good reproducibility.

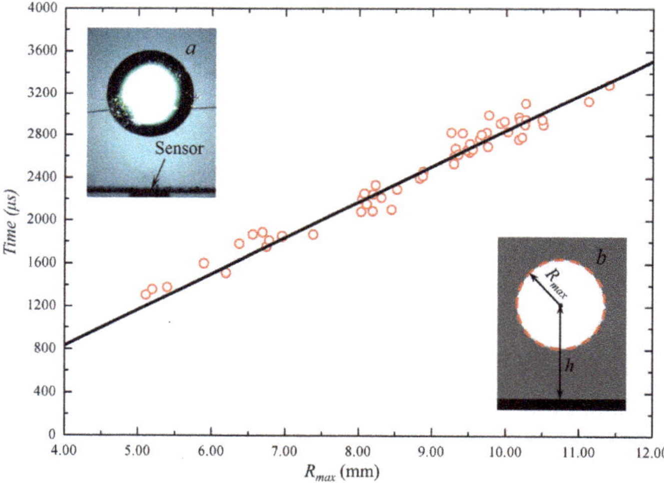

Figure 2. Processing method of characteristic parameters. (**a**) Cavitation bubble photograph; (**b**) characteristic parameters.

3. Results and Discussion

3.1. Influence of Distance Between the Cavitation Bubble and Wall on the Development of Micro-Jetting

A selected set of high-speed photographic images of the cavitation bubbles in the collapse stage for different γ (dimensionless distances between the cavitation bubble and the wall) are shown in Figure 3. Therein, the wall surface is located at the bottom of the image. In the three sets of tests, the dimensionless distances γs are 2.91, 1.91 and 0.83, respectively, the maximum values R_{max} of the equivalent radius R_{eq} are 9.40 mm, 9.31 mm and 9.47 mm respectively, and the distances h from the center of the cavitation bubbles to the wall's surface when the cavitation bubbles expand to their maximum radii are 27.35 mm, 17.76 mm and 7.86 mm, respectively.

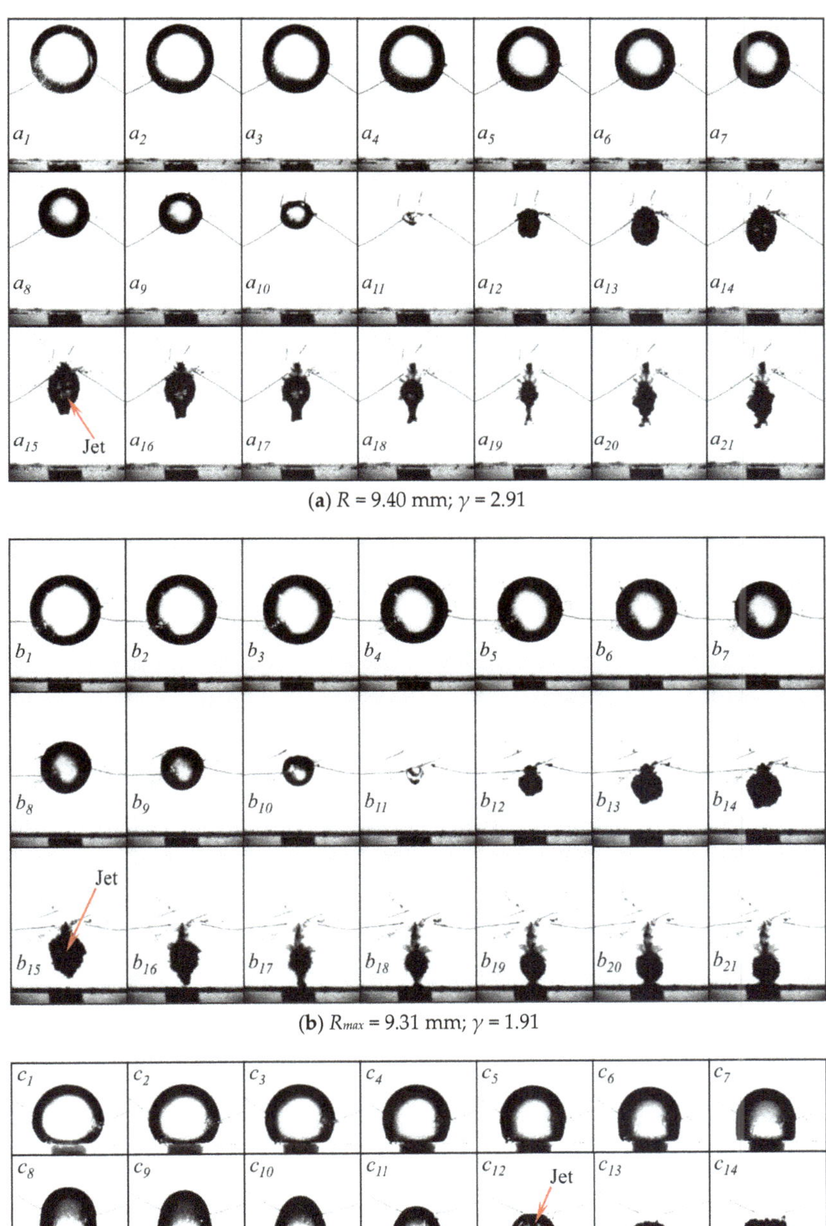

(**a**) R = 9.40 mm; γ = 2.91

(**b**) R_{max} = 9.31 mm; γ = 1.91

(**c**) R_{max} = 9.47 mm; γ = 0.83

Figure 3. High-speed photographic images of the cavitation bubbles in the contraction collapse stage under the conditions of different distances γ (frame-rate: 150,000 fps; exposure time: 5.06 μs; frame width: 32.00 mm).

When the cavitation bubble expands to the maximum radius in the testing, as shown in Figure 3 (a_1,b_1,c_1), the cavitation bubbles begin the contraction stage through an action of the external water body. With the gradual contraction of the cavitation bubble, due to the existence of the wall under the cavitation bubble, the surrounding water begins to fill the surrounding space released by the cavitation bubbles in their contraction process, and the upper and lower surfaces of the cavitation bubble begin to form asymmetrical shapes gradually. The wall surface under the cavitation bubble blocks the filling of water, resulting in a slow contraction speed of the lower surface of the cavitation bubble, while the upper surface belongs to the unbounded domain and contracts quickly. Thus, this asymmetric contraction gradually begins to form from the upper surface (as shown in a_9, b_9 and c_{11} in Figure 3). The development of the asymmetric collapse of the surface has been accompanied by the shrinkage of cavitation bubbles to the minimum volume (as shown in a_{11}, b_{11} and c_{13} in Figure 3). The cavitation bubbles will rebound–regenerate after they shrink to the minimum volume, and as shown in Figure 3, the surface of the rebound cavitation bubble is not very smooth (as shown in a_{12} and b_{12} in Figure 3), but just as in the first cycle, the cavitation bubble will again undergo the expansion–contraction process. In the second expansion–contraction process, the surface is not smooth enough, but this non-smoothness is not enough to change the collapse direction of cavitation bubbles. In Figure 3b, the rebound cavitation bubbles move quickly toward the wall surface in the expansion–collapse process and impact onto the wall in the second collapse.

From the above three tests, the following can be concluded: when the dimensionless distance γ is small, the micro-jet directly impacting the wall surface can be formed during the first asymmetric contraction of the cavitation bubble; as the dimensionless distance γ has been increasing continuously, the cavitation bubbles will move to the wall surface quickly and will eventually impact on the wall surface. With the further increase of γ, cavitation bubbles do not impact on the wall surface through two or three times of contraction. Therefore, the collapse of cavitation bubbles near the wall surface can be divided into main impact area collapse, secondary impact area collapse and slow release area collapse as per the parameters of the distance between the cavitation bubble and wall.

3.2. Impact of Cavitation Bubble Collapse on Wall

The collapse position of cavitation bubbles near the wall surface can be obtained directly through high-speed photography for the development of the micro-jet under the above different γ conditions, but the impact strength of cavitation bubbles on the wall surface during the first collapse, whether the impact can reach the wall surface during the second collapse, and the impact strength of the second collapse on the wall surface cannot be ascertained. With these questions, the high-speed dynamic collection and analysis system and transient pressure testing system are combined in this part, obtaining the impact process of the first and second cavitation bubble collapses on the wall surface (that is, the main impact area and secondary impact area). Figure 4 shows the impact process of the two groups of cavitation bubble collapses on the wall surface when the dimensionless distances γ are 0.91 and 1.79 respectively. The wall surface is located at the bottom of the image, and the pressure sensor is placed inside of the wall surface.

The maximum radius of the cavitation bubble in Figure 4a is 9.98 mm, and when the cavitation bubble reaches the maximum volume, the distance from the center to the wall surface is 9.08 mm. When the electric spark discharges into the water during the induction of the cavitation bubble, the pressure value appears to fluctuate slightly, and the pressure returns to a normal condition after discharge. The subsequently generated cavitation bubbles will begin the expansion–contraction–collapse–rebound stage. In the cavitation bubble development process, the pressure on the wall surface is stable; when the cavitation bubble contracts to the minimum volume (the first expansion-contraction cycle is 2.91 ms), the pressure on the wall surface sharply increases, abd from the contents of the previous section it can be seen that because of the short dimensionless distance γ, the collapse generated due to asymmetrical cavitation bubble contraction directly forms the micro-jet impacting the wall surface (shown as the image in Figure 4a), and the maximum peak value of impact pressure of the first cavitation bubble collapse on the wall surface is 19.37 MPa. After that, the peak pressure gradually decreases when the cavitation bubble is in the rebound–regeneration stage, the pressure on the wall surface is relatively stable for the whole rebound regeneration stage, and the pressure borne on the wall surface greatly increases again when the cavitation bubble contracts to the minimum volume again. The time period of the whole regeneration rebound is 1.23 ms; the wall surface bears the collapse impact of the rebound cavitation bubble again, and by this time, the maximum pressure on the wall surface is 6.50 MPa. It can be seen from the whole impact process of the cavitation bubbles on the wall surfaces in the condition of $\gamma = 0.91$ that the strength of the two impacts of cavitation bubbles on the wall surfaces will gradually reduce from 19.37 MPa in the first impact to 6.50 MPa in the second impact.

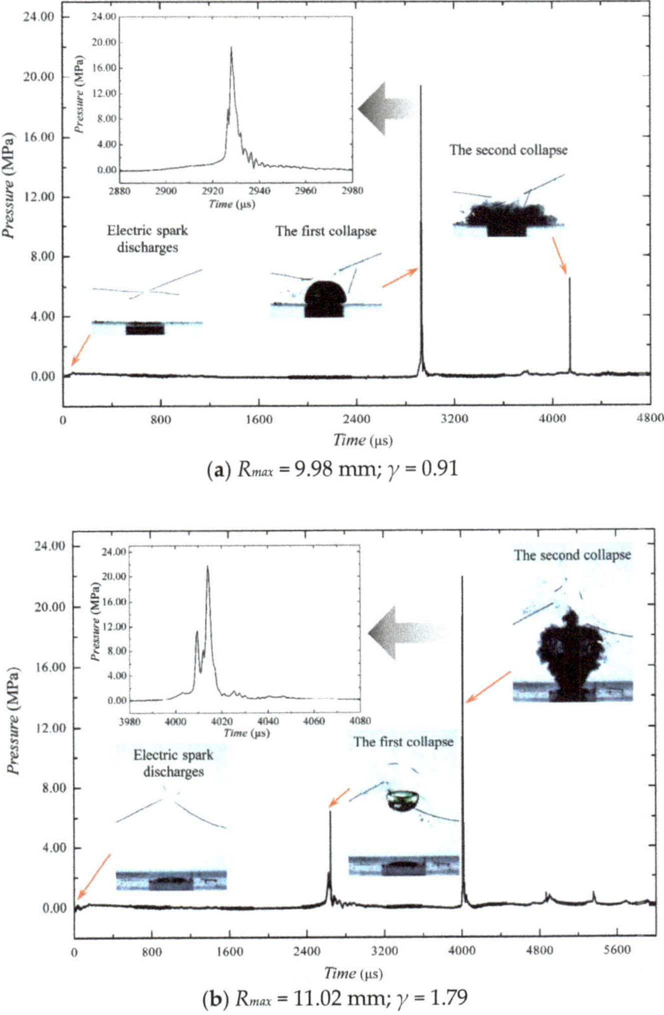

Figure 4. The impulsive process generated by cavitation bubble collapse for (**a**) $\gamma = 0.91$; (**b**) $\gamma = 1.79$.

The maximum radius of the cavitation bubble in Figure 4b is 11.02 mm, and the distance from the center to the wall surface is 19.73 mm when the cavitation bubble reaches the maximum radius.

The first peak and the second peak are the impact on the wall surface during the first collapse and rebound collapse of the cavitation bubble, with the peak pressure values of 6.38 MPa (2.64 ms) and 21.89 MPa (4.01 ms), respectively. The minimum distance to the wall surface when the cavitation bubble shrinks to the minimum volume for the first time is 13.81 mm, and at this moment, the dimensionless distance from the center of the cavitation bubble to the wall surface is 5.06. The cavitation bubble will quickly move to the wall surface in the rebound evolution process; when the cavitation bubble contracts to the minimum volume once more, the cavitation bubble is close to the wall surface, which is as shown in Figure 4b.

It can be clearly seen from Figure 4 that when the dimensionless distance γ is small, the impact of the first collapse of the cavitation bubble on the wall surface is greater than the impact during rebound collapse; when γ can meet a certain condition and the cavitation bubble completes the impact of the first collapse on the wall surface, the rebound cavitation bubble will quickly move to the wall surface, and the impact pressure greater than that of the first collapse will act on the wall surface once more.

Shaw et al. [6] used laser-induced cavitation bubbles and a pressure (voltage) transducer to obtain the waveform of the impact of the first collapse of cavitation bubbles on the wall when γ is between 0.56 and 1.5. The waveform of the impact of the first collapse of cavitation bubbles on the wall reported in the literature is very similar to that in Figure 4a. The similarities between the two are as follows: (1) there are three peaks in the pressure rise. Except the maximum peak, the other two are distributed on both sides of the maximum pressure value; (2) the time interval between the two peaks before the pressure in the experiment increases to the maximum value is 8 μs, and the time interval during the pressure drop is 3 μs; (3) the time of the pressure rise is slightly greater than that of the pressure drop. The differences between the two are as follows: (1) the time span between several peaks is different. The time span in the experiment in this paper is greater than the corresponding value in the literature; (2) the time span of the pressure rise and that of the pressure drop differ. The time of the pressure rise in this experiment is about 13 μs, and that of the pressure drop about 6 μs, which are both greater than that in the literature. The above phenomenon is mainly caused by the influence of the size of the cavitation bubbles. The maximum radius of the corresponding cavitation bubble is 1.17 mm when γ is 0.89 in the literature, and 9.98 mm when γ is 0.91 in this paper.

The impact of the jet on the wall will cause a water hammer effect. The water hammer pressure keeps a linear relation with the jet velocity and is calculated as follows [3]:

$$P_{wh} = \frac{\rho_w c_w \rho_s c_s}{\rho_w c_w + \rho_s c_s} v_t \tag{1}$$

where ρ_w and c_w are the density and sound velocity of the jet medium, respectively; ρ_s and c_s are the density and sound velocity of the solid wall, respectively; and v_t is the velocity of the jet as it impacts the wall. The wall used in this paper is made of plexiglass, with a density and sound velocity of 2700 kg/m^3 and 2692 m/s, respectively. The density and sound velocity of the jet medium are 1000 kg/m^3 and 1435 m/s, respectively.

Figure 5 is a high-speed photographic image of the collapse of cavitation bubbles in Figure 4b, where v is the micro-jet velocity. The frame-rate used in the experiment was 150,000 frames per second, and the pictures in Figure 5 were acquired every 15 frames based on the experimental rate. The micro-jet was formed during the first collapse of the cavitation bubble when the velocity of the micro-jet was approximately 74.96 m/s. When the cavitation bubble entered the rebound stage, a micro-jet was again formed with impact on the wall. The micro-jet developed gradually and eventually imposed an impact on the wall, with its speed decreasing to approximately 13.00 m/s when arriving at the wall. According to the formula for the water hammer pressure, the impact pressure of the micro-jet formed from the rebound cavitation bubbles against the wall is estimated to be 21.36 MPa, and the pressure peak measured by the pressure sensor is 21.89 MPa, both of which are very close. It can be seen that the impact of the micro-jet formed from the rebound cavitation bubbles against the wall mainly comes from the impact of the micro-jet. For the impact of the micro-jet from the first collapse on the wall when the distance between the bubble and the wall is small, due to the limited speed of the high-speed camera used in this experiment, no clear shock wave can be obtained. Therefore, it is currently impossible to tell whether the main reason for the pressure peak of the impact of the cavitation bubble on the wall from the first collapse is the shock wave or the micro-jet.

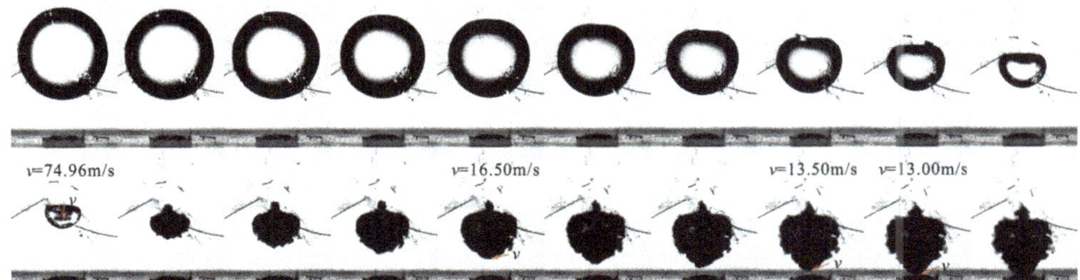

Figure 5. Process of the impact of micro-jet of rebound cavitation bubbles on a wall (frame-rate: 150,000 fps; exposure time: 5.06 µs; R_{max} = 11.02 mm, γ = 1.79).

3.3. Influence of the Distance Between the Cavitation Bubble and Wall on the Impact Strength of the Wall Surface

In the whole process of cavitation bubble evolution, when the dimensionless distance γ meets a certain condition, the second collapse impact strength on the wall surface is greater than the first collapse impact strength. Thus, in this section, the dimensionless distance γ and the impact strength on the wall surface are systematically studied to gain the relationship of the peak impact pressure on the wall surface at cavitation bubble collapse stages under different γ conditions.

In order to obtain the evolution process and impact process of the cavitation bubble with different dimensionless distances γ on the wall surface, the distance from the discharge electrode to the wall surface (with pressure sensor buried inside) will be slightly adjusted in the testing, and the characteristic radius R_{max} of the cavitation bubble will be changed under each h condition. For the peak value of the impact strength of the cavitation bubble on the wall surface, the maximum pressure of the first impact and second impact during the impact of the cavitation bubble on the wall surface is selected in this section to reflect the impact strength of the cavitation bubble on the wall surface.

Figure 6 shows the relationship between the dimensionless distance γ and the wall peak pressure; therein, the horizontal axis indicates the dimensionless distance γ and the vertical axis indicates the corresponding pressure peaks on the wall surface during the first and second collapses of the cavitation bubble. It can be seen from Figure 6 that for the first collapse of the cavitation bubble, the relationship between the dimensionless distance γ and pressure peak on the wall surface shows an exponential-type distribution overall: with the gradual decrease of γ, the maximum value of the impact on the wall surface rapidly increases. Then, the peak value will increase first and then rapidly decrease as a whole during the impact of the second collapse of the cavitation bubble on the wall surface. When the distance between the cavitation bubble and the wall increases, the peak pressure on the wall can be divided into the following three parts: when γ is less than 1.33, the impact of the first collapse of the cavitation bubble on the wall surface is greater than the impact of the second collapse on the wall surface; when γ between the cavitation bubble and wall is greater than 1.33 and less than 2.37, the impact strength on the wall surface caused by the second collapse of the cavitation bubble will be greater than the impact strength of the first collapse on the wall surface; when γ is greater than 2.37, the impact of the first collapse of the cavitation bubble on the wall surface is greater than the impact of the second collapse on the wall surface, and the maximum of the first impact peak is less than 4 MPa.

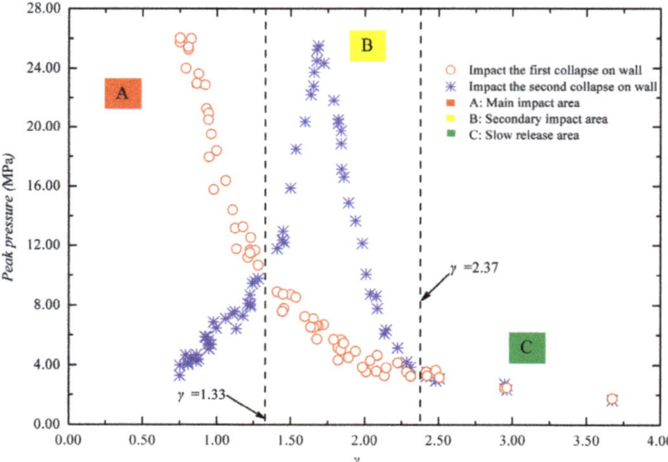

Figure 6. The relationship between the dimensionless distance (γ) and the wall peak pressure.

It can be seen from the above analysis that for the collapse of cavitation bubbles near the wall surface, when the dimensionless distance γ is less than 2.37, the impact of the first and second collapses of the cavitation bubble on the wall surface is huge; when γ is greater than 1.33 and less than 2.37, the impact on the wall surface caused due to the second collapse of the cavitation bubble is greater than the impact on the wall surface caused due to the first collapse. The reason why the above impact characteristics are seen can be analyzed from the following two aspects: on the one hand, when the bubble is very close to the wall surface, during the first bubble-asymmetric shrinkage, a micro-jet with a direct impact on the wall is formed. When the cavitation bubble is a little further away from the wall surface, the impact of the wall on the first shrinkage form of the cavitation bubble is weakened, and a micro-jet is formed during the shrinkage of the cavitation bubble. The micro-jet towards the wall surface drives the cavitation bubble, and finally, during the second shrinkage, the micro-jet is formed. This micro-jet can impose an impact on the wall. Although the energy of the cavitation bubble itself is attenuated after the first cycle of evolution, the impact of the micro-jet from the second shrinkage on the wall is still considerably large. As the distance between the cavitation bubble and the wall surface further increases, the bubble during the second shrinkage obviously moves toward the wall surface, but the resulting micro-jet from the second collapse is not sufficient to impose an impact on the wall, and then the bubble enters the next expansion and contraction stage. On the other hand, the cavitation bubble has a certain energy after the birth. The energy is dissipated due to the viscosity of the water body and other factors after the first and second expansion. After the two-fold expansion and shrinkage, the energy is significantly reduced. As a result, less obvious movement toward the wall is observed and the impact on the wall is reduced greatly in the subsequent evolution of the cavitation bubble.

Philipp and Lauterborn [3] studied the deterioration of the wall under different bubble–wall conditions when the cavitation bubble's R_{max} = 1.45 mm with the laser-induced cavitation technology. There were over 100 groups of samples. By analyzing the deterioration and bubble–wall distance γ, it is found that when $\gamma \geq 2.2$, no deterioration was found on the solid wall; when $2.2 \geq \gamma \geq 1.5$, the area of deterioration increased with the decrease of γ. By comparing the deterioration in the literature and the impact of the cavitation bubble on the wall obtained in this paper, it can be found that the critical values are basically consistent. Since a low-pressure discharge-induced cavitation bubble is used here, it is more difficult to achieve cavitation under the condition of smaller values of γ. Furthermore, the cavitation bubble was larger with the low-voltage underwater discharge technology used in this paper, and the size in the literature is smaller, which may cause the slight difference in the critical value between the paper and the literature. This difference, however, does not affect the zoning characteristics of the impact of the cavitation bubble on the wall.

4. Conclusions

Through the combination of the high-speed dynamic collection and analysis system and transient pressure testing system, the impact characteristics of the collapse of the low-voltage spark-discharge cavitation bubble on the wall surface are studied in this paper, drawing the following important conclusions:

(1) The dimensionless distance between the cavitation bubble and wall (γ) has an important influence on the formation of the micro-jet in the late evolution stage of the cavitation bubble. When the dimensionless distance γ is less than 1.33, a micro-jet directly impacting the wall surface can be formed during the first contraction of the cavitation bubble; subsequently, the collapse of the main impact area will occur. When γ is within 1.33~2.37, the second collapse of the cavitation bubble after rebound will cause a greater impact on the wall surface, then the collapse of the secondary impact area will occur. When γ is greater than 2.37, the influence of wall on the development of a micro-jet is not obvious, that is, collapse in the slow release area occurs.

(2) Concerning the peak value of the impact pressure of the cavitation bubble collapse on the wall surface, the impact of cavitation bubbles on the wall surface during the main impact gradually decreases with the increase of the distance between the cavitation bubble and the wall, and the impact of the cavitation bubble on the wall surface during the secondary impact increases first and then decreases. More specifically, when the dimensionless distance γ is less than 1.33, the impact on the wall surface due to cavitation bubble collapse is mainly from the first collapse, and with increasing γ, the impact of the second collapse on the wall surface increases. When γ is within 1.33~2.37, the impact of the cavitation bubble on the wall surface during the first collapse will be less than the impact strength of the re-collapse after rebound on the wall surface with increase of γ. When γ is greater than 2.37, the impact of the first and second collapses of the cavitation bubble on the wall surface will decrease considerably.

The above conclusions have potential theoretical value for the anti-erosion design of hydraulic engineering, the erosion resistance of blades in hydraulic machinery and the better exertion of the cavitation effect in ultrasonic cleaning technology.

Author Contributions: Conceptualization, W.X. and J.L.; Methodology, J.D. and J.L.; Software, J.L.; Validation, W.X. and J.D.; Formal Analysis, J.L.; Investigation, J.L.; Resources, J.L.; Data Curation, Y.Z. and Q.Z.; Writing-Original Draft Preparation, J.L.; Writing-Review & Editing, J.D.; Visualization, J.L.; Supervision, W.X.; Project Administration, W.X.; Funding Acquisition, J.D.

Funding: This research was funded by the National Key R&D Program of China (Grant No.2016YFC0401901) and the National Natural Science Foundation of China (Grant No.51409180).

Conflicts of Interest: The authors declare no conflict of interest.

References

1. Shima, A.; Takayama, K.; Tomita, Y.; Ohsawa, N. Mechanism of impact pressure generation from spark-generated bubble collapse near a wall. *Aiaa J.* **1983**, *21*, 55–59. [CrossRef]
2. Tomita, Y.; Shima, A. Mechanisms of impulsive pressure generation and damage pit formation by bubble collapse. *J. Fluid Mech.* **1986**, *169*, 535–564. [CrossRef]
3. Philipp, A.; Lauterborn, W. Cavitation erosion by single laser-produced bubbles. *J. Fluid Mech.* **1998**, *361*, 75–116. [CrossRef]
4. Mitelea, I.; Bordeaşu, I.; Pelle, M.; Crăciunescu, C. Ultrasonic cavitation erosion of nodular cast iron with ferrite–pearlite microstructure. *Ultrason. Sonochem.* **2015**, *23*, 385–390. [CrossRef] [PubMed]
5. Dong, F.; Li, X.; Zhang, L.; Ma, L.; Li, R. Cavitation erosion mechanism of titanium alloy radiation rods in aluminum melt. *Ultrason. Sonochem.* **2016**, *31*, 150–156. [CrossRef] [PubMed]

6. Shaw, S.J.; Schiffers, W.P.; Emmony, D.C. Experimental observations of the stress experienced by a solid surface when a laser-created bubble oscillates in its vicinity. *J. Acoust. Soc. Am.* **2001**, *110*, 1822. [CrossRef] [PubMed]

7. Lindau, O.; Lauterborn, W. Cinematographic observation of the collapse and rebound of a laser-produced cavitation bubble near a wall. *J. Fluid Mech.* **2003**, *479*, 327–348. [CrossRef]

8. Liu, X.M.; He, J.; Lu, J.; Ni, X.W. Effect of surface tension on a liquid-jet produced by the collapse of a laser-induced bubble against a rigid boundary. *Opt. Laser Technol.* **2009**, *41*, 21–24. [CrossRef]

9. Xu, R.; Zhao, R.; Cui, Y.; Lu, J.; Ni, X. The collapse and rebound of gas-vapor cavity on metal surface. *Opt. Int. J. Light Electron. Opt.* **2009**, *120*, 115–120. [CrossRef]

10. Ren, X.D.; He, H.; Tong, Y.Q.; Ren, Y.P.; Yuan, S.Q.; Liu, R.; Zou, C.Y.; Wu, K.; Sui, S.; Wang, D.S. Experimental investigation on dynamic characteristics and strengthening mechanism of laser-induced cavitation bubbles. *Ultrason. Sonochem.* **2016**, *32*, 218–223. [CrossRef] [PubMed]

11. Rochester, M.C.; Brunton, J.H. Surface pressure distribution during drop impingement. In Proceedings of the Fourth International Conference on Rain Erosion and Related Phenomena, Neues Schloss, Meersburg, Germany, 8–10 May 1974; pp. 371–393.

12. Klaseboer, E.; Fong, S.W.; Turangan, C.K.; Khoo, B.C.; Szeri, A.J.; Calvisi, M.L.; Sankin, G.N.; Zhong, P. Interaction of lithotripter shockwaves with single inertial cavitation bubbles. *J. Fluid Mech.* **2007**, *593*, 33–56. [CrossRef] [PubMed]

13. Lauterborn, S.; Urban, W. Ultrasonic cleaning of submerged membranes for drinking water applications. *J. Acoust. Soc. Am.* **2008**, *123*, 3291. [CrossRef]

14. Verhaagen, B.; Fernández, R.D. Measuring cavitation and its cleaning effect. *Ultrason. Sonochem.* **2016**, *29*, 619–628. [CrossRef] [PubMed]

15. Fong, S.W.; Adhikari, D.; Klaseboer, E.; Khoo, B.C. Interactions of multiple spark-generated bubbles with phase differences. *Exp. Fluids* **2009**, *46*, 705–724. [CrossRef]

16. Pain, A.; Terence Goh, B.H.; Klaseboer, E.; Ohl, S.W.; Cheong Khoo, B. Jets in quiescent bubbles caused by a nearby oscillating bubble. *J. Appl. Phys.* **2012**, *111*, 137. [CrossRef]

17. Luo, J.; Xu, W.L.; Niu, Z.P.; Luo, S.J.; Zheng, Q.W. Experimental study of the interaction between the spark-induced cavitation bubble and the air bubble. *J. Hydrodyn.* **2013**, *25*, 895–902. [CrossRef]

18. Luo, J.; Xu, W.L.; Li, R. High-speed photographic observation of collapse of two cavitation bubbles. *Sci. Chin. Technol. Sci.* **2016**, *59*, 1707–1716. [CrossRef]

19. Goh, B.H.T.; Ohl, S.W.; Klaseboer, E.; Khoo, B.C. Jet orientation of a collapsing bubble near a solid wall with an attached air bubble. *Phys. Fluids* **2014**, *26*, 221–240. [CrossRef]

20. Xu, W.L.; Zhang, Y.L.; Luo, J.; Arong; Zhang, Q.; Zhai, Y.W. The impact of particles on the collapse characteristics of cavitation bubbles. *Ocean Eng.* **2017**, *131*, 15–24. [CrossRef]

21. Bing, H.T.G.; Shi, W.G.; Ohl, S.W.; Khoo, B.C. Spark-generated bubble near an elastic sphere. *Int. J. Multiphase Flow* **2016**, *90*, 156–166. [CrossRef]

22. Cui, P.; Zhang, A.M.; Wang, S.; Khoo, B.C. Ice breaking by a collapsing bubble. *J. Fluid Mech.* **2018**, *841*, 287–309. [CrossRef]

23. Duncan, J.H.; Zhang, S. On the interaction of a collapsing cavity and a compliant wall. *J. Fluid Mech.* **1991**, *226*, 401–423. [CrossRef]

24. Chahine, G.L.; Kalumuck, K.M.; Duraiswami, R. Bubble dynamics fluid-structure interaction simulation by coupling fluid BEM and structural FEM codes. *Int. J. Multiphase Flow* **1995**, *9*, 861–883. [CrossRef]

25. Chahine, G.L.; Kalumuck, K.M.; Hsiao, C.T. Simulation of surface piercing body coupled response to underwater bubble dynamics utilizing 3dynafs, a three-dimensional bem code. *Comput. Mech.* **2003**, *32*, 319–326. [CrossRef]

26. Zhang, Y.L.; Yeo, K.S.; Khoo, B.C.; Wang, C. 3d jet impact and toroidal bubbles. *J. Comput. Phys.* **2001**, *166*, 336–360. [CrossRef]

27. Wang, C.; Khoo, B.C. An indirect boundary element method for three-dimensional explosion bubbles. *J. Comput. Phys.* **2004**, *194*, 451–480. [CrossRef]

28. Best, J.P. The formation of toroidal bubbles upon the collapse of transient cavities. *J. Fluid Mech.* **1993**, *251*, 79–107. [CrossRef]

29. Klaseboer, E.; Hung, K.C.; Wang, C.; Wang, C.W.; Khoo, B.C.; Boyce, P.; Debono, S.; Charlier, H. Experimental and numerical investigation of the dynamics of an underwater explosion bubble near a resilient/rigid structure. *J. Fluid Mech.* **2005**, *537*, 387–413. [CrossRef]
30. Goh, B.H.; Oh, Y.D.; Klaseboer, E.; Ohl, S.W.; Khoo, B.C. A low-voltage spark-discharge method for generation of consistent oscillating bubbles. *Rev. Sci. Instrum.* **2013**, *84*, 014705. [CrossRef] [PubMed]

Article

Numerical and Experimental Comparative Study on the Flow-Induced Vibration of a Plane Gate

Chunying Shen [1,2], Wei Wang [1,*], Shihua He [2] and Yimin Xu [2]

[1]   State Key Laboratory of Hydraulics and Mountain River Engineering, Sichuan University, Chengdu 610065, China; shenchunying@kmust.edu.cn
[2]   Faculty of Electrical Power Engineering, Kunming University of Science and Technology, Kunming 650500, China; hoxiwa@aliyun.com (S.H.); yiminxu@sina.com (Y.X.)
*   Correspondence: wangwei@scu.edu.cn; Tel.: +86-136-1801-0451

Received: 13 September 2018; Accepted: 29 October 2018; Published: 31 October 2018

Abstract: A numerical method is applied here to simulate the unstable flow and the vibration of a plane gate. A combination of the large eddy simulation (LES) method and the volume of fluid (VOF) model is used to predict the three-dimensional flow field in the vicinity of a plane gate with submerged discharge. The water surface profile, the streamline diagrams, the distribution of turbulent kinetic energy, the power spectrum density curve of the fluctuating pressure coefficient at typical points underneath the gate, and the complete vortex distribution around the gate are obtained by LES-VOF numerical calculation. The vibration parameters of the gate are calculated by the fluid-structure coupling interface transferring the hydrodynamic load. A simultaneous sampling experiment is performed to verify the validity of the algorithm. The calculated results are then compared with experimental data. The difference between the two is acceptable and the conclusions are consistent. In addition, the influence of the vortex in the slot on the flow field and the vibration of the gate are investigated. It is feasible to replace the experiment with the fluid-structure coupling computational method, which is useful for studying the flow-induced vibration mechanism of plane gates.

Keywords: LES-VOF method; flow-induced vibration; plane gate; numerical analysis

---

1. Introduction

With the wide application of a high head gate in hydraulic engineering, the vibration problem of the plane gate is increasingly prominent. Essentially, the flow-induced vibration of the plane gate is a complex fluid-structure interaction phenomenon. The time varying hydrodynamic load that is caused by the unstable flow around the gate is the main excitation source. The coexistence of the vortices at the gate's bottom edge, in the gate slot and in the gate downstream, further complicates the flow structure of the sluice flow. However, it is difficult to capture these vortices simultaneously and to catch the comprehensive information of the vortex-induced vibration by using an experimental approach. Researchers, such as Hardwick [1], Kolkman [2], Jongeling [3], and Ishii [4], have been making efforts to explore the mechanism of gate vibration. Hardwick conducted a model test investigation on the plane gate vibration. It is believed that this nonlinear vertical vibration response is caused by the resonance between the fluid shear force acting at the bottom edge of the gate and the gate. Kolkman et al. believed that the flow fluctuation and the variation of the pressure due to the inertia effect of sluice flow might be the main reason for the vertical vibration of the gate. Jongeling proved that the streamwise vibration of the gate is induced by the instability of the bottom edge shear layer. Ishii considered that the vortices at the bottom edge of the gate and the interaction with the structure lead to the streamwise vibration of the gate. These research results are mostly limited to the qualitative description of the physical process of the gate vibration, and tend to separate the vertical vibration and the streamwise flow vibration.

Therefore, it is especially important to analyze the vibration of the gate combining the vertical and the streamwise vibration. Simultaneously, these researches mainly focus on investigating the vibration mechanism that resulted from the vortex at the bottom edge of the gate.

With the rapid development of computational fluid dynamics (CFD) and computer technology, the accuracy of numerical calculation of flow field and the complexity of solving problems have been greatly improved, and it has been possible to numerically simulate the sluice flow in some extent. The method adopted is mainly based on the finite volume scheme. The SIMPLE series algorithm is used in the calculation of pressure field and velocity field. When dealing with turbulence problems, the turbulence model is employed to solve the Reynolds time-averaged Navier-Stokes equation. Pani et al. [5] used a finite element technique to investigate the fluid-structure interaction effect on the hydrodynamic pressure. Erdbrink et al. [6] also assessed the local velocities and pressures of the two-dimensional field by using the finite element method. Liu et al. [7] adopted the two-dimensional (2-D) mathematical model of Reynolds-averaged equations to study the turbulent flow behind a sluice gate. Kostecki [8–10] predicted the two-dimensional flow field in the vicinity of an underflow vertical lift gate by using a combinative numerical model of the vortex method and the boundary element method. Kazemzadeh [11] applied the smoothed fixed grid finite element method to simulate free surface flow in gated tunnels. In recent years, some novel computational methods and models have been developed. For example, Zhu et al. [12,13] proposed a new slip model, which was used in the study of lid-driven cavity flows in both continuum and transition flow regimes. However, there is still a lack of theories and methods to analyze effectively and deal with the complex vibration phenomenon of plane gates.

The development of parallel algorithms and supercomputers has strongly promoted the progress of direct numerical simulation (DNS) and large eddy simulation (LES) methods. Analyzing the pressure pulsation of turbulent flow using LES method has received more and more attention. However, the application of CFD technology, especially simulating pressure pulsation, in the sluice flow is still preliminary.

Different from the arc gate, the side walls of a plane gate have the gate slots that are used for the lifting movement of gate. In the case of the submerged sluice flow, the random movement of multiple vortices behind the gate, in the gate slot and at the gate's bottom edge will aggravate the pulsation of velocity and the pressure. When water flow crosses the slot, such as that which exists in the hydraulic plane gate, the vortices in the slot will interact with the separated flow near the downstream corner of the slot due to local boundary mutation in the slot area, which can easily lead to flow cavitation and even endanger the safe operation of hydraulic projects [14]. The vortex-induced vibration of a blunt body or other solid boundary caused by vortex shedding has been studied most [15–18], while studies of the gate vibration induced by the vortex in the gate slot has been reported rarely.

Based on the theory of fluid-structure interaction, the present investigation studies the characteristics of the flow-induced vibration of the hydraulic plane gate with a submerged discharge. The three-dimensional numerical simulation method with a large eddy simulation turbulence model can obtain a complete flow structure, including a funnel vortex in the slot. The vibration parameters of the gate are acquired by the fluid-structure coupling interface transferring the hydrodynamic load. The fluid-structure coupling method is validated by the experimental evidence.

2. Computational Modeling

2.1. Turbulence Model

The large eddy simulation (LES) method uses instantaneous flow control equations to directly simulate the large scale vortex in a turbulent flow field. After processing by using the filtering function, the instantaneous Navier-Stokes equation and continuity equation of control fluid flow become:

$$\frac{\partial}{\partial t}(\rho \bar{u}_i) + \frac{\partial}{\partial x_j}(\rho \bar{u}_i \bar{u}_j) = -\frac{\partial \bar{p}}{\partial x_i} + \rho g_i + \frac{\partial}{\partial x_j}\left[\mu\left(\frac{\partial \bar{u}_i}{\partial x_j} + \frac{\partial \bar{u}_j}{\partial x_i}\right) - \rho \tau_{ij}\right] \tag{1}$$

$$\frac{\partial \rho}{\partial t} + \frac{\partial}{\partial x_i}(\rho \bar{u}_i) = 0 \tag{2}$$

where $i, j = 1, 2, 3$; ρ is the specific mass; μ is the dynamic viscosity; g_i is the mass force; and, $\bar{p}$ and $\bar{u}_i$ are the filtered pressure and velocity components, respectively. $\tau_{ij} = \overline{u_i u_j} - \bar{u}_i \bar{u}_j$ are the components of the sub grid scale (SGS) stress tensor representing the effect of the small scale motion on the large scale motion. According to the basic SGS model, Smagorinsky [19] assumes the SGS stress in the following form:

$$\tau_{ij} - \frac{1}{3}\tau_{kk}\delta_{ij} = -2\mu_t \bar{R}_{ij} \tag{3}$$

The Smagorinsky-Lilly model is used in this paper. In the Smagorinsky-Lilly [20] model, the eddy-viscosity is modeled by:

$$\mu_t = (C_s \Delta)|\bar{R}| \tag{4}$$

where $|\bar{R}| = \sqrt{2\bar{R}_{ij}\bar{R}_{ij}}$, $\bar{R}_{ij} = \frac{1}{2}\left(\frac{\partial \bar{u}_i}{\partial x_j} + \frac{\partial \bar{u}_j}{\partial x_i}\right)$, C_s is the Smagorinsky constant; and, Δ is the local grid scale, which is computed according to the volume of the computational cell using $\Delta = V^{1/3}$.

It is more difficult to simulate the free surface accurately, and the volume of fluid (VOF) method is more advantageous to track the complex free surface in the multiphase flow models. Due to the downstream flow directly in contact with the atmosphere, the VOF model is used to track the air-water surface in the present investigation.

The tracking of the interface between the phases is accomplished by solving the continuity equation of the volume fraction of one or more phases. The volume fraction of water is a function of time and space, so the VOF model must be solved by transient. Assuming that α_w denotes the volume fraction for water phase, $\alpha_w = 0$ indicates the exclusion of the water phase in the region, $\alpha_w = 1$ indicates that the region is full of water phase, and $0 < \alpha_w < 1$ indicates the interface for water phase and other phases.

The tracking equations of the interface are as follows:

$$\frac{\partial \alpha_w}{\partial t} + u_i \frac{\partial \alpha_w}{\partial x_i} = 0 \tag{5}$$

2.2. Structural Model

The gate structure is assumed to consist of linear elastic small deformation material, which is made of organic glass. Under the Lagrangian system, structural vibration control equations with a small deformation assumption are as follows:

$$\rho^s \frac{\partial^2 u^s}{\partial t^2} = \text{div}\left(J\sigma^s F^{-T}\right) + \rho^s g \tag{6}$$

$$\sigma^s = J^{-1}F(\lambda^s(\text{tr}E)I + 2\mu^s E)F^T \tag{7}$$

The F, E, and J are defined as:

$$F = I + \nabla u^s \tag{8}$$

$$E = \frac{1}{2}\left(F^T F - I\right) \tag{9}$$

$$J = \det F \tag{10}$$

where ρ^s is the density of structure, u^s is the vibration displacement, σ^s is Cauchy's stress tensor of the gate structure, $\lambda^s = vE^s/((1+v)(1-2v))$ and $\mu^s = E^s/(2(1+v))$ are the constants of the gate material, E^s is material elastic modulus, and v is Poisson's ratio.

For complete fluid and structure coupling, the geometry coordination condition and the force equilibrium condition on the fluid-structure coupling interface should be satisfied.

The geometry coordination condition on the fluid-structure coupling interface $S^{fs} = \Omega^f \cap \Omega^s$ is:

$$u_i^f = u_i^s \text{ on } S^{fs} \tag{11}$$

The force equilibrium condition on the fluid-structure coupling surface $S^{fs} = \Omega^f \cap \Omega^s$ is:

$$\sigma_{ij}^f n_j^{fs} = \sigma_{ij}^s n_j^{fs} \text{ on } S^{fs} \tag{12}$$

where n_j^{fs} is the normal unit vector on the fluid-structure interface, Ω^f is the fluid region, and Ω^s is the structure region. Due to the water flow effect, the gate structure contacting with the fluid vibrates, which leads to flow field changes. The changing flow field acts on the gate structure again, which forms fluid-structure interaction problems.

2.3. Meshing and Numerical Approach

The direction of flow is the x direction, the depth direction of flow is the y direction, and the direction across the flume is the z direction. Submerged discharge occurs downstream of the gate. The flow inlet is the speed boundary, and the flow outlet is the pressure boundary converted by the head. The top of the downstream flow is the air boundary, and the relative pressure is zero. The flume bottom and the sidewalls are no-slip solid boundary conditions, and the given normal velocity is zero. The method of the wall function is used for processing the near-wall viscous sublayer.

Due to the symmetry of the gate and the flume, half the model is selected to be meshed to reduce the calculating work. In this paper, the cases of the calculation are shown in Table 1, where e is the opening of the gate and H is the upstream water head. The underflow ratio U, is used to access the characteristics of the gate vibration caused by the submerged flow. Comprehensively considering the upstream and the downstream water head, underflow ratio U is defined, as follows:

$$U = \frac{h_t - h_c''}{H_0 - h_c''} \tag{13}$$

where h_t is the downstream water depth, h_c is the contraction section depth, h_c'' is the conjugate water depth of the contraction section, and H_0 is the total head.

Table 1. A list of the cases of calculation.

Case	e/H	U
1	0.049	0.059
2	0.197	0.355
3	0.375	0.730

For $U = 0.059$, the half grid of fluid with tetrahedral mesh has 252,758 units and a total of 49,112 nodes (Figure 1). The time step length of unsteady flow is 0.001 s, and the maximum number of iterations of each step is 20 times. At the same time, the numerical simulation of turbulent flow under the unslotted condition is carried out to calculate the vortex effect of the gate slot. The half grid of the plane gate is shown in Figure 2a, and the size of the plane gate is shown in Figure 2b.

The ANSYS Worckbench17.0 is used to solve the system of dependent variables. After constructing a three-dimensional fluid-structure coupling model, the flow field in the vicinity of the lift gate with a submerged discharge condition is simulated by the finite volume method. A combination of the $k - \epsilon$ double-equation turbulence model and the VOF model of the multiphase flow is used to calculate the flow field before 10.5 s. With the result as the initial conditions of the Smagorinsky-Lilly subgrid model and the LES method after 10.5 s, the VOF model is still adopted to determine of the water-air interface. Velocity pressure coupling by the PISO algorithm and the momentum equation in discrete form by the second-order windward scheme are used to obtain the hydrodynamic pressure distribution that acts on the gate surface. Under the action of the surface pressure load, the vibration response of the gate structure is analyzed by the finite element method. To analyze the calculation results visually, the magnitude and the materials of the calculation model are consistent with the synchronous experiment model. The density, elastic modulus, and the Poisson's ratio of the gate material are 1400 kg/m^3, 3×10^9 N/m^2, and 0.4, respectively. Constraints are imposed on the top and sides of the gate. The width-to-depth ratio of the gate slot is 1.78.

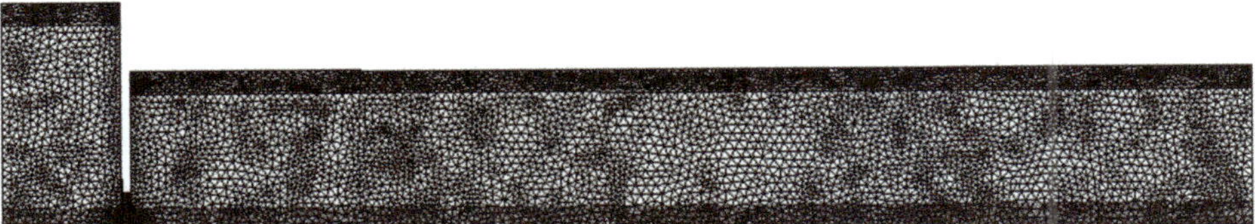

Figure 1. The grid of fluid.

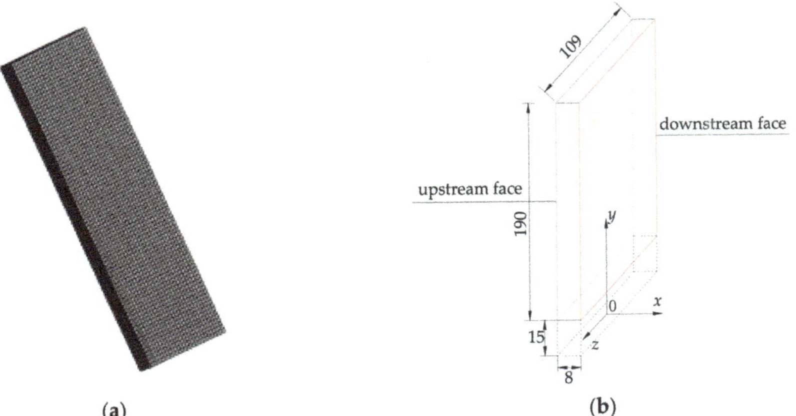

(a)   (b)

Figure 2. The plane gate: (**a**) the grid; and, (**b**) the size (unit: mm).

3. The Results of Numerical Simulation

To verify the validity of the numerical simulation method, the numerical results are compared with the experimental data. The flow field behind the plane gate with submerged discharge, the hydrodynamic pressure process at typical points underneath the gate and the vibration parameters of the gate are acquired by the combination experiment [21] of three-dimensional particle image velocimetry (3D-PIV) made by TSI Company, the digital pressure sensor with high precision and the multichannel vibrating data acquisition system made by Lance Measurement Technologies Company, respectively.

The PIV test uses the Insight3G software to start the laser, and simultaneously captures the flow field particle images with two PIV-specific cross-frame CCD cameras. The autocorrelation or cross-correlation principle is used to extract the image of flow characteristics. Finally vector diagram of transient flow velocity in the measured range is obtained by professional post-processing software, such as Tecplot and Matlab. The maximum emission frequency of the laser is set to 14.5 Hz. In this test, a certain concentration of SiO_2 is selected as the tracer particle, and the particle size is 10–15 µm. The three-axis accelerometer is directly attached to the upstream surface of the gate. The accuracy of the pressure sensor and accelerometer both are ±0.1%FS, and the sampling interval of the pulsating pressure and the vibration displacement are 0.01 s and 0.001 s, respectively. Three instruments are tested simultaneously.

The size of the organic glass flume is 4880 mm × 100 mm × 190 mm, and the length behind the plane gate is approximately 3000 mm. The calculations and the experimental cases are consistent with each other. The general arrangement of the experiment is shown in Figure 3.

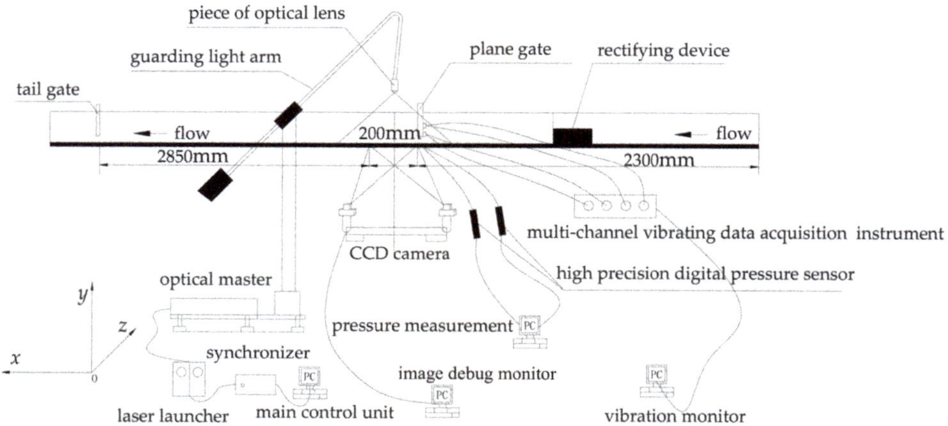

Figure 3. General arrangement of the experiment.

3.1. Flow Field Distribution

The calculation length after the lift gate is 920 mm and width B of the flume is 100 mm. The computation results of the time-averaged surface profiles with the whole flow passage of the modeling after the gate in $U = 0.059$ are shown in Figure 4. Blue represents the water phase and red represents the gas phase.

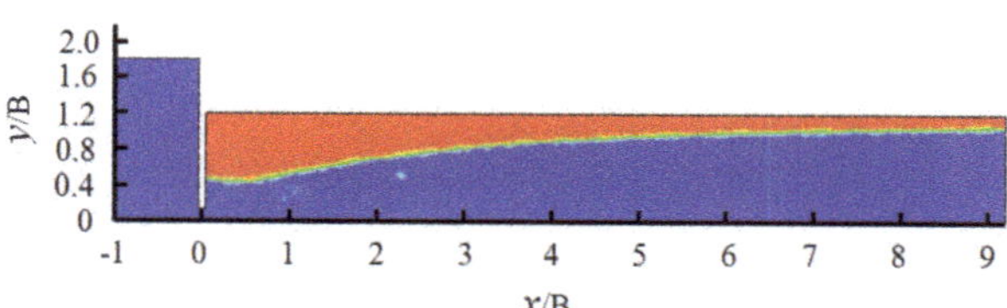

Figure 4. The color nephogram of the time-averaged flow profile.

Figure 5 shows the time-averaged flow profile comparison behind the gate of section $z = 0$ mm for the calculation and experiment of $U = 0.059$. The maximum and minimum of relative difference of the height between the calculation and the test values of the flow profile are 4.8% and 0.16%, respectively, which shows that the VOF method can simulate the water-air interface well.

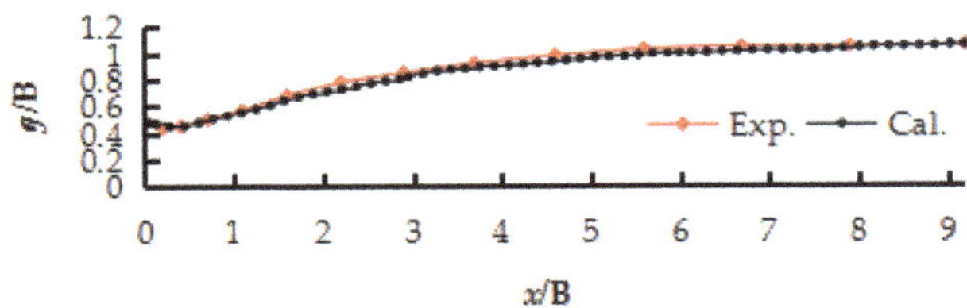

Figure 5. The time-averaged flow profile.

The calculational time-averaged streamlines on sections $z = 0$ mm and $z = 35$ mm in different cases are presented in Figure 6. According to (a), (b), and (c) in Figure 6, there is only a vortex on section $z = 35$ mm in the time-averaged flow field behind the plane gate in the x direction, but there are two vortices on section $z = 0$ mm that are within the 300 mm range. At different sections in the same case, the vertical distances of the vortex center from the x-axis are approximately the same. With the e/H increasing, the vertical distance of the vortex center from the x-axis increases gradually.

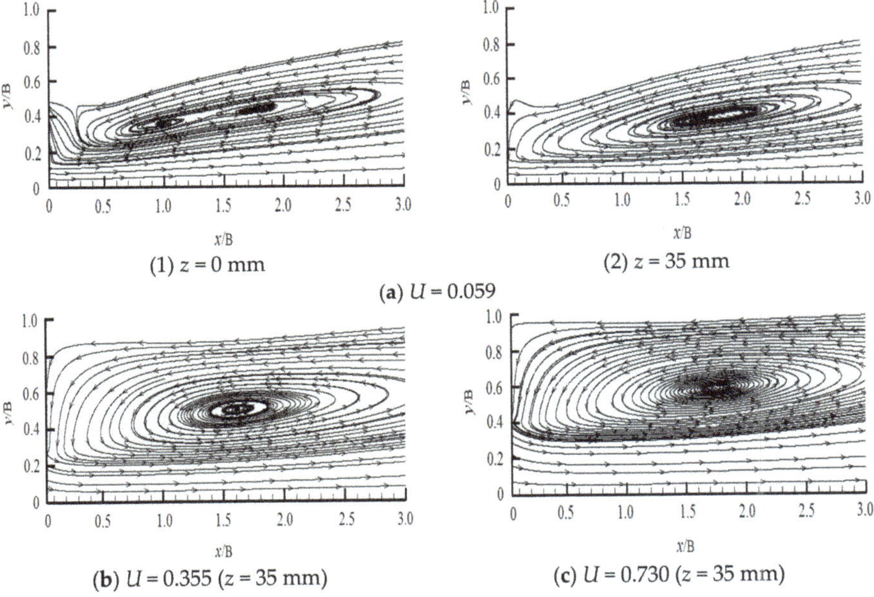

(1) $z = 0$ mm

(2) $z = 35$ mm

(a) $U = 0.059$

(b) $U = 0.355$ ($z = 35$ mm)

(c) $U = 0.730$ ($z = 35$ mm)

Figure 6. The calculational time-averaged streamlines.

The size of the PIV calibration target is 200 mm × 200 mm. Due to the size limitation, it is impossible to test the vortices in the slot and at the bottom edge of the gate by PIV, so the computational results of only the vortices behind the gate are compared carefully with those of the three-dimensional PIV test. Both the simulated and the experimental velocity vector diagrams are achieved at 20.6 s.

Figure 7 shows the instantaneous streamlines behind the gate on the same section $z = 0$ mm for the calculation and the experiment. The numerical simulation can make up for the drawback of losing the velocity vector by 3D-PIV positioning. When compared with the experimental results, the vortex center, which is closest to the gate obtained by the simulation, is lower in the y direction under $U = 0.059$, but this distance is basically same for the other two cases. On the whole, the numerical simulation still obtained the most ideal vortex information.

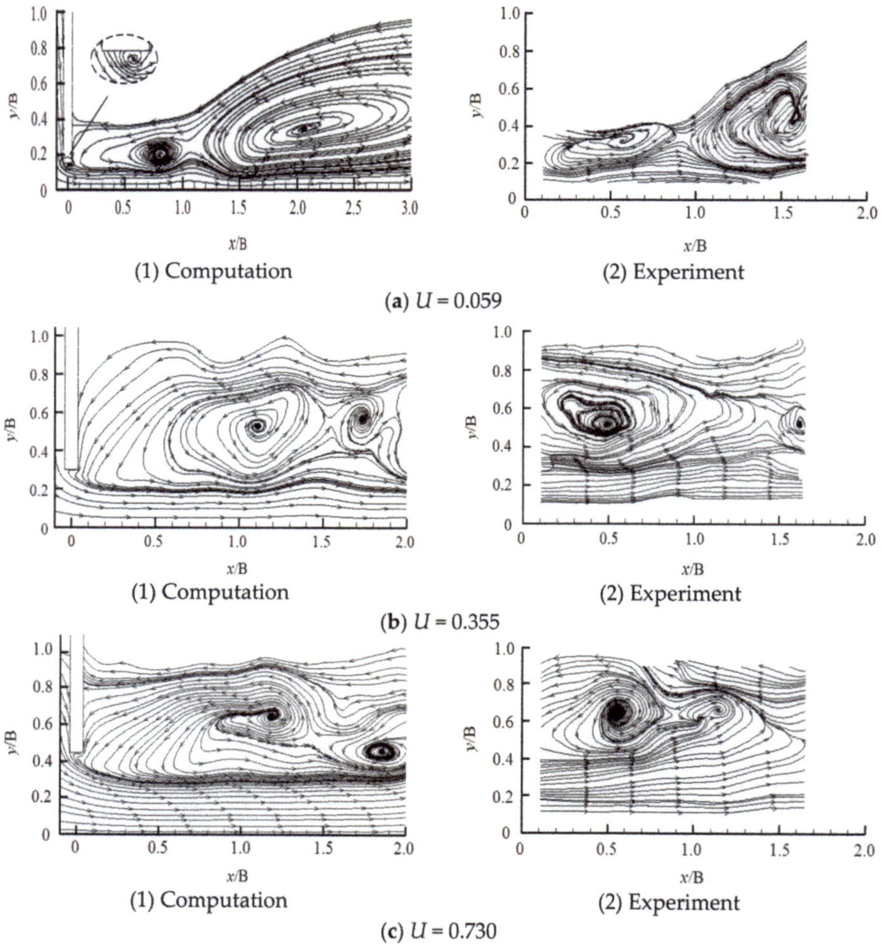

(1) Computation (2) Experiment

(**a**) $U = 0.059$

(1) Computation (2) Experiment

(**b**) $U = 0.355$

(1) Computation (2) Experiment

(**c**) $U = 0.730$

Figure 7. The instantaneous streamlines ($z = 0$ mm).

It can be seen from the instantaneous streamlines in Figure 7a–c that there are the same number of vortexes on section $z = 0$ mm between the computation and the experiment. Under $U = 0.059$, the numerical simulation captures the phenomenon of the vortex detachment from the gate's bottom edge, but there is no such phenomenon when $U = 0.355$ or $U = 0.730$. Therefore, when the e/H and U reach a certain extent, the vortex shedding phenomenon can be generated from the bottom edge of the gate.

Normalizing the turbulent kinetic energy by the square of the mean flow velocity of the gate downstream section, Figure 8 presents the computational normalized turbulent kinetic energy distribution on section $z = 0$ mm and section $z = 35$ mm. Turbulent kinetic energy under the gate on section $z = 0$ mm is higher than the value on section $z = 35$ mm. The different magnitude of turbulent kinetic energy is one of the significant factors affecting the gate vibration extremum. The experiment proved that the extremum of vibration displacement on section $z = 0$ mm is larger than that on section $z = 35$ mm.

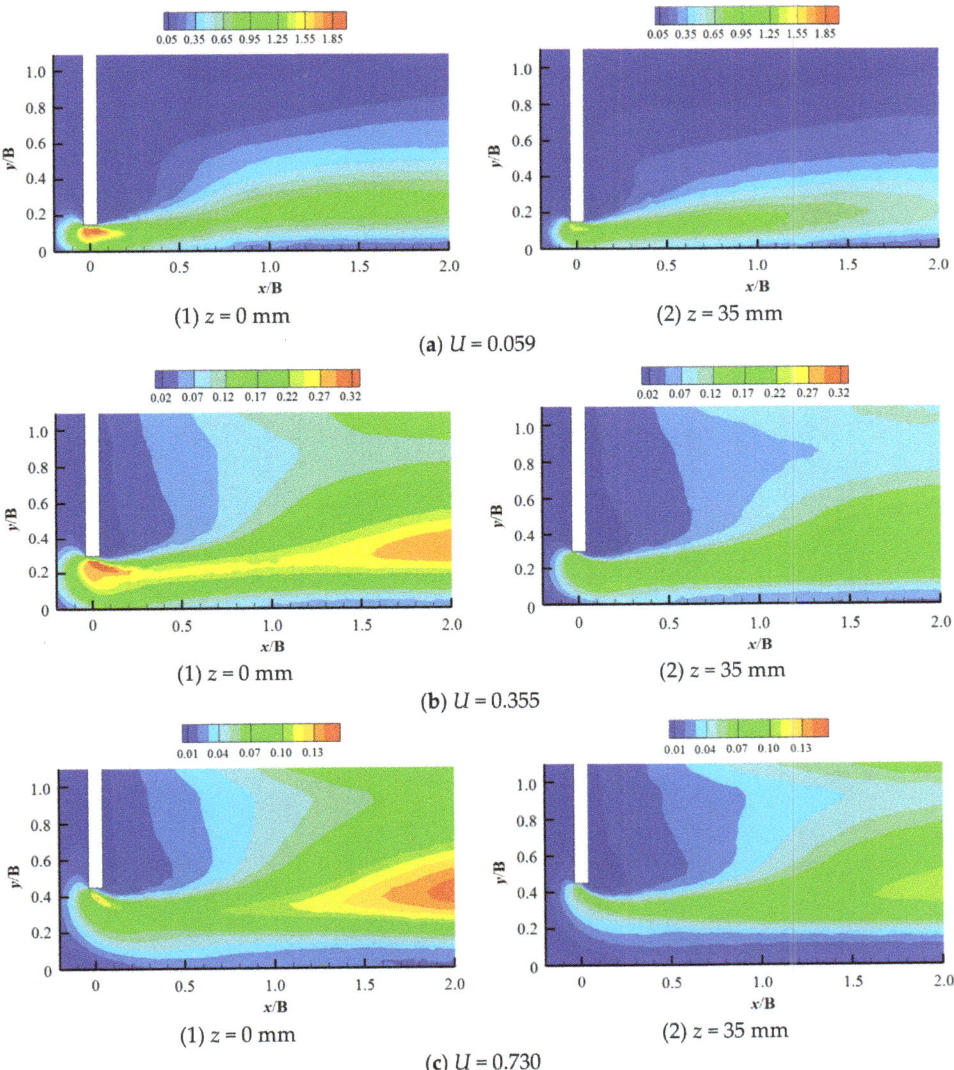

Figure 8. Normalized turbulent kinetic energy distribution.

3.2. Hydrodynamic Pressure

The measuring points of hydrodynamic pressure just below the gate are point 1 (0, 0, 0 mm) and point 2 (0, 0, 35 mm). The digital pressure sensor with high precision is used to test the hydrodynamic pressure with a sampling frequency of 0.01 s, while the monitoring points are also set up in the numerical calculation.

Figure 9 shows the time history curves of the fluctuating pressure coefficient of points 1 and 2 in 10 s. $C_p = (p - \overline{p})/0.5\rho v_2$, where C_p is the fluctuating pressure coefficient, p is the instantaneous pressure, $\overline{p}$ is the average pressure, ρ is the water flow density, and v is the average velocity of the exit section. Under the same case, the fluctuating pressure of point 1 is stronger than that of point 2. The time to reach the extreme of the two points is out of synchronization.

Figure 10 shows the normalized power spectrum density curve of the pressure for point 1 in different cases. *St* is the Strouhal number and $St = f \cdot L/\bar{v}$, Where *f* is the frequency of the pressure pulsating, *L* is the characteristic length and taken as the gate opening, and $\bar{v}$ is the mean velocity of section just below the gate. With the *e/H* or *U* increasing, the energy of pulsating pressure of point 1 decreases. The computational Strouhal numbers for dominant frequencies of the pressure are 0.332×10^{-3}, 2.217×10^{-3}, 1.545×10^{-3} with the corresponding cases of *U* = 0.059, 0.355, 0.730. Accordingly, the experimental Strouhal numbers for dominant frequencies of the pressure are 0.337×10^{-3}, 2.242×10^{-3}, 1.552×10^{-3} in corresponding cases. The Strouhal number difference for main frequency of the pressure between the computational and the experimental results is very small.

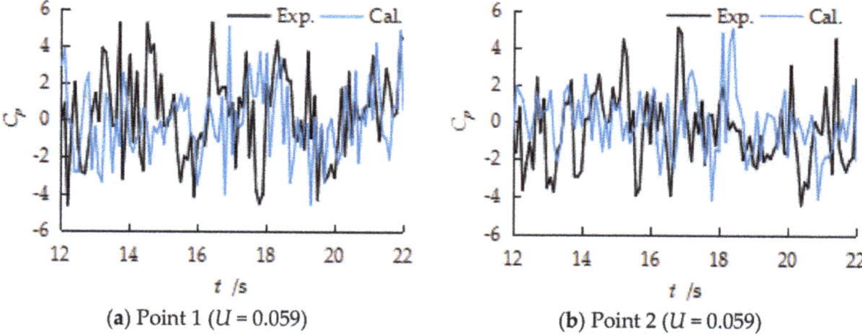

(a) Point 1 (*U* = 0.059)   (b) Point 2 (*U* = 0.059)

Figure 9. The time history curve of the pressure pulsating coefficient for different point.

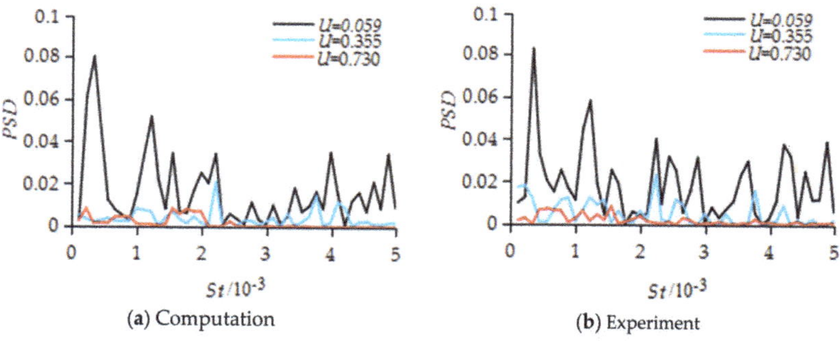

(a) Computation   (b) Experiment

Figure 10. The power spectrum density (PSD) curve of the pressure for point 1 in different cases.

In order to illustrate the influence of the gate slot on the flow field and the gate vibration, both calculation and experiment are carried out on the filling of the slots underneath the gate (unslotted). Figure 11 shows the time history curve of the slotted and unslotted pressure pulsating coefficients in *U* = 0.059. It can be found that the maximum values of the slotted pressure pulsation coefficients for both points 1 and 2 were larger than those of the without a slot. The characteristic values of the pressure pulsating coefficients (slotted and unslotted) for both the experiment and for the calculation in *U* = 0.059 are presented in Figure 12. When comparing the value of the computational and the synchronous experimental results, the relative difference is within 10%, which indicates that the numerical simulation results meet the basic requirements.

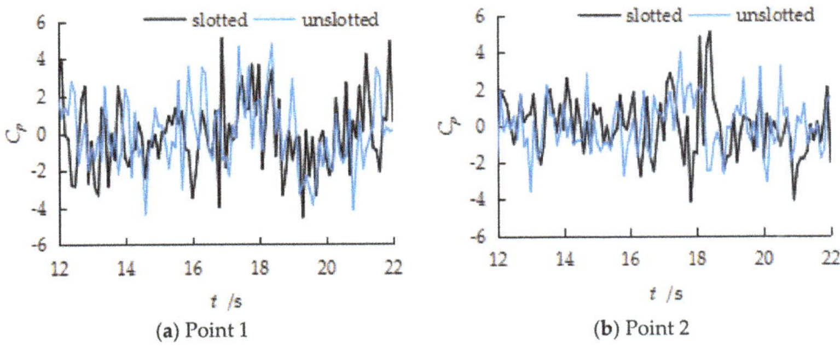

Figure 11. The time history curve of the pressure pulsating coefficients (slotted and unslotted).

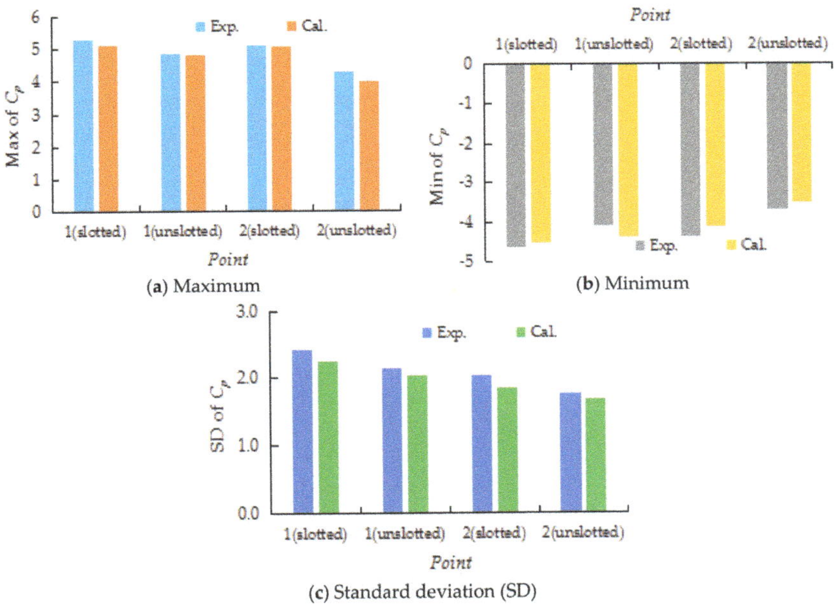

Figure 12. The comparison of characteristic value of the pressure pulsating coefficient.

In the experiment, the dominant frequency of the pressure at the measuring points underneath the gate are only 0.025~1.245 Hz, and the dominant frequency of the pressure at the measuring points behind the gate are only 0.183~14.63 Hz. The natural frequencies of the gate vibration by the modal analysis are 247.43 Hz, 529.85 Hz, 1086.6 Hz, 1280.6 Hz, 1605.1 Hz, and 1963.1 Hz, respectively. The natural frequency of the gate vibration is far from the dominant frequency of the flow pressure, so it would not have produced resonance.

Figure 13a presents the computational pressure distribution and streamlines in the gate slot for different level cross sections ($y/e = 0.067$, $y/e = 0.2$, $y/e = 0.333$, $y/e = 0.467$, $y/e = 0.6$, $y/e = 0.733$) from the bottom to the top in y direction for $U = 0.059$. It can be seen from the superimposed vortex diagram that the low-pressure zone of the vortex gradually enlarges from the bottom to the top, which is shaped like a funnel and it shifted to the downstream. The pressure is lowest in the vortex center where it is also most prone to generate cavitation in the gate slot. Due to the water impacting the slot from the left to the right, the pressure is highest at the right of the slot. The closer to the bottom the place is, the higher the pressure is. Under $U = 0.355$ (Figure 13b) and $U = 0.730$ (Figure 13c), the range of the low pressure and the high pressure zones in the gate slot is much smaller than that in $U = 0.059$.

The funnel vortex of $U = 0.355$ still exists, but the center of the vortex is shifted to the upstream from the bottom to the top, and the center pressure of the bottom vortex is relatively large. For $U = 0.730$, the centerline of the vortex is similar to the spiral line, and the funnel vortex is not obvious.

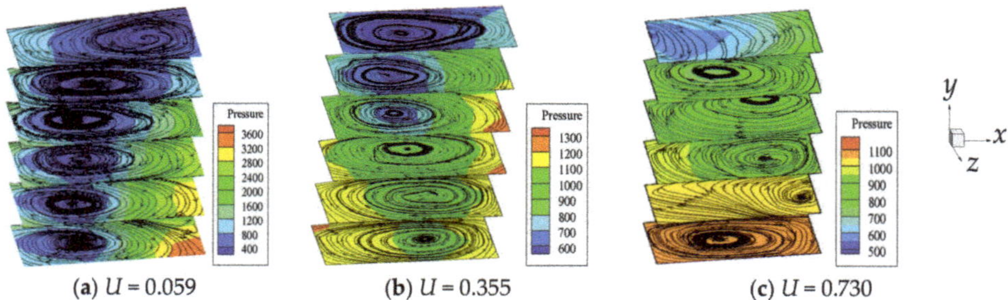

(a) $U = 0.059$ (b) $U = 0.355$ (c) $U = 0.730$

Figure 13. Pressure (pa) distribution and streamlines at 22.0 s in the gate slot.

The maximum vibration displacement of the gate with a submerged condition in the x direction is not due to the resonance, but it is due to the mixed vortex-induced process, including the vortices behind the gate, in the gate slot and at the gate's bottom edge. It can be seen from the above analysis that the numerical simulation method in this paper can obtain not only the pressure fluctuation distribution, but also a more complete flow structure, such as multiple vortices of the sluice flow, and it forms the excitation force of the plane gate vibration.

3.3. Vibration of the Gate

Define the non-dimensional coefficient $K_d = (d - \bar{d})/\sigma_d$, where d is vibration displacement, $\bar{d}$ is the average vibration displacement, and σ_d is the vibration standard deviation. Point A, which is located on the central axis of the upstream surface of the gate and 10 mm away from the bottom edge, is used to study the response of the gate vibration.

The normalized power spectrum density curve of vibration displacement for point A in $U = 0.059$, which is obtained by the computation and the experiment, is shown in Figure 14. In the experiment, the acceleration sensor and the multichannel data acquisition instrument were used to monitor the acceleration of the gate vibration in the x direction and y direction. The vibration displacement is achieved by the second integral of the acceleration. The computational Strouhal number for dominant frequency of vibration displacement in x direction is 2.582 with the corresponding experiment of 2.660. In addition, the computational Strouhal number for dominant frequency of vibration displacement in y direction is 3.724 with the corresponding experiment of 3.768. The Strouhal number for main frequency of the vibration displacement between the computational and the experimental results is closer.

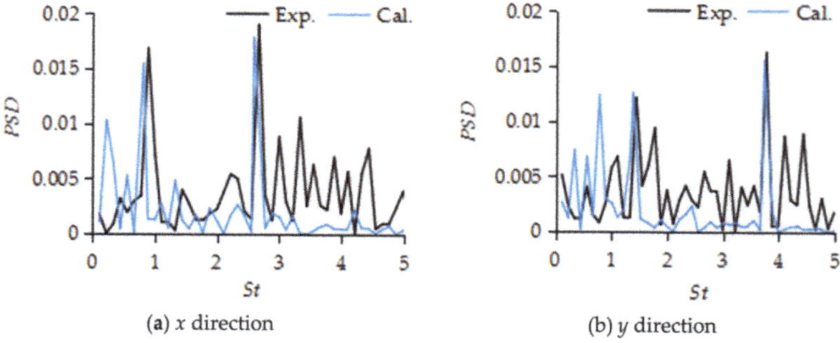

(a) x direction   (b) y direction

Figure 14. The power spectrum density (PSD) curve of vibration displacement for point A.

Figure 15 plots the time-history curve of the vibration displacement coefficient K_d in the x direction and in the y direction of point A, with and without slots, by using the numerical simulation. It can be seen that the maximum value of the vibration displacement coefficient of the plane gate with a slot is larger than that of the plane gate without a slot in both the x and y directions. When there is no slot, the maximum value of the x direction vibration displacement coefficient appears at 12.7 s and the minimum value appears at 21.8 s. The maximum value in the y direction appears at 14.7 s and the minimum value appears at 17.6 s. The moment when the vibration displacement coefficient of the x direction and the y direction appears is out of sync.

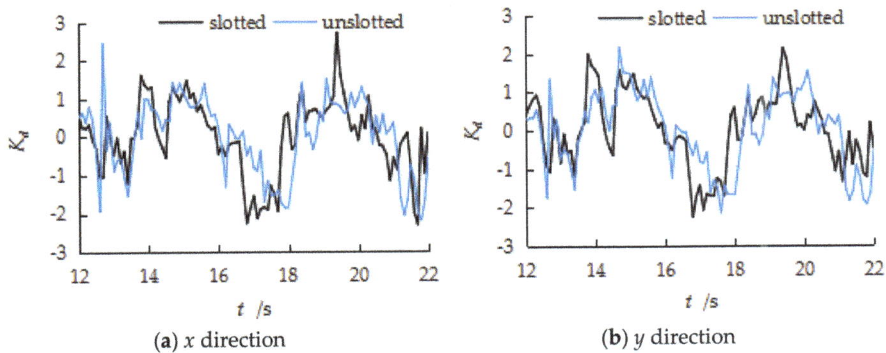

Figure 15. The time history curve of the vibration displacement coefficient K_d of point A.

4. Conclusions

In this study, numerical simulations were performed to compute the flow field and the water surface profile for a plane gate with a submerged discharge by the combination of the LES and the VOF methods. Overall, the numbers and locations of the vortex are consistent with the synchronous experimental results. When compared with the experimental values, the relative difference of both the standard deviation pressure and the maximum pressure of the calculation is within 10%, which verifies the validity of the numerical simulation methods that were adopted in the present work. Simultaneously, the maximum vibration displacement coefficient was acquired from a calculation of the vibration displacement process of Point A, which was very close to that of the experiment. So, the fluid-structure coupling computational method used in certain conditions can replace the complex experiment and make up for the lack of experiment.

The turbulent kinetic energy under the gate on section $z = 0$ mm was higher than the value on section $z = 35$ mm. The change trend of the turbulent kinetic energy was consistent with the amplitude of the gate vibration displacement, which shows that the turbulent kinetic energy under the gate plays an important role in the amplitude of the gate vibration. The maximum value of the vibration displacement coefficient of the plane gate with a slot was larger than that of the plane gate without a slot in both x and y directions. Thus, the vortex in the gate slot aggravates the vibration of the gate. Turbulence pressure pulsation and multiple vortices in the flow field are the main causes of gate vibration. The mixed vortex-induced process includes the vortices behind the gate, in the gate slot, and at the gate's bottom edge.

Author Contributions: C.S. and W.W. conceived and designed the model; C.S. and S.H. set up and debugged the programmes; C.S. analysed the results and wrote the paper; C.S. and Y.X. provided editorial improvements to the paper.

Acknowledgments: Financial support provided by the National Key Research and Development Program (Grant No. 2016YFC0401603) and the National Natural Science Foundation of China (Grant No. 51369013) are gratefully acknowledged.

Conflicts of Interest: The authors declare no conflict of interest.

References

1. Hardwick, J.D. Flow-induced vibration of vertical lift gate. *J. Hydraul. Div.* **1974**, *100*, 631–644.
2. Kolkman, P.A.; Vrijer, A. Gate edge suction as a cause of self-exciting vertical vibrations. In Proceedings of the 17th IAHR Congress, Baden, Germany, 15–19 August 1977; Volume 17, pp. 425–437.
3. Jongeling, T.H.G. Flow-induced Self-excited in Flow Vibrations of Gate Plates. *J. Fluids Struct.* **1988**, *2*, 541–566. [CrossRef]
4. Ishii, N. Flow-induced vibration of long span gates, Part I: Model development. *J. Fluids Struct.* **1992**, *6*, 539–562. [CrossRef]
5. Pani, P.K.; Bhattacharyya, S.K. Hydrodynamic pressure on a vertical gate considering fluid-structure interaction. *Finite Elem. Anal. Des.* **2008**, *44*, 759–766. [CrossRef]
6. Erdbrink, C.D.; Krzhizhanovskaya, V.V.; Sloot, P.M.A. Reducing cross-flow vibrations of underflow gates: Experiments and numerical studies. *J. Fluids Struct.* **2014**, *50*, 25–48. [CrossRef]
7. Liu, S.; Liao, T.; Luo, Q. Numerical simulation of turbulent flow behind sluice gate under submerged discharge conditions. *J. Hydrodyn.* **2015**, *27*, 257–263. [CrossRef]
8. Kotecki, S.W. Numerical modeling of flow through moving water-control gates by vortex method. Part I—Problem formulation. *Arch. Civ. Mech. Eng.* **2008**, *8*, 73–89. [CrossRef]
9. Kotecki, S.W. Numerical modeling of flow through moving water-control gates by vortex method. Part II—Calculation result. *Arch. Civ. Mech. Eng.* **2008**, *8*, 39–49. [CrossRef]
10. Kotecki, S.W. Numerical analysis of hydrodynamic forces due to flow instability at lift gate. *Arch. Civ. Mech. Eng.* **2011**, *11*, 943–961. [CrossRef]
11. Kazemzadeh-Parsi, M.J. Numerical flow simulation in gated hydraulic structures using smoothed fixed grid finite element method. *Appl. Math. Comput.* **2014**, *246*, 447–459. [CrossRef]
12. Zhu, T.; Ye, W. Theoretical and numerical studies of noncontinuum gas-phase heat conduction in micro/nano devices. *Numer. Heat Transf. Part B Fundam.* **2010**, *57*, 203–226. [CrossRef]
13. Liu, H.; Xu, K.; Zhu, T.; Ye, W. Multiple temperature kinetic model and its applications to micro-scale gas flows. *Comput. Fluids* **2012**, *67*, 115–122. [CrossRef]
14. He, S.; Zhang, L. Large Eddy Simulation of Effect of Geometric Parameters on Low Pressure of Turbulent Flow over Gate Slot. *J. Syst. Simul.* **2015**, *27*, 1081–1086.
15. Leclercq, T.; de Langre, E. Vortex-induced vibrations of cylinders bent by the flow. *J. Fluids Struct.* **2018**, *80*, 77–93. [CrossRef]
16. Chizfahm, A.; Yazdi, E.A.; Eghtesad, M. Dynamic modeling of vortex induced vibration wind turbines. *Renew. Energy* **2018**, *121*, 632–643. [CrossRef]
17. Postnikov, A.; Pavlovskaia, E.; Wiercigroch, M. 2DOF CFD calibrated wake oscillator model to investigate vortex-induced vibrations. *Int. J. Mech. Sci.* **2017**, *127*, 176–190. [CrossRef]
18. Bouratsis, P.; Diplas, P.; Dancey, C.L.; Apsilidis, N. Quantitative Spatio-Temporal Characterization of Scour at the Base of a Cylinder. *Water* **2017**, *9*, 227. [CrossRef]
19. Smagorinsky, J. General circulation experiment with primitive equations. *Mon. Weather Rev.* **1963**, *91*, 99–164. [CrossRef]
20. Lilly, D.K. Aproposed modification of the Germa no subgrid-scale closure method. *Phys. Fluid A* **1992**, *4*, 633–635. [CrossRef]
21. Shen, C.; He, S.; Yang, T.; Wang, W. Model tests for synchronous measurement of fluid-structure interaction vibration of a plane vertical lift gate. *J. Vib. Shock* **2016**, *35*, 219–224.

Article

Experimental Optimization of Gate-Opening Modes to Minimize Near-Field Vibrations in Hydropower Stations

Yong Peng [1], Jianmin Zhang [1,*], Weilin Xu [1] and Matteo Rubinato [2]

[1]   State Key Laboratory of Hydraulics and Mountain River Engineering, Sichuan University, Chengdu 610065, China; pengyongscu@foxmail.com (Y.P.); xuwl@scu.edu.cn (W.X.)

[2]   Civil and Structural Engineering Department, University of Sheffield, Sheffield S1 3JD, UK; m.rubinato@sheffield.ac.uk

*   Correspondence: zhangjianmin@scu.edu.cn; Tel.: +86-28-8540-5706

Received: 26 July 2018; Accepted: 10 October 2018; Published: 12 October 2018

Abstract: Multi-Horizontal-Submerged Jets are successfully applied to dissipate energy within a large-scale hydropower station. However, notable near-field vibrations are generated when releasing high discharges through the gates, which is generally typical in a flooding case scenario. Under these conditions, the magnitude of the vibrations varies when applying different gate-opening modes. To investigate and find optimized gate-opening modes to reduce the near-field vibration, multiple combinations were tested by varying gate-opening modes and hydraulic conditions. For each of the tests conducted, fluctuating pressures acting on side-walls and bottoms of a stilling basin were measured. The collected datasets were used to determine the maximum and minimum fluctuating pressure values associated with the correspondent gate-opening mode and a detailed comparison between each of the gate-opening modes was completed. The paper presents the quantitative analysis of the discharge ratio's effect on fluctuating pressures. It also investigates the influence of different gate-opening modes by including side to middle spillways and upper to lower spillways configurations. The flow pattern evolutions triggered by each different gate-opening mode are discussed and optimal configurations that minimize near-field vibrations at high discharges are recommended to support both the design of new systems and assessment of the performance of existing ones.

Keywords: multi-horizontal-submerged jets; energy dissipation; near-field vibration; fluctuating pressure; gate-opening modes

---

1. Introduction

Since 2000, over 300 hydropower projects have been completed or are under construction in China [1]. Flood discharge and energy dissipation are crucial to maintaining the safety of these hydropower installations. Furthermore, lessons must be drawn from the past dam disasters, e.g., Vajont Dam [2,3] and Banqiao Dam [4]. Decades ago, hydropower stations were not constructed in proximity to residential areas, and as a result, research at that time mainly focused on the stability and the functionality of the hydraulic structures. However, in recent years, due to increasing urbanization [5], some large hydropower stations had to be located in the proximity of residential areas. Moreover, due to the climatic scenario [6], it is likely that in the future there will be numerous and more frequent intense rainfall events, and consequently large volumes of water may need to be released from dam reservoirs. Therefore, it is essential to provide solutions to a need for increasing the stability of these hydraulic structures and surrounding areas, minimizing environmental impacts due to extreme flow discharges [7].

The purpose of this work is to optimize a recently developed novel technique for energy dissipation, Multi-Horizontal-Submerged Jets (simplified as MHSJ). MHSJ have been studied and improved by many researchers in recent years [8–17]. It has been demonstrated that MSHJ possesses three main advantages when compared with the hydraulic jump, which is the traditional energy dissipater: (1) lower atomization; (2) a higher rate of energy dissipation; (3) more flexibility of gate-opening modes [12–14]. Thus, it seems feasible to implement MHSJ to tackle the problem of energy dissipation within large hydropower stations that are typically characterized by higher water heads and larger flowrates. The adoption of MHSJ was tested by Deng et al. [12] and successfully applied as energy dissipation in large-scale hydropower stations [14].

However, despite positive progress on solving energy dissipation and erosion problems, there is still a lack of studies on the considerable flow-induced vibrations that occur in this kind of structure and propagate towards the near-field when releasing the greater flowrates. These near-field vibrations can be considered a serious threat for the stability of the structure, and hence this phenomenon needs to be further investigated. Yin and Zhang [18] and Yin et al. [19] studied the relationship between the discharge flow and the induced vibrations and simulated the results with a finite element method. Yin et al. [19] found that at higher flowrates during flooding situations, the flow is complex and unsteady, and this is the main cause of near-field vibrations. Therefore, if the vibrations are caused by releasing higher flowrates, optimizing gate-opening modes could reduce this effect and near-field vibrations could decrease correspondingly. To date, this issue has not been investigated in detail and to address this gap, this paper presents an experimental study to explore the relationship between complex flow scenarios and gate-opening modes in order to recommend an optimal solution to minimize the formation of vibrations.

2. Methods

The experimental facility, shown in Figure 1, was constructed based on a large-scale hydropower station (Xiangjiaba hydropower station. It sits on the Jinsha River, a tributary of the Yangtze River in Yunnan Province and Sichuan Province, Southwest China). The experimental facility follows Froude similarity, i.e., the ratio between inertial and gravitational forces remains constant, with the scale of 1:80. The experimental model is composed of the upper and lower reaches of the river, a dam, mid-level channels, spillways and two stilling basins with identical layout (the difference between two stilling basins is the their location on the dam, being parallels). The single stilling basin includes 5 mid-level channels and 6 spillways. Both of mid-level channels and spillways were arranged alternately and division piers were located between them.

In the model, the outlets of spillways are horizontal with symmetrical contraction of 1 m and width of 8 m. Also, the mid-level channels' outlets are horizontal, without any contraction, and their width is 6 m. The division piers are 3 m wide. The offset of mid-level channels, spillways, and division piers from the bottom of the stilling basin are 8 m, 16 m, 26 m, respectively. The test section (L = 15 m, W = 2 m and H = 1 m) is made of plexiglass. The modelled stilling basin is 3 m long and a gate is adopted to control tail-water level downstream of the model. To measure the fluctuating pressure within the system, 135 measurement taps for pressure transducers are located in the right-side wall and bottom of the stilling basin, as well as in the right-side wall and bottom of spillway (1# spillway in Figure 1) and mid-level channel (1# mid-level channel in Figure 1). For each test, the fluctuating pressure was measured synchronously by sensors for pressure transducers at each measurement point. The accuracy of the pressure sensors utilized in this study is 0.1%, with a 0–50 kPa working range and a frequency response of 100 Hz.

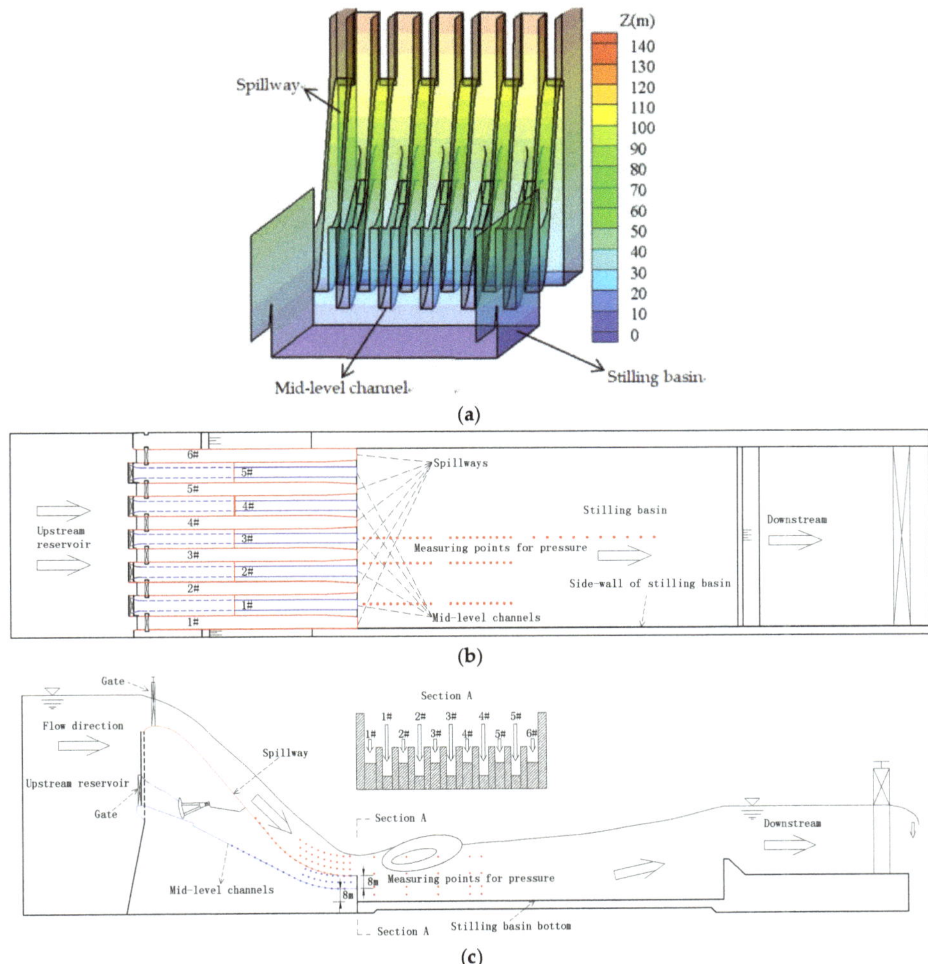

Figure 1. Sketch of experimental arrangement (single stilling basin). (**a**) 3D view of gate session; (**b**) plan view; (**c**) side view.

Multiple experimental flooding scenarios were completed within the experimental facility, varying flowrate and changing gate-opening modes. The area under investigation includes the: (i) stilling basin bottom (BB); (ii) stilling basin side-wall (BS); and (iii) piers, spillways and mid-level channels (PSO). Discharges considered varied gradually between 800 and 24,500 m^3/s. The downstream water elevation was measured within the model for each test.

3. Results and Discussion

3.1. Relation between Fluctuating Pressure and Flood Discharge

After calculating the root-mean-square (RMS) of the fluctuating pressure measured for each hydraulic condition tested, the arithmetic mean of RMS values obtained for every part of the BB, BS and PSO were calculated individually and these final values are used in the following sections.

Figure 2 shows (a) the maximum and minimum RMS of fluctuating pressure on BB against the flood discharge scenario and (b) the gate-opening modes considered. In Figure 2a, the best fitting line and its correspondent equation are displayed. The configuration of gate-opening modes has a significant impact on the magnitude of the fluctuating load acting on the BB during the release of flows in the range of 2000 to 12,000 m^3/s. This is a clear indicator that the fluctuating load can be reduced effectively by optimizing gate-opening modes. As a consequence, the near-field vibrations induced by releasing higher flow associated with flooding scenario can also be reduced. However, this cannot be confirmed when the flood discharge increases to 15,000 m^3/s into a single stilling basin (as shown in Figure 2a) due to the limited adjustment of the gate-opening modes available within this experimental facility. Additionally, Figure 2b highlights that the use of side spillways seems favorable to reduce fluctuating pressure on the BB.

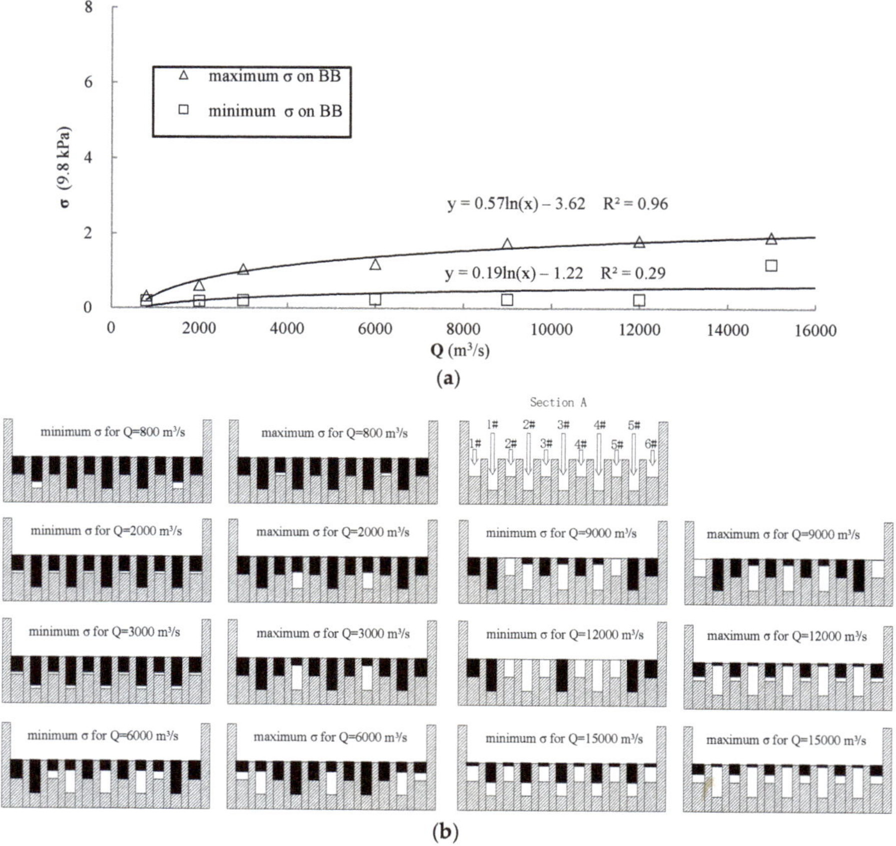

Figure 2. (a) Maximum and minimum RMS of fluctuating pressure on BB and (b) corresponding gate opening modes with flood discharge. (Black indicates closed and white indicates open).

Figure 3 displays the maximum and minimum RMS of fluctuating pressure on the BS during flood discharge scenarios, similar to Figure 2. Outcomes displayed in Figure 3 are clear indicators that it is beneficial to reduce pressure values in the BS by reducing the discharge associated with the side spillways as they are directly connected with BS, and hence cause a higher impact. For example, the maximum is more than 7 times higher than the minimum on BS, with Q of 12,000 m^3/s.

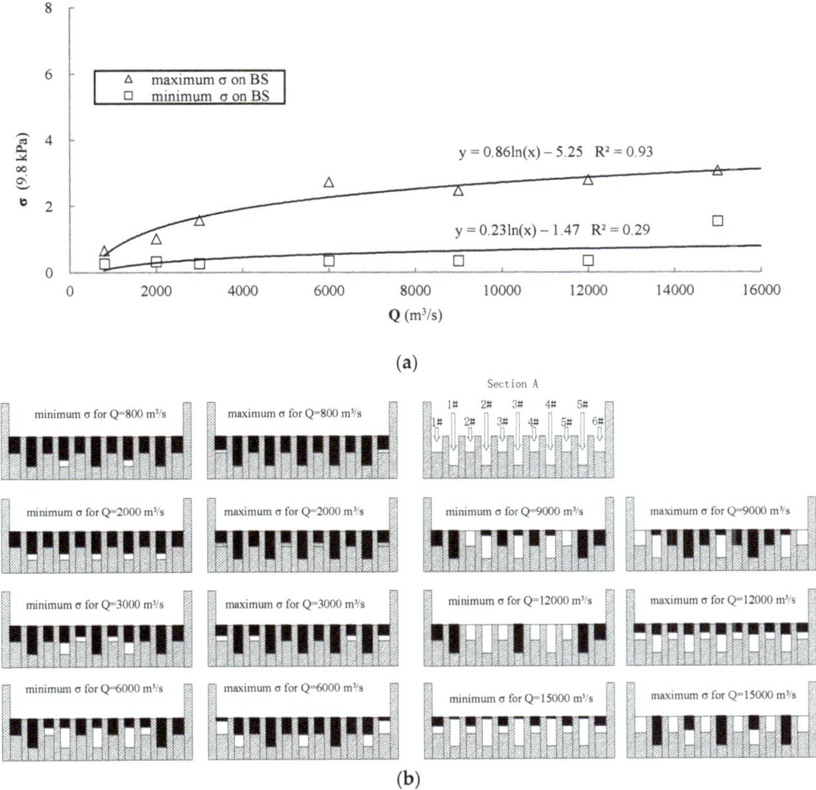

Figure 3. (**a**) Maximum and minimum RMS of fluctuating pressure on BS and (**b**) corresponding gate-opening modes with flood discharge. (Black indicates closed and white indicates open).

Figure 4 indicates maximum and minimum RMS of fluctuating pressure on the PSO and corresponding gate opening modes. Although fluctuating pressure on the PSO is larger than those on the BB and BS, the main sources of near-field vibration are the BB and BS. The effect of PSO is relatively small according to Yin and Zhang [18] and Yin et al. [19].

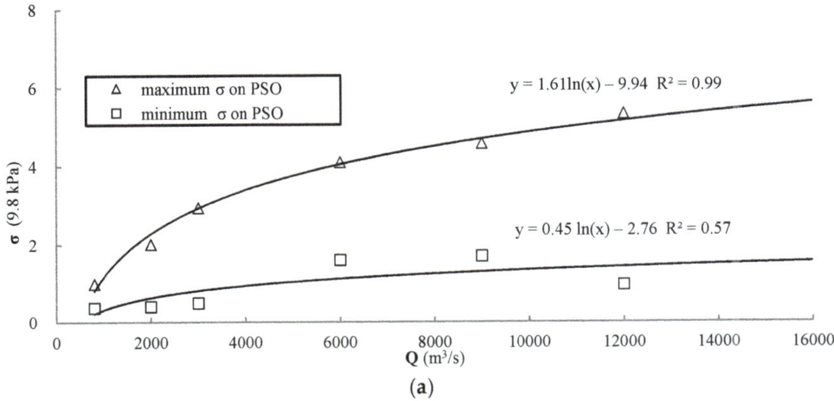

Figure 4. *Cont.*

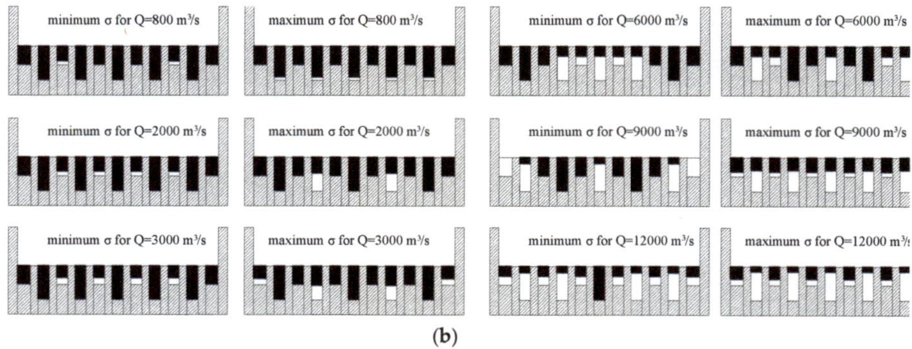

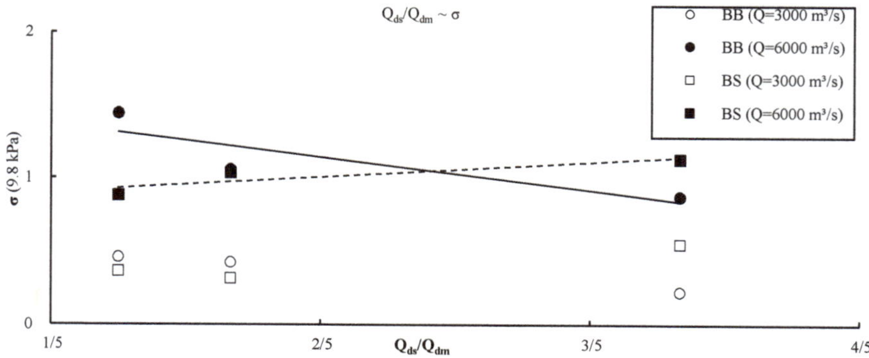

(b)

Figure 4. (**a**) Maximum and minimum RMS of fluctuating pressure on PSO and (**b**) corresponding gate-opening modes with flood discharge. (Black indicates closed and white indicates open).

3.2. Effect of Flow Discharge Ratio and Configuration Adopted to Release the Flow on Fluctuating Pressure

To quantitatively study the effect of different flowrates, more experiments were carried out. Specifically, three kinds of gate-opening modes were tested at higher flowrates typical of flooding scenarios: (i) mid-level channels separately, (ii) spillways separately, and (iii) combining both previous cases (i) and (ii). Based on measured fluctuating pressure and flow pattern evolution of these three different configurations, recommended gate-opening modes are proposed for various flood discharge rates.

The following notations are used to simplify the understanding of outcomes described in the upcoming sections: Q is total flood discharge for single stilling basin (m³/s); Q_u and Q_d are the total discharge of spillways and mid-level channels, respectively (m³/s); Q_{ds} and Q_{dm} are the total discharge of two side mid-level channels (1# and 5# as shown in Figure 1) and three middle ones (2–4#), respectively (m³/s); Q_{us} and Q_{um} are total flood discharge of two side spillways (1# and 6#) (m³/s) and four middle ones (2–5#) (m³/s).

Figure 5 illustrates the effect of the ratio Q_{ds}/Q_{dm} on fluctuating pressure on the BB and BS. It indicates that fluctuating pressure tends to increase on the BS and to reduce on the BB as Q_{ds} increases slightly if the mid-level channels are used separately. This result suggests that mid-level channels should be opened uniformly to avoid this disruption.

Figure 5. Effect of the ratio Q_{ds}/Q_{dm} on RMS of fluctuating pressure on BB and BS.

Figure 6 shows the effect of the ratio Q_{us}/Q_{um} on fluctuating pressure on BB and BS. By slightly increasing Q_{us}, if spillways are used separately, Figure 6 shows that the corresponding fluctuating pressure generated increases on the BS and reduces on the BB. This effect is consistent if applied to the mid-level channels configuration. Therefore, the same as the mid-level channels, spillways should be opened uniformly to reduce the magnitude of pressure on the BB.

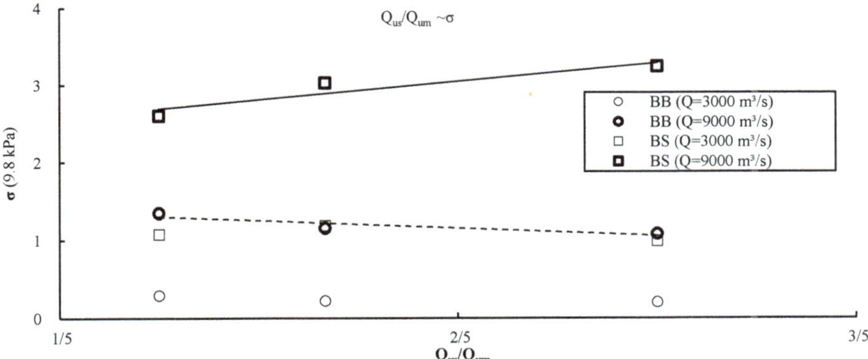

Figure 6. Effect of the ratio Q_{us}/Q_{um} on RMS of fluctuating pressure on BB and BS.

Figure 7 displays the effect of the ratio Q_u/Q_d on fluctuating pressure on the PSO, BB, and BS. When combination of the spillways and mid-level channels is adopted, fluctuating pressure increases on the BS and reduces on the BB and PSO as Q_u increases maintaining constant Q. Additionally, by applying this configuration, fluctuating load varies more rapidly on the PSO, and this effect is clearer as Q increases. When the ratio Q_u/Q_d is between 2 and 3, values of fluctuating pressure in all of three areas of study are relatively small.

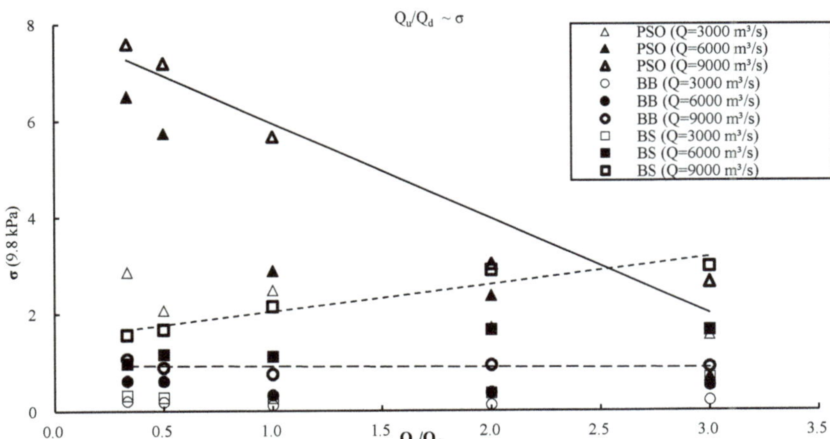

Figure 7. Effect of the ratio Q_u/Q_d on RMS of fluctuating pressure on BB, BS and PSO.

3.3. Effect of Flow Discharge Ratio and Configuration Adopted to Release the Flow on Flow Patterns in the Stilling Basin

In this section, the flow patterns in the stilling basin generated by the different hydraulic conditions and gate-opening configurations tested are discussed. As observed by previous studies [10], there is a correlation between the flow fluctuations and the flow patterns in the stilling basin. This explains why there is a significant difference, considering the same discharge flowrate, between fluctuating pressure on the PSO, on BB, and on BS. Figure 8 displays some examples of flow patterns observed in the stilling basin when the mid-level channels were used separately for diverse discharge flowrates. This is an indicator that various jets tend to quickly merge together and are likely to swing in the stilling basin if two or three of the middle mid-level channels are opened symmetrically. The formation of these flow patterns generates an increase in fluctuating pressure on the side-walls of the stilling basin. This phenomenon is mainly due to the outlet' width of each mid-level channel that is much smaller than the width of the stilling basin. When replicating this hydraulic condition of using mid-level channels separately, a backflow characterized by an "S" shape emerges in the stilling basin. The flow stability is relative good in the case of combination of 1#, 3#, 5# orifices (Figure 8c), or 1#, 2#, 4#, 5# orifices (Figure 8d), but the flow is not stable when all 1–5# orifices (Figure 8f) are fully open. This result suggests that mid-level channels can only be used individually for small flowrates during flooding events, and mid-level channels should only be opened symmetrically.

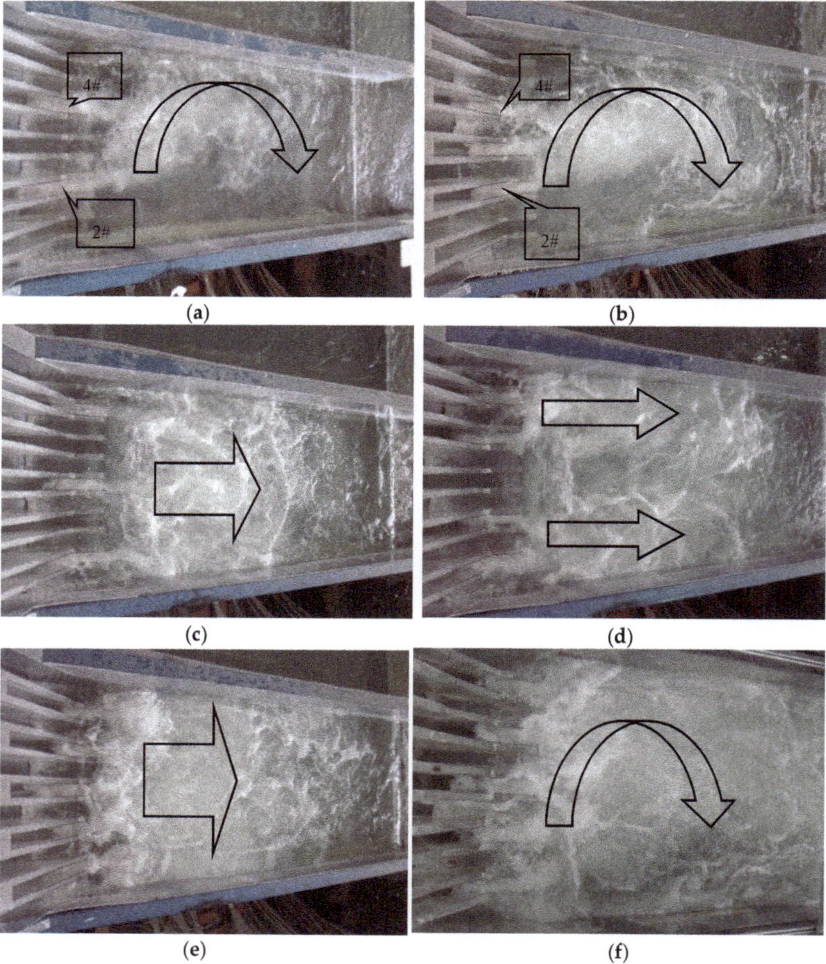

Figure 8. Flow pattern evolution in the stilling basin with opening of mid-level channels separately for different flow rates. (**a**) Opening partly of 2# and 4# Q = 2000 m^3/s; (**b**) Opening partly of 2#, 3# and 4# Q = 4000 m^3/s; (**c**) Opening fully of 1#, 3# and 5# Q = 5300 m^3/s; (**d**) Opening fully of 1#, 2#, 4# and 5# Q = 7000 m^3/s; (**e**) Opening partly of 1–5#, Q = 7000 m^3/s; (**f**) Opening fully of 1–5#, Q = 8874 m^3/s.

Figure 9 displays flow patterns that tend to develop when spillways are opened separately. It shows that the flow pattern is stable when the spillways are uniformly opened and separately regardless of the gate-opening height. Moreover, flow patterns are still stable if the middle spillways are opened symmetrically by closing side spillways. Finally, Figure 10 shows flow patterns generated when combining discharging orifices and spillways. In this case, flows discharged and flow in the stilling basin fully mix, and the flow patterns produced confirm the stability of the flow in the stilling basin. This stability continues until the outlet section of the stilling basin.

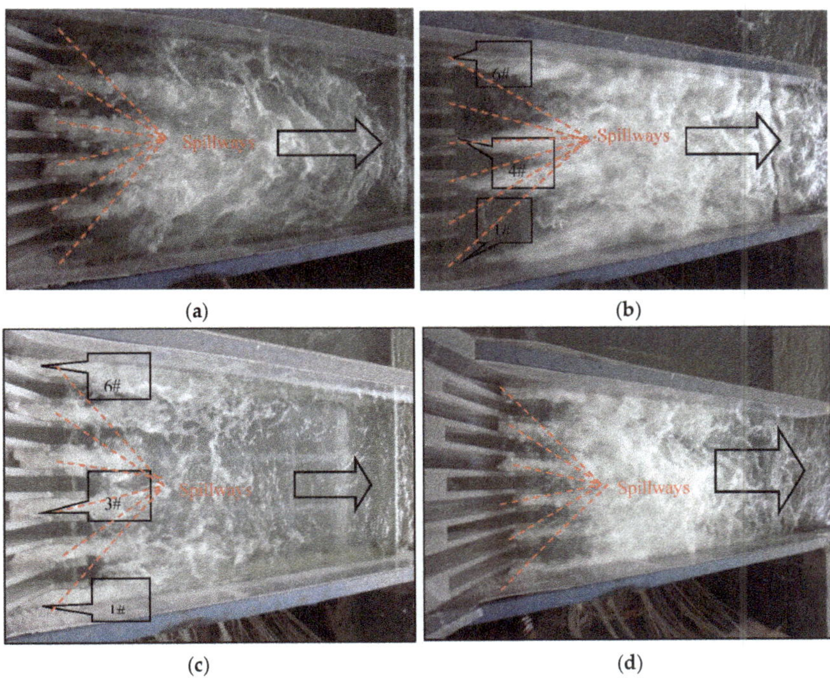

Figure 9. Flow pattern evolution in the stilling basin with opening of spillways separately for different flowrates. (**a**) Opening fully of 2–5#, $Q = 4000$ m^3/s; (**b**) Opening fully of 1#, 3#, 4#, 6#, $Q = 8634$ m^3/s; (**c**) Opening partly of 1–6#, $Q = 2000$ m^3/s; (**d**) Opening fully of 1–6#, $Q = 5868$ m^3/s.

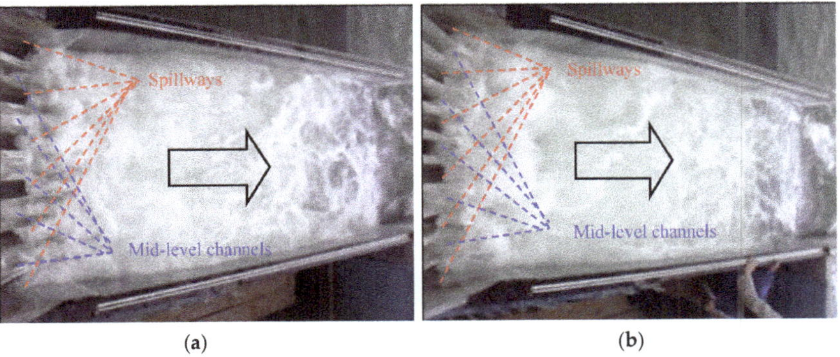

Figure 10. Flow pattern evolution in the stilling basin for combining mid-level channels with spillways for different flowrates. (**a**) Opening fully of 1–6# spillways and partly of 1–5# mid-level channels $Q = 17,800$ m^3/s, (**b**) Opening fully of 1–6# spillways and 1–5# mid-level channels $Q = 24,200$ m^3/s.

To summarize, flow patterns are stable in the stilling basin when combinations of mid-level channels and spillways are applied. The lack of clear and repeatable flow patterns in the stilling basin generated under different scenarios when mid-level channels are opened separately suggest that such a scenario should be avoided. Despite this, there is a consistent good agreement between the flow patterns and the measured fluctuating pressure values acting on the overflow walls.

4. Conclusions

According to the experimental results, the relationship between gate-opening modes and fluctuating pressure on the overflowing walls was explored and optimal gate-opening modes were recommended. Based on the above study, the following conclusions can be stated:

(1) Gate-opening modes are very flexible for MHSJ. Moreover, proper and strategical adjustment of gate-opening modes can be very effective in reducing fluctuating loads on overflow walls (e.g., fluctuating pressure can be reduced by nearly 50% for the same flood discharge by optimizing the opening mode of the gates).

(2) To improve the overall performance and stability of the hydraulic structure, both spillways and mid-level channels should be opened uniformly, but side spillways should be opened slightly less to reduce fluctuating pressure on the side-wall of the stilling basin. The discharge ratio Q_u/Q_d (the ratio of spillway discharge to mid-level channel discharge) should be kept between 2–3 when side spillways and mid-level channels are used together.

(3) The flow patterns in the stilling basin show remarkable differences depending on gate-opening modes. No repeatable shape was observed when trying different configurations. It is recommended to open the mid-level channels partially and uniformly when discharging orifices are used separately. The flow patterns are always stable when spillways are used separately regardless of full or partial opening. The flow patterns are relatively stable in the stilling basin when a combination of spillways and mid-level channels are used to release a high flowrate typical of flooding scenarios.

(4) It is proposed to use spillways and mid-level channels separately for middle orifices and small flowrates released. The combination of mid-level channels and spillways can be used for large discharge.

Author Contributions: J.Z. and W.X. conceived and designed the experiments; Y.P. performed the experiments; J.Z. and Y.P. analyzed the data; Y.P. and M.R. wrote the paper.

Funding: This research was funded by the National Natural Science Foundation of China (51579166) and the National Key Research and Development Program of China (2016YFC0401705).

Conflicts of Interest: The authors declare no conflict of interest.

Abbreviations

BB	stilling basin bottom
BS	stilling basin side-wall
H	height dimension (m)
L	length dimension (m)
MHSJ	Multi-Horizontal-Submerged Jets
PSO	piers, spillways and mid-level channels
RMS	root-mean-square of fluctuating pressure
W	width dimension (m)
Q	discharge (m^3/s)
Q_d	total discharge of mid-level channels (m^3/s)
Q_{ds}	total discharge of two side mid-level channels(1# and 5#) (m^3/s)
Q_{dm}	total discharge of three middle mid-level channels (2–4#) (m^3/s)
Q_u	total discharge of spillways (m^3/s)
Q_{um}	total flood discharge of four middle spillways (2–5#) (m^3/s)
Q_{us}	total flood discharge of two side spillways (1# and 6#) (m^3/s)
σ	RMS of fluctuating pressure (9.81 kPa)

References

1. Wu, J.H.; Ai, W.Z.; Zhou, Q. Head loss coefficient of orifice plate energy dissipater. *J. Hydraul. Res.* **2010**, *48*, 526–530.
2. Manenti, S.; Pierobon, E.; Gallati, M.; Sibilla, S.; D'Alpaos, L.; Macchi, E.; Todeschini, S. Vajont Disaster: Smoothed Particle Hydrodynamics Modeling of the Postevent 2D Experiments. *J. Hydraul. Eng.* **2015**, *142*, 05015007. [CrossRef]
3. Vacondio, R.; Mignosa, P.; Pagani, S. 3D SPH numerical simulation of the wave generated by the Vajont rockslide. *Adv. Water Resour.* **2013**, *59*, 146–156. [CrossRef]
4. Xu, Y.; Zhang, L.; Jia, J. Lessons from Catastrophic Dam Failures in August 1975 in Zhumadian. *China Geocongr.* **2008**, *178*, 162–169.
5. Tan, Y.T.; Xu, H.; Zhang, X.L. Sustainable urbanization in China: A comprehensive literature review. *Cities* **2016**, *55*, 82–93. [CrossRef]
6. Intergovernmental Panel on Climate Change (IPCC). *Summary for Policymakers, Climate Change 2014: Impacts, Adaptation and Vulnerability—Contributions of the Working Group II to the Fifth Assessment Report*; IPCC: Geneva, Switzerland, 2014; pp. 1–32.
7. Zhang, J.M.; Peng, Y.; Xu, W.L. *Study on the Cause of Vibration Induced by Flood Discharge and Its Countermeasures in Xiangjiaba Hydropower Station*; Technical Report; Sichuan University: Chengdu, China, 2013.
8. Zhang, J.M.; Li, Y.L.; Yang, Y.Q.; Xu, W.L.; Zeng, X.H.; Cheng, H. Investigation on energy dissipation of multiple submerged horizontal jets with pressure flow. *Water Resour. Hydropower Eng.* **2004**, *35*, 30–33. (In Chinese)
9. Zhang, J.M.; Chen, J.G.; Xu, W.L.; Peng, Y. Characteristics of the vortex structure in multi-horizontal submerged jets stilling basin. *Proc. Inst. Civ. Eng.-Water Manag.* **2013**, *167*, 322–333. [CrossRef]
10. Li, Y.L.; Hua, G.C.; Zhang, J.M.; Yang, Y.Q.; Hu, Y.H. Analysis on energy dissipation of the single-lever with multi-strand and multi-lever with multi-strand horizontal submerged jets. *Chin. J. Hydrodyn.* **2006**, *1*, 26–31. (In Chinese)
11. Li, Y.L.; Hua, G.C.; Zhang, J.M.; Yang, Y.Q. Factors affecting the hydraulic characteristics of horizontal submerged jets. *Adv. Water Sci.* **2006**, *17*, 761–766. (In Chinese)
12. Deng, J.; Xu, W.L.; Zhang, J.M.; Qu, J.; Yang, Y. A new type of plunge pool—Multi-horizontal submerged jets. *Sci. China (E Technol. Sci.)* **2008**, *51*, 2128–2141. [CrossRef]
13. Deng, J.; Xu, W.L.; Zhang, J.M.; Qu, J.X.; Yang, Y.Q. A new type of stilling pool—Multi-strand horizontal submerged jets. *Sci. China (E Technol. Sci.)* **2009**, *39*, 29–38. (In Chinese)
14. Huang, Q.J.; Feng, S.R.; Li, Y.N.; Wu, J.H. Experimental study on energy dissipation characteristics of multi-horizontal submerged jets. *Chin. J. Hydrodyn.* **2008**, *23*, 694–701. (In Chinese)
15. Chen, J.G.; Zhang, J.M.; Xu, W.L.; Li, S.; He, X.L. Particle image velocimetry measurements of vortex structures in stilling basin of multi-horizontal submerged jets. *J. Hydrodyn. Ser. B* **2013**, *25*, 556–563. [CrossRef]
16. Chen, J.G.; Zhang, J.M.; Xu, W.L.; Wang, Y.R. Scale effects of air-water flows in stilling basin of multi-horizontal submerged jets. *J. Hydrodyn. Ser. B* **2010**, *22*, 788–795. [CrossRef]
17. Chen, J.G.; Zhang, J.M.; Xu, W.L.; Wang, Y.R. Numerical simulation investigation on the energy dissipation characteristics in stilling basin of multi-horizontal submerged jets. *J. Hydrodyn.* **2010**, *22*, 732–741. [CrossRef]
18. Yin, R.G.; Zhang, J.H. Vibration source and shock absorption scheme research of near-field vibration caused by flood discharge and energy dissipation of a stilling pool. *Geo Shanghai Int. Congr.* **2014**, *237*, 107–116.
19. Yin, R.G.; Zhang, J.H.; Liu, X.K. 3D dynamic FEM analysis of near-field vibration caused by flood discharge and energy dissipation of a stilling pool. *J. Water Resour. Arch. Eng.* **2014**, *12*, 72–76. (In Chinese)

Article

River Bathymetry Model Based on Floodplain Topography

Ludek Bures [1],*, Petra Sychova [1], Petr Maca [1], Radek Roub [1] and Stepan Marval [2]

[1] Department of Water Resources and Environmental Modeling, Faculty of Environmental Sciences, Czech University of Life Sciences Prague, Kamycka 1176, 165 21 Prague 6, Suchdol, Czech Republic; sychova@fzp.czu.cz (P.S.); maca@fzp.czu.cz (P.M.); roub@fzp.czu.cz (R.R.)

[2] Research Institute for Soil and water Conservation, Zabovreska 250, 156 27 Prague 5, Czech Republic; marval.stepan@vumop.cz

* Correspondence: Buresl@fzp.czu.cz; Tel.: +420-22438-2153

Received: 16 April 2019; Accepted: 17 June 2019; Published: 20 June 2019

Abstract: An appropriate digital elevation model (DEM) is required for purposes of hydrodynamic modelling of floods. Such a DEM describes a river's bathymetry (bed topography) as well as its surrounding area. Extensive measurements for creating accurate bathymetry are time-consuming and expensive. Mathematical modelling can provide an alternative way for representing river bathymetry. This study explores new possibilities in mathematical depiction of river bathymetry. A new bathymetric model (Bathy-supp) is proposed, and the model's ability to represent actual bathymetry is assessed. Three statistical methods for the determination of model parameters were evaluated. The best results were achieved by the random forest (RF) method. A two-dimensional (2D) hydrodynamic model was used to evaluate the influence of the Bathy-supp model on the hydrodynamic modelling results. Also presented is a comparison of the proposed model with another state-of-the-art bathymetric model. The study was carried out on a reach of the Otava River in the Czech Republic. The results show that the proposed model's ability to represent river bathymetry exceeds that of his current competitor. Use of the bathymetric model may have a significant impact on improving the hydrodynamic model results.

Keywords: DEM; hydrodynamic modelling; river bathymetry; floods; Bathy-supp

---

1. Introduction

Knowledge of terrain morphology is crucial for the hydrodynamic modelling of floods. The accuracy and applicability of hydrodynamic models is driven by the nature, availability, and accuracy of source topographic data [1–3].

The digital elevation models (DEMs) are required as a main input for hydrodynamic modelling. Topographic mapping is conventionally conducted by ground surveying. The main advantage of such a method is its high accuracy. Among the major limitations of measured data acquisition are its high cost and time-consuming data collection. Therefore, ground mapping is increasingly being replaced by remote-sensing methods. Radar and laser altimetry are among the most commonly used remote-sensing techniques [4–6].

A description of the river channel and its surrounding area is necessary to create a DEM for purposes of hydrodynamic modelling. The DEM must have precise vertical accuracy and spatial resolution [1,7]. DEMs obtained from satellites are commonly used on a global scale, but the spatial resolution of these models often does not meet the requirements of precise hydrodynamic modelling. The most commonly cited DEMs used for hydrodynamic modelling are the Advanced Spaceborne Thermal Emission and Reflection Radiometer (ASTER), and the Shuttle Radar Topography Mission (SRTM) [8–11].

Aerial laser scanning (ALS) can be another source of input data for a DEM intended for hydrodynamic modelling. This data source can produce DEMs with high spatial resolution [4]. The method is based on Light Detection and Ranging (LiDAR) technology. ALS methods usually use an infrared laser beam, which is unable to penetrate the water surface or scan the river bed [12]. This problem can be solved by using dual LiDAR (DiAL) technology, which uses a combination of two laser beams [13–15]. However, in some cases, this technology fails because the green laser beam (used for scanning water areas) is not able to penetrate an aquatic environment characterized by high turbidity [16–18].

The most commonly used method for creating a DEM with correct bathymetry is to merge topographic (e.g., ALS) data with another source of bathymetric data. Often used for this purpose are acoustic Doppler current profilers (ADCP), Sound Navigation and Ranging (SONAR) techniques, theodolites, and Global Positioning System (GPS) stations [3,19,20]. The main advantage of these methods is their high accuracy of the acquired data. The main disadvantage lies in the cost of the data's acquisition.

Interpolation approaches have been developed in an attempt to reduce the volume of measured data needed to represent river bathymetry [21,22]. Some interpolation techniques for the creation of bathymetry use cross-sections with various spacings as input data [21,23,24]. Nevertheless, measured data are still required.

Another option for representing river bathymetry can be to use mathematical modelling methods. In works of Dutch and Wang [25] and James [26], the cross-section shape of fluvial sediment deposits was estimated using analytical curves. Moramarco et al. [27] have used entropy-based methods for estimation of the cross-section. Roub et al. [28] introduced a bathymetric model, which estimates the river channel on the basis of hydraulic parameters. The river channel is thereby schematized into a trapezoidal shape. The flow area of this channel is determined using Chezy's equation.

This paper explores new possibilities in mathematical representation of river bathymetry. A new model of river bathymetry based on topographic data describing the area surrounding the river channel is proposed. In this work, the model's ability to represent actual bathymetry is evaluated. A two-dimensional (2D) hydrodynamic model is used to evaluate the influence of the bathymetric model on the results of hydrodynamic modelling. Also presented is a comparison of the proposed model with another state-of-the-art model, the Cross-section Solver ToolBox (CroSolver) [28].

2. Materials and Methods

The proposed bathymetry model is based on analytical curves. The curves are bent into the shapes of the cross-sections. The most precise values of the model parameters are needed in order to obtain the best description of the river bathymetry.

2.1. Input Data for the Bathy-Supp Model

Three types of input data are needed for a river bathymetric model. The first is a DEM, describing the floodplain topography. Also extracted from the DEM are the terrain characteristics, the altitudes for height adjustment of the new cross-sections, and a definition for the aquifer area of the river channel. The second type of input data is the design flow rate. The design flow rate is the flow rate at the time of elevation data acquisition. The last, but not least, type of input data is Manning's roughness coefficient for the river channel.

2.2. Construction of the Bathy-Supp Model

The proposed bathymetric model is constructed in four main steps:

1. The user defines the number and location of the new cross-sections (location of cross-section endpoints) from which a new bathymetry model will be composed.

2. Computation of the spatial terrain characteristics (predictor variables) derived from the floodplain DEM, and estimation of the model parameters m_1 and m_2.
3. Cross-section construction and transformation.
4. River bed reconstruction.

2.2.1. Location of the New Cross-Sections

After displaying the input DEM, it is possible to identify the position of the river channel itself as a no-data region, which thereby divides the model into two (or more) parts. In this no-data region, the user defines the number of new cross-sections and their locations. The distance between the first and last cross-section defines that part of the river for which new bathymetry will be estimated. Each cross-section is defined by two endpoints. The distance between these endpoints defines the width of the channel. The endpoints have coordinates X and Y. A DEM point with the lowest altitude is searched in the circular space around each endpoint. The altitude of this lowest point is used as the Z coordinate of the endpoint. The radius of the circular space is called the lowest point search radius. An example of the cross-section location can be seen in Figure 1.

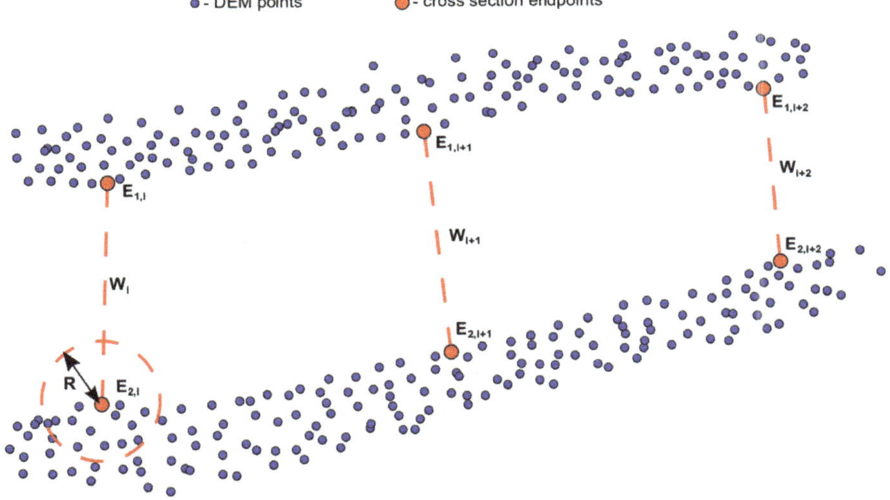

Figure 1. Defining the position of the new cross-sections. (width (W), endpoint (E), lowest point search radius (R)).

2.2.2. Explaining Model Parameters

Usually, the best-fit model parameters (m_1, m_2) are unknown. To establish them (without using an inverse problem-solving method), it is necessary to find the relationship between the search parameters and other explanatory variables. Spatial terrain characteristics can be used as possible explanatory variables [21,26]. The overall curvature, planar curvature, profile curvature, overall slope, slope in the x-direction, and slope in the y-direction were, in the present case, selected for predicting individual model parameters. The characteristics were determined as described by Zevenbergen and Thorne [29]. The source of terrain characteristics was the DEM input. Terrain characteristics were calculated for the all raster cells located behind the bank lines around the river. The terrain characteristics of the nearest raster cell were used to estimate the parameters of a given endpoint. As explanatory variables, the left bank terrain characteristics for model parameter m_1 and the right bank terrain characteristics for parameter m_2 were used.

Statistical learning methods were used for finding the dependence between model variables and terrain characteristics. Those statistical learning methods studied were multiple linear regression, extended linear regression, and random forest (RF).

A simple linear model (LM) is used for finding a linear relationship between a response and its predictor, and, in the case of multiple linear regressions, this relationship is based upon more than a single predictor. The LM is described in Equation (1):

$$y = \beta_0 + \sum \beta_i x_i + \varepsilon_i, \ \varepsilon_i \sim N(0, \sigma^2), \ i = 1, \ldots, M, \tag{1}$$

where βx are the linear parameters, x_i are predictor variables, ε_i is the error term, and M is the number of predictor variables. A mixed selection procedure, as described by Gareth et al. [30], was adopted for choosing the optimal number of variables.

An extended linear model with no random effects (GLS) was also used. This method extends linear regression with an ability to fit models with heteroscedastic and correlated within-group errors, but with no random effects [31]. The extended formula of the LM is described in Equation (2):

$$y = \beta_0 + \sum \beta_i x_i + \varepsilon_i, \ \varepsilon_i \sim N(0, \sigma^2 A^i), \ i = 1, \ldots, M, \tag{2}$$

where A_i are positive-definite matrices composed using variance and covariance matrices, β_x are the linear parameters, x_i are predictor variables, ε_i is the error term, and M is the number of predictor variables. Again, the mixed selection procedure, as described by Gareth et al. [30], was adopted for choosing the optimal number of variables.

Random forest (RF) is a combined machine learning method for classification and regression. This method is based on an ensemble of a regression tree (RT) algorithm. RT deals with tree structure by dividing the dataset into homogenous groups. That division is driven by some classification criterion, such as minimizing the variance of a given set of variables. In the case of RF, a dataset is divided into multiple sub-datasets by a bootstrap aggregating algorithm. For each sub-dataset, an RF of its own is constructed. This creates a group of random trees, termed RF. For each predictor variable, a measure of variable importance can be determined. Based on variable importance values, it is possible to decide which variables have significant impact for the response and which can be omitted [32].

All statistical analyses in this work were performed using statistical software R. The extended linear model with no random effects was applied using the package nlme [33], and the package randomForest [34] was used for random forests.

The coefficient of determination (R^2) was used to evaluate the reliability of model parameters. As a second quality assessment, vertical differences between models based on the best model parameters and a model based on estimated parameters were calculated. For this comparison, a similar approach is used in Section 2.1.

2.2.3. Cross-Section Construction and Transformation

Once the model parameters are estimated, new cross-sections can be constructed. A tested theoretical cross-section model Equation (3) is able to estimate the natural river cross-section on the basis of estimated parameters [35]. The studied theoretical model of the river cross-section is explained as:

$$z(d) = m_1 m_2 d^{m_1 - 1} (1 - d^{m_1})^{m_2 - 1}, \tag{3}$$

where $z(d)$ is the depth of water at a distance d from the left endpoint of the cross-section, while m_1 and m_2 are theoretical model parameters that are unique for each river cross-section. An example of the estimated shapes of the cross-section is shown in Figure 2.

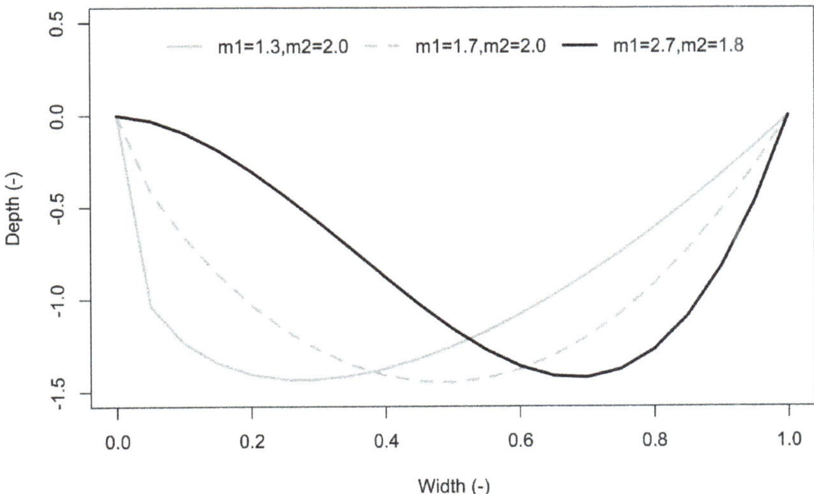

Figure 2. Example showing ability of the proposed bathymetric model to schematize a cross-section.

Due to the mathematical nature of the proposed model, a new cross-section width of 0–1 [35] must be selected in the first step. Note that the value 0 represents the left side of the cross-section. Additional points are inserted between the points 0 and 1. The user decides upon the number of points to insert. Once new stationing is defined, the depth value for each cross-section point is computed by applying Equation (3). New cross-sections produced by the proposed bathymetric model are in normative state (width and flow area are equal to (1)). Width transformation is simple. Stationing of the new cross-section is multiplied by the distance (W) between its endpoints (Figure 1). For the flow area transformation, an adequate flow area must be identified. This adequate flow area defines the flow area of the river channel required for the transfer of the design flow. The adequate flow area is determined using Chezy's equation on the basis of Manning's roughness coefficient, water surface slope, and design flow rate. This adequate flow area is compared with the area of the new cross-section. If the adequate flow area is smaller than the cross-sectional area, then the depths of the cross-section points are multiplied by the area multiplication parameter. This step is repeated until the adequate flow area is equal to or less than the area of the cross-section. Manning's coefficient, design flow rate, and area multiplication parameter are the input parameters. The water longitudinal surface slope and water surface elevation of each cross-section are extracted from the DEM. A similar approach had been used in the work of Roub et al. [28].

The final transformation step is to add the new cross-section into the coordinate system used. The XY coordinates of the first and last point of the cross-section correspond to the coordinates of the endpoints E_1 and E_2, for which the cross-section has been created. All internal points are placed on the line connecting the endpoints. Therefore, the coordinates of the internal points can be calculated based on the coordinates of the endpoints and its station value. The method of calculating the coordinates of internal points may vary depending on the coordinate system used. The lower of the altitudes of both endpoints is determined as the water level of this cross-section. The altitude (Z value) of each point is obtained by subtracting the water level and its depth. All cross-section points have coordinates X, Y, and Z and stationing after transformation.

2.2.4. River Bed Reconstruction

To create a three dimensional (3D) bathymetric model composed of isolated linear structures, such as cross-sections, spatial interpolation between these cross-sections must be used. Many different bed reconstruction algorithms are available in the literature [23,24,36,37]. In this contribution, we adopted the approach of Caviedes–Voullième [23]. This bed reconstruction algorithm is based upon cubic Hermite splines (CHS). Spline trajectory connects two depth points in two consecutive cross-sections, and it is driven by their normal vectors. The spatial interpolation (X, Y) of new bathymetric points is performed using CHS, and linear interpolation is performed for the height (Z). For a more detailed description, see the work of Caviedes–Voullième [23].

2.3. Model Suitability

Global optimization methods are used for estimating the best model parameters. The global optimization schemes are based on heuristics inspired by natural processes. Methods of differential evolution [38] are used for determining the solutions of inverse problems related to the parameters of a mathematical model of river bed surfaces. Differential evolution is a population-based stochastic optimization search algorithm that iteratively estimates a candidate solution with regard to a given measure of quality [38]. The analyzed objective function is a least squares method, determining the differences between the depths of the real cross-sections and the modelled cross-sections. We employed the *best1bin* algorithm in this work [38]. Analysis of the model's suitability was made in the C++ programming environment.

Model Suitability Evaluation

For evaluating the model's suitability, a vertical differences comparison between the model cross-sections (with the best model parameters) and the measured cross-sections (see Section 2.5.2) was made. The root mean square error (RMSE) and the mean absolute error (MAE) were calculated for this purpose. The equations for these evaluation criteria are as follows:

$$RMSE = \sqrt{\frac{1}{N}\sum_{i=1}^{N}(Elev_{MOD} - Elev_{REF})^2}, \tag{4}$$

$$MAE = \frac{1}{N}\sum_{i=1}^{N}|Elev_{MOD} - Elev_{REF}|, \tag{5}$$

where $Elev_{MOD}$ represents the elevation (m) obtained from each model cross-section and $Elev_{REF}$ represents the equivalent reference point obtained from the measured cross-section (see Section 2.5.2). N represents the number of the cross-section points.

2.4. Case Study Area

A part of the Otava River in the Czech Republic was chosen as an area of interest (Figure 3). The full length of the Otava River is 111.7 km with the basin area of 3840 km^2. The studied river reach is located into the lower part of the river (between 22.83 and 24.58 river km) and is 1.75 km long. The average depth of the river reach fluctuates around 1 m. The average bankfull width is between 22.8 and 52.7 m. The average annual flow rate is 23.4 m^3/s. The average water level is 354.84 m above sea level. The flow rates for N–year floods in the Otava River reach are shown in Table 1. $Q_{a.a.}$ denotes the average annual discharge.

Table 1. The flow rates for N-year floods in the Otava River reach [39].

N-year Floods	$Q_{a.a.}$	Q_1	Q_5	Q_{10}	Q_{50}	Q_{100}
Flow rate (m^3/s)	23.4	146	300	394	680	837

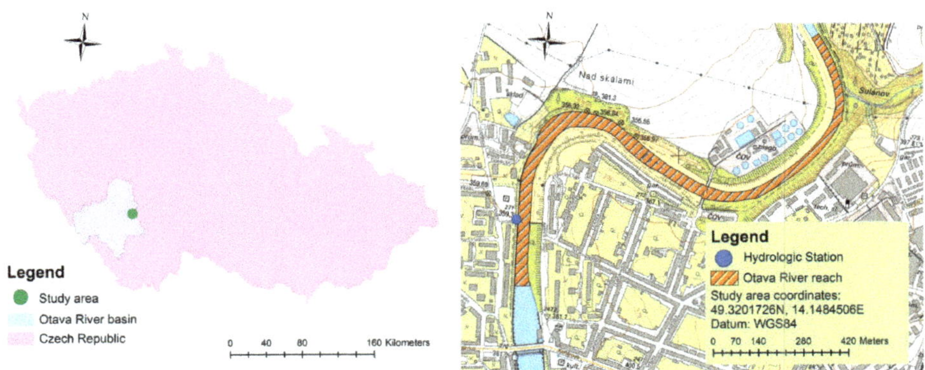

Figure 3. The study area: Otava River reach.

2.5. Ground and Bathymetry Data

This section of the paper describes the topographic source data and bathymetry data source used for identification of the model parameters and creating the DEMs.

2.5.1. Aerial Laser Scanning Data

The technology used for ALS data collection in this study was aerial laser scanning with one infrared laser beam. These ALS data were provided by the Czech Office for Surveying, Mapping, and Cadastre. Provided data were in the form of the list of elevation points. The point density per square meter varies depending on the slope of the terrain described. The point density starts with one point (flat area) and ends with dozens of points (steep area). Vertical data accuracy is +/−0.18 m in open areas and +/−0.3 m in forest areas. [40].

2.5.2. Acoustic Doppler Current Profiler Data

The data obtained by using an ADCP measurement device have the depth range 0.2–80 m and accuracy of +/−1% of the depth. The ADCP technology was successfully applied in recent works [18,41,42]. The data were sampled as a set of single cross-sections. The distance between these cross-sections was approximately 5 m. The distance between points within the cross-sections was less than 0.5 m. A total of 375 cross-sections were measured. These data were used for testing the suitability of the bathymetric model as well as for determining the best model parameters.

2.5.3. Compared DEMs

A total of four DEMs, based upon different sources of bathymetric data, were compared in this practical demonstration. The first DEM (ALS) was created only from the ALS data. It contains no bathymetric data and represents models of that sort produced by remote-sensing techniques.

The second DEM (CRO) was composed using ALS data, which describe the floodplain, and data from the Cross-section Solver ToolBox (CroSolver) bathymetric model. CroSolver is a software tool for improving an existing DEM in the riverbed area. The newly formed channel is schematized as trapezoidal or rectangular. The Chezy equation is used to determine the channel flow area for the design flow rate. The depth of the new channel is determined on the base of the size of the flow area, channel width, and selected channel schematization [28]. Table 2 provides the settings overview for the CroSolver model.

Table 2. The Cross-section Solver ToolBox (CroSolver) software specific settings.

Parameter	Value
Calculation method	Longitudinal gradient
Lowest point search radius (m)	5
Manning's roughness (s/m$^{1/3}$)	0.031
Bank slope 1:m	2
Flow rate (m^3/s)	36.8
Water surface slope calculation (m)	100

The third DEM (BAT) was composed using ALS data, which describes the floodplain and the Bathy-supp model proposed in this paper. The cross-section distance for the bathymetry creation was about 100 m. Table 3 provides the settings overview for the proposed model.

Table 3. Bathy-supp model specific settings.

Parameter	Value
Lowest point search radius (m)	5
Manning's roughness (s/m$^{1/3}$)	0.031
Flow rate (m^3/s)	36.8

The last DEM (ADP) is a model composed using ALS data and ADCP data. This model is used as a reference model because it is based on measured data (i.e., on data with assured accuracy). Table 4 provides background information on the DEMs being compared.

Table 4. Overview of compared digital elevation models (DEMs).

Digital Elevation Model	Floodplain Data	Bathymetry Data	Resolution (m)
ALS	LiDAR	–	0.5
CRO	LiDAR	CroSolver	0.5
BAT	LiDAR	Bathy-supp	0.5
ADP	LiDAR	ADCP	0.5

2.5.4. DEM evaluation

For evaluation of the DEM quality, the vertical differences between compared DEMs (ALS, CRO, BAT), and the reference DEM (ADP) were assessed. This comparison was made for all corresponding cells in the river channel area. Comparison of the models in the floodplain is not relevant. Identical data were used for its description. RMSE Equation (4) and MAE Equation (5) values were employed for evaluation.

2.6. Hydrodynamic Modelling

All hydrodynamic simulations were performed using a 2D HEC-RAS model [43]. HEC-RAS 2D allows computations for steady and unsteady flow conditions. The hydrodynamic results are determined on the base of solving of 2D Saint–Venant equations. Many authors successfully employed HEC-RAS 2D for hydrodynamic simulations [44,45].

In this study, hydrodynamic simulations for four different DEMs were made (see Section 2.5.3). For each DEM, four simulations were made with chosen N-year flow rates. Manning's roughness coefficients were set separately for each land cover category. Values in the range 0.035–0.10 s/m$^{1/3}$ were determined for inundations and the value for the main channel was 0.031 s/m$^{1/3}$. Selection of the Manning's values was verified by a calibration-verification process. The validation of model parameters was based on a flood event from December 2012, where, for discharges of 143 m^3/s, the recorded water level (hydrological station located in the model river reach) was equal to 356.43 m. The normal depth was used as the downstream boundary condition. The results of the basic hydrodynamic model (topography source was ADP DEM) for the given discharge provided a difference of 2 cm in the water level. That verified the accuracy of the model setup. All simulations began from a minimal start discharge, which then grew until eventually becoming steady [18,46,47]. Table 1 shows the selected N-year floods that were used as boundary conditions.

Water surface elevation (WSE) and Inundation areas (IAs) were evaluated to determine the influence of channel bathymetry representation. WSE was evaluated by comparing the differences between vertical measurements for the raster-based WSE (ALS, CRO, BAT) and those of the reference WSE (ADP). The RMSE (Equation (4) and MAE (Equation (5) values were used in the evaluation.

For IAs, the following compliance criterion was used:

$$IA_{dif} = \frac{|IA_{DEM} - IA_{REF}|}{IA_{REF}} \times 100 \tag{6}$$

where IA_{dif} represents the difference in inundation areas (%), IA_{DEM} represents the inundation area (km^2) of compared models (ALS, CRO, BAT), and IA_{REF} represents the inundation area of the ADP model (km^2).

3. Results and Discussion

3.1. Bathymetric Model Suitability

The ability of the proposed bathymetric model to schematize the measured cross-section was evaluated. The best model parameters were used for this purpose. The parameter range m_1 ranged from 1.0202 to 1.8219, and for the m_2 parameter from 1.0929 to 2.0376. Overall, 375 measured cross-sections were evaluated. The mean RMSE and MAE values were 0.16 m and 0.11 m, respectively. The variability of the RMSE and MAE values is shown in Figure 4.

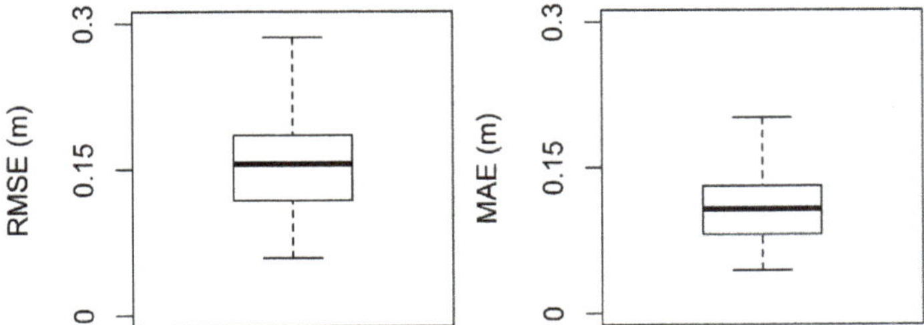

Figure 4. Comparing ability of the bathymetric model (with ideal parameters) to represent a measured cross-section. Shown are variances of root mean square error (RMSE) and mean absolute error (MAE) values.

3.2. Explaining Bathymetric Model Parameters

High-quality coefficient estimation is a prerequisite for correct bathymetric model creation. Therefore, various methods for its estimation were evaluated. For LM and GLS methods, overall curvature, planar curvature, overall slope, slope in the x direction, and slope in the y direction were considered as predictor variables. For the RF method, overall curvature, planar curvature, profile curvature, overall slope, slope in the x direction, and slope in the y direction were considered. LM and GLS methods provided similarly poor results. In contrast, the RF method produced a good result. Table 5 provides an overview of determination coefficients for the compared techniques. In view of these findings, the RT method was adopted for creating the bathymetric model.

Table 5. Coefficients of determination for the best-fit model parameters in comparing the estimation techniques linear model (LM), the extended linear model with no random effects (GLS), and the random forest (RF).

	LM	GLS	RF
R^2 (m_1)	0.145	0.161	0.918
R^2 (m_2)	0.085	0.118	0.914

Differences in parameter estimation quality may be due to nonlinear relationships in the data structure. More detailed analysis of the suitable regression structures and evaluation of its results are planned in follow-up research.

3.3. DEM Comparison

This comparison was made for cross-sections extracted from compared DEMs. The smallest divergence from the reference model (ADP) was achieved by the BAT model. The CRO model is unable to closely simulate the depth variability across the cross-section. This is due to the trapezoidal schematization of the channel. [28]. The ALS model almost completely neglects the riverbed area, which is specific for the sampling method used [4,12,40,48]. A graphical comparison is shown in Figure 5.

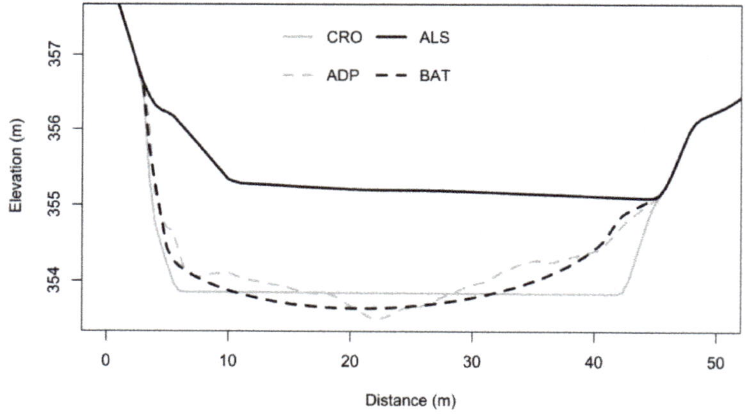

Figure 5. Visual comparison: Random cross-section derived from evaluated digital elevation models.

Table 6 describes the RMSE and MAE values achieved for the river channel (raster comparison). Both evaluated errors for the ALS model are >1. Note that the mean water depth in the channel at the time of acquiring ALS data was around 1 m. This suggests that the ALS model neglected almost the entire flow area of the river channel [16,20,28,42]. The RMSE and MAE values for the CRO model were 0.46 and 0.36 m, respectively. The smallest error values, RMSE 0.30 and MAE 0.23 m, were achieved by the BAT model. The raster cell difference variability is shown in Figure 6. Greater difference variability was achieved by the ALS model, and the smallest variability was achieved by the BAT model.

Table 6. RMSE and MAE errors for the compared DEMs (channel comparison).

	ALS	CRO	BAT
RMSE (m)	1.19	0.46	0.30
MAE (m)	1.06	0.36	0.23

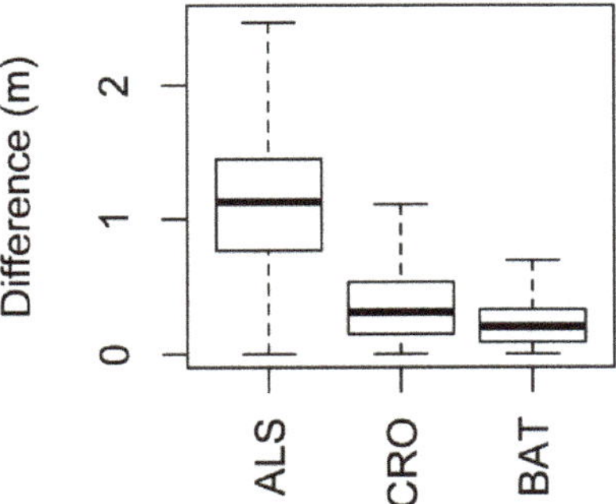

Figure 6. Variance of raster cell differences among compared digital elevation models.

Channel difference maps are shown in Figure 7. The biggest differences can be seen for the ALS model. The largest difference values are located along the thalweg (or stream centerline) location. Indeed, description of the river bottom was unrealistic. The error map presented for the CRO model identifies some areas with high cell difference values. These areas are located near embankment areas. The differences would probably be less significant if the cross-section were actually to have a trapezoidal shape (reflecting technical modification of the channel). For a natural river, the differences may be more significant. The BAT difference map provides the best results. Even here, however, it is possible to identify areas with a poor match. In this case, there was an excessive recess of one cross-section in the BAT model. The water surface slope derived from inundation data was slightly underestimated. This may be due to the accuracy of the data used to describe the floodplain [40].

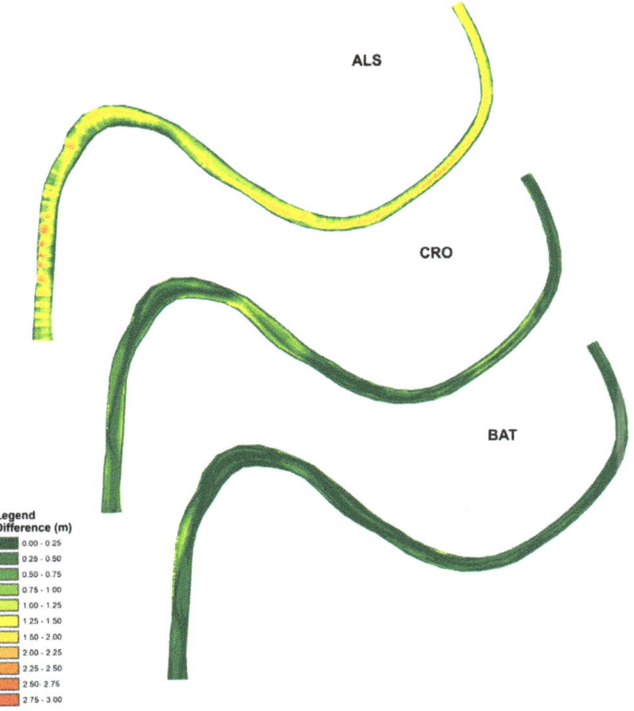

Figure 7. Error maps comparing evaluated digital elevation models (ALS, CRO, BAT) and the reference model (ADP).

3.4. Thalweg Comparison

The ALS model provided a poor match with the ADP model. Thalweg derived from the ALS model was shifted toward the water surface, where it fluctuated. Thus, it is evident that almost the entire flow area was neglected [16,20]. The CRO model had higher deviations relative to the BAT model. These deviations may occur in places where the channel narrows or expands locally. This is due to the fact that the local changes in flow velocity are not reflected in the CroSolver software [28]. The graphical comparison is shown in Figure 8, and the results of the numerical comparison of thalwegs are shown in Table 7.

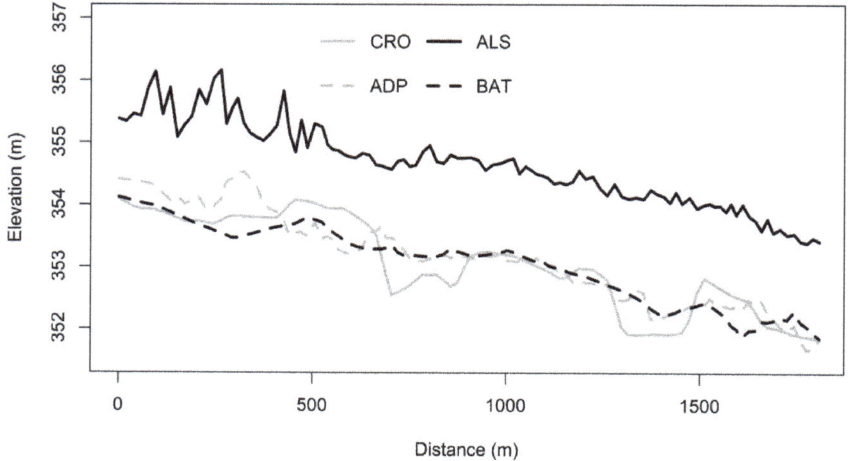

Figure 8. Visual comparison of thalwegs.

Table 7. RMSE and MAE values for the compared DEMs (thalweg comparison).

	ALS	CRO	BAT
RMSE (m)	1.52	0.37	0.30
MAE (m)	1.49	0.31	0.21

3.5. Water Surface Elevations Comparison

Figure 9 presents the variability in WSE errors between the compared DEMs (ALS, CRO, BAT) and the ADP DEM. The greatest deviations in WSE were seen in the ALS model, which manifested significant overestimation of WSE at all modelled flow rates. At flow $Q_{a.a.}$, the RMSE was more than 1.2 m, although the WSE variability was comparable to that for other models. For models CRO and BAT, the medians of the differences, as well as their variability, decreased with the increasing flow rate. The BAT model provided the smallest median values and the smallest variance for all rated flows. Table 8 presents RMSE and MAE values for WSE comparison.

Table 8. RMSE and MAE values for WSE comparison.

	Model	$Q_{a.a.}$	Q_1	Q_{10}	Q_{100}
	ALS	1.24	0.87	0.80	0.85
RMSE (m)	CRO	0.15	0.14	0.10	0.07
	BAT	0.13	0.06	0.04	0.03
	ALS	1.24	0.87	0.79	0.84
MAE (m)	CRO	0.13	0.12	0.09	0.05
	BAT	0.10	0.04	0.03	0.02

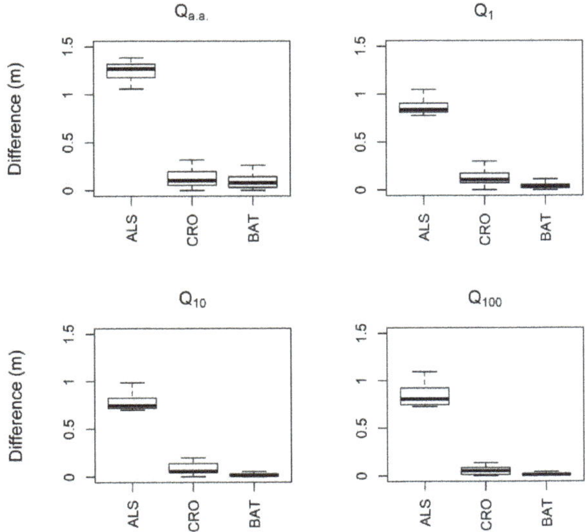

Figure 9. Variance of raster cell differences in water surface elevation between the compared digital elevation models (ALS, CRO, BAT) and the reference model (ADP).

3.6. Inundation Areas Comparison

Table 9 presents a comparison of inundation areas. The ALS model in all cases overestimated inundation vis-à-vis the reference ADP model. This overestimation was >50% in the case of $Q_{a.a.}$. These differences were diminishing with increasing flow, but nevertheless were >20% for the Q_{100} flow rate. Similar results had been presented in the works of Bures et al. [18] and Roub et al. [42]. In both those cases, software-modified DEMs (CRO, BAT) provided better results than the ALS model. The CRO model underestimated the ADP model values by as much as 4.5%. The BAT model underestimated them by as much as 3.8%. The fact that CRO and BAT consistently underestimate the results may reflect their similar model settings. The flow area for model cross-section transformation was set the same for the two models.

Table 9. Inundation area (IA) differences for compared DEMs.

	Model	$Q_{a.a.}$	Q_1	Q_{10}	Q_{100}
Inundation	ADP	0.0540	0.0911	0.1288	0.1608
Area	ALS	0.0830	0.1182	0.1415	0.1953
(km^2)	CRO	0.0536	0.0870	0.1277	0.1587
	BAT	0.0519	0.0898	0.1283	0.1598
Difference	ADP	-	-	-	-
in Area	ALS	53.83	29.73	9.84	21.49
(%)	CRO	0.6	4.5	0.87	1.27
	BAT	3.76	1.44	0.35	0.62

Overall, the ALS model showed the poorest results. That is because the technology used is unable to capture terrain under water. It was a LiDAR technology that, in our case, used an infrared laser beam [40]. Similar poor results could be expected from other remote-sensing methods [8–11]. The error produced by these models depends on the size of the neglected flow area of the river channel [18,42]. Although differential absorption LiDAR (DiAL) technologies may constitute exceptions to this general case [14,15], even using DiAL technology does not ensure any increase in accuracy. That is particularly the case when a high-turbidity river channel is measured [16–18,20].

The ALS model used in this study is a basic DEM, representative of the group of other remote-sensing models.

In the BAT model, the longitudinal water surface slope and water surface elevations were determined for all constructed cross-sections. Both of these parameters are derived from DEM inputs. The uncertainty in determining these parameters depends upon the accuracy of the data from which they are derived. It is possible, however, to assume that the development of remote-sensing techniques will reduce the uncertainty in determining water surface slope and elevations. Uncertainty in determining roughness parameters is the same for both models (CRO, BAT) and depends upon user experience. Further research in the field of model parameter estimation may bring improvement in this area. More appropriate regression relationships can be found, as can other explanatory variables. These variables might, for example, be the curvature of the flow centerline [21,38], mutual tilt of consecutive cross-sections [23], or ratio of the bank line distances.

In comparing the quality of the DEMs, we can see that the BAT model produced better results than the CRO model. The extent of the differences between the CRO and BAT models will probably depend upon the nature of the channel described. With respect to its adaptability, it can be assumed that the BAT model will produce better results in the case of natural channels. If the river channel has been technically adjusted (into trapezoidal shape), it can be assumed that the BAT model will produce similar results to a CRO model.

In the present bathymetric model, the interpolation mechanism introduced by Caviedes–Voullième [23] was used. Several interpolation methods can alternatively be used for this step [24,37,38]. Which of these methods will yield more reliable results remains an open question.

In the presented study, the importance of river bathymetry was evaluated mainly from the perspective of flood modeling. However, the proposed bathymetric model can find its place in other scientific disciplines, such as understanding the morphological changes of the river segment or understanding the hydrodynamic behavior of the river.

4. Conclusions

In this study, the new theoretical bathymetric model Bathy-supp was presented. The model is based on the analytical curves, which schematizes the river cross-sections. The shape of the analytical curves is driven by the floodplain topography. In the case study, the practical usability of the model was evaluated.

The vertical differences in the model's ability to represent the cross-sections were 0.16 m for RMSE and 0.11 m for MAE. For estimation of the model parameters, the three regression methods were used. The best parameter estimates were provided by the RF method. However, the regression relationships between model parameters and topographical characteristics need to be further investigated.

The study's results also show that, in the DEM quality assessment, the BAT model provides the best results in the river channel and thalweg representation. When assessing the WSE and IA, the BAT model provided better results than the CRO model, especially for high design flows rates.

Both DEMs created by merging ALS data with mathematical bathymetric models (CRO and BAT) provided significantly better results. The extent of this improvement was directly proportional to the size of the flow area neglected by using the ALS data itself.

The newly proposed Bathy-supp model was also compared with the state-of-the-art CroSolver model. The results showed that the Bathy-supp model describes the river bathymetry more accurately than the CroSolver model. This is because the analytical curves used in the Bathy-supp tool are more able to simulate the individual cross-sections of the channel than the trapezoidal shapes used by the CroSolver tool. Another advantage of the Bathy-supp model is in determining the flow area for each single cross-section from which the model is composed. This allows the model to take into account local narrows or expansions of the river channel.

The merging of bathymetric models with other types of remote-sensing data can significantly improve the DEMs, thereby enhancing their practical usefulness. Hence, the methods for mathematical schematizing of riverbeds should be further studied and improved.

The Bathy-supp model as an appropriate mechanism to improve DEM quality when other bathymetry measurements that are not available can be used. However, it will never be able to fully replace large bathymetric measurements.

Author Contributions: Conceptualization, L.B. and P.M.; methodology, L.B., P.M., and R.R.; validation, L.B., P.S., and R.R.; investigation, P.S.; data curation, L.B.; writing—original draft preparation, L.B.; writing—review and editing, P.S., S.M.; visualization, L.B. and S.M.; supervision, P.M. and R.R.; funding acquisition, R.R.

Funding: This research was supported from the European Union by the Operational Programme Prague–Growth Pole of the Czech Republic, project No. CZ.07.1.02/0.0/0.0/17_049/0000842, Tools for effective and safe management of rainwater in Prague city–RainPRAGUE and the Internal Grant Agency (IGA) of the Faculty of Environmental Sciences (CULS) (IGA/20164233).

Conflicts of Interest: The authors declare no conflict of interest.

References

1. Horritt, M.S.; Bates, P.D. Effects of spatial resolution on a raster based model of flood flow. *J. Hydrol.* **2001**, *253*, 239–249. [CrossRef]
2. Horritt, M.S.; Bates, P.D.; Mattinson, M.J. Effects of mesh resolution and topographic representation in 2D finite volume models of shallow water fluvial flow. *J. Hydrol.* **2006**, *329*, 306–314. [CrossRef]
3. Merwade, V.; Cook, A.; Coonrod, J. GIS techniques for creating river terrain models for hydrodynamic modeling and flood inundation mapping. *Environ. Model. Softw.* **2008**, *23*, 1300–1311. [CrossRef]
4. Baltsavias, E.P. Airborne laser scanning: Basic relations and formulas. *ISPRS J. Photogramm. Remote Sens.* **1999**, *54*, 199–214. [CrossRef]
5. Lyzenga, D.R.; Malinas, N.R.; Tanis, F.J. Multispectral bathymetry using a simple physically based algorithm. *IEEE Trans. Geosci. Remote Sens.* **2006**, *44*, 2251–2259. [CrossRef]
6. Gao, J. Bathymetric mapping by means of remote sensing: Methods, accuracy and limitations. *Prog. Phys. Geogr.* **2009**, *33*, 103–116. [CrossRef]
7. Bates, P.D.; De Roo, A.P.J. A simple raster-based model for flood inundation simulation. *J. Hydrol.* **2000**, *236*, 54–77. [CrossRef]
8. Schumann, G.; Matgen, P.; Cutler, M.E.J.; Black, A.; Hoffmann, L.; Pfister, L. Comparison of remotely sensed water stages from LiDAR, topographic contours and SRTM. *ISPRS J. Photogramm. Remote Sens.* **2008**, *63*, 283–296. [CrossRef]
9. Ali, A.M.; Solomatine, D.P.; Di Baldassarre, G. Assessing the impact of different sources of topographic data on 1-D hydraulic modelling of floods. *Hydrol. Earth Syst. Sci.* **2015**, *19*, 631–643. [CrossRef]
10. Domeneghetti, A. On the use of SRTM and altimetry data for flood modeling in data-sparse regions. *Water Resour. Res.* **2016**, *52*, 2901–2918. [CrossRef]
11. Yan, K.; Di Baldassarre, G.; Solomatine, D.P.; Schumann, G.J.-P. A review of low-cost space-borne data for flood modelling: Topography, flood extent and water level. *Hydrol. Process.* **2015**, *29*, 3368–3387. [CrossRef]
12. Casas, A.; Benito, G.; Thorndycraft, V.R.; Rico, M. The topographic data source of digital terrain models as a key element in the accuracy of hydraulic flood modelling. *Earth. Surf. Proc. Landf.* **2006**, *31*, 444–456. [CrossRef]
13. Irish, J.L.; Lillycrop, W.J. Scanning laser mapping of the coastal zone: The SHOALS system. *ISPRS J. Photogramm. Remote Sens.* **1999**, *54*, 123–129. [CrossRef]
14. Hilldale, R.C.; Raff, D. Assessing the ability of airborne LiDAR to map river bathymetry. *Earth. Surf. Proc. Landf.* **2008**, *33*, 773–783. [CrossRef]
15. Mandlburger, G.; Hauer, C.; Wieser, M.; Pfeifer, N. Topo-Bathymetric LiDAR for Monitoring River Morphodynamics and Instream Habitats-A Case Study at the Pielach River. *Remote Sens.* **2015**, *7*, 6160–6195. [CrossRef]
16. Skinner, K.D. *Evaluation of Lidar-Acquired Bathymetric and Topographic Data Accuracy in Various Hydrogeomorphic Settings in the Lower Boise River, Southwestern Idaho*; US Geological Survey: Reston, VA, USA, 2007.

17. Bailly, J.S.; Le Coarer, Y.; Languille, P.; Stigermark, C.J.; Allouis, T. Geostatistical estimations of bathymetric LiDAR errors on rivers. *Earth. Surf. Proc. Landf.* **2010**, *35*, 1199–1210. [CrossRef]

18. Bures, L.; Roub, R.; Sychova, P.; Gdulova, K.; Doubalova, J. Comparison of bathymetric data sources used in hydraulic modelling of floods. *J. Flood Risk Manag.* **2018**, e12495. [CrossRef]

19. Allouis, T.; Bailly, J.S.; Pastol, Y.; Le Roux, C. Comparison of LiDAR waveform processing methods for very shallow water bathymetry using Ra-man, near-infrared and green signals. *Earth. Surf. Proc. Landf.* **2010**, *35*, 640–650. [CrossRef]

20. Laks, I.; Sojka, M.; Walczak, Z.; Wrozynski, R. Possibilities of Using Low Quality Digital Elevation Models of Floodplains in Hydraulic Numerical Models. *Water* **2017**, *9*, 283. [CrossRef]

21. Merwade, V.M.; Maidment, D.R.; Goff, J.A. Anisotropic considerations while interpolating river channel bathymetry. *J. Hydrol.* **2006**, *331*, 731–741. [CrossRef]

22. Jha, S.K.; Mariethoz, G.; Kelly, B.F. Bathymetry fusion using multiple-point geostatistics: Novelty and challenges in representing non-stationary bedforms. *Environ. Model. Softw.* **2013**, *50*, 66–76. [CrossRef]

23. Caviedes-Voullième, D.; Morales-Hernández, M.; López-Marijuan, I.; García-Navarro, P. Reconstruction of 2D river beds by appropriate interpolation of 1D cross-sectional information for flood simulation. *Environ. Model. Softw.* **2014**, *61*, 206–228. [CrossRef]

24. Schäppi, B.; Perona, P.; Schneider, P.; Burlando, P. Integrating river cross-section measurements with digital terrain models for improved flow modelling applications. *Comput. Geosci.* **2010**, *36*, 707–716. [CrossRef]

25. Deutsch, C.V.; Wang, L. Hierarchical object-based stochastic modeling of fluvial reservoirs. *Math. Geol.* **1996**, *28*, 857–880. [CrossRef]

26. James, L.A. Polynomial and power functions for glacial valley cross-section morphology. *Earth. Surf. Proc. Landf.* **1996**, *21*, 413–432. [CrossRef]

27. Moramarco, T.; Corato, G.; Melone, F.; Singh, V.P. An entropy-based method for determining the flow depth distribution in natural channels. *J. Hydrol.* **2013**, *497*, 176–188. [CrossRef]

28. Roub, R.; Hejduk, T.; Novák, P. Automating the creation of channel cross-section data from aerial laser scanning and hydrological surveying for modeling flood events. *J. Hydrol. Hydromech.* **2012**, *60*, 227–241. [CrossRef]

29. Zevenbergen, L.W.; Thorne, C.R. Quantitative analysis of land surface topography. *Earth. Surf. Proc. Landf.* **1987**, *12*, 47–56. [CrossRef]

30. Gareth, J. *An Introduction to Statistical Learning: With Applications in R*; Springer: New York, NY, USA, 2010.

31. Pinheiro, J.; Bates, D. *Mixed-Effects Models in S and S-PLUS*; Springer: New York, NY, USA, 2006.

32. Breiman, L. Random forests. *Mach. Learn.* **2010**, *45*, 5–32. [CrossRef]

33. Pinheiro, J.; Bates, D.; DebRoy, S.; Sarkar, D.; Package'nlme'. Linear and Nonlinear Mixed Effects Models. Available online: http://www.cran.r-project.org/web/packages/nlme/nlme.pdf (accessed on 18 June 2018).

34. Liaw, A.; Wiener, M.; Package'randomForest'. Breiman and Cutlers Random Forests for Classification and Regression; Version 4.6–12. Available online: http://www.cran.r-project.org/web/packages/randomForest/randomForest.pdf (accessed on 18 June 2018).

35. Kumaraswamy, P. A generalized probability density function for double-bounded random processes. *J. Hydrol.* **1980**, *46*, 79–88. [CrossRef]

36. Merwade, V. *Creating River Bathymetry Mesh from Cross-Sections*; School of Civil Engineering, Purdue University: West Lafayette, Indiana, 2017. Available online: https://web.ics.purdue.edu/~vmerwade/research.html#river (accessed on 10 February 2019).

37. Dysarz, T. Development of RiverBox—An ArcGIS Toolbox for River Bathymetry Reconstruction. *Water* **2018**, *10*, 1266. [CrossRef]

38. Storn, R.; Price, K. Differential evolution—A simple and eficient heuristic for global optimization over continuous spaces. *J. Global. Optim.* **1997**, *11*, 341–359. [CrossRef]

39. CHMI. Evidence Card for Profile No. 127. Available online: http://hydro.chmi.cz/hpps/popup_hpps_prfdyn.php?seq=307049 (accessed on 8 February 2019).

40. Brazdil, K.; Belka, L.; Dusanek, P.; Fiala, R.; Gamrat, J.; Kafka, O. *The Technical Report to the Digital Elevation Model 5th Generation DMR 5G*; VGHMÚř Dobruška: ZÚ Pardubice, Czech Republilc, 2012.

41. Nihei, Y.; Kimizu, A. A new monitoring system for river discharge with horizontal acoustic Doppler current profiler measurements and river flow simulation. *Water Resour. Res.* **2008**, *44*, 1–15. [CrossRef]

42. Roub, R.; Kurkova, M.; Hejduk, T.; Novak, P.; Bures, L. Comparing a hydrodynamic model from fifth generation dtm data and a model from data modified by means of crosolver tool. *AUC Geogr.* **2016**, *51*, 29–39. [CrossRef]

43. Brunner, G.W. *HEC-RAS River Analysis System 2D Modeling User's Manual*; US Army Corps of Engineers—Hydrologic Engineering Center: Davis, CA, USA, 2016. Available online: https://www.hec.usace.army.mil/software/hec-ras/downloads.aspx (accessed on 10 February 2019).

44. Quirogaa, V.M.; Kurea, S.; Udoa, K.; Manoa, A. Application of 2D numerical simulation for the analysis of the February 2014 Bolivian Amazonia flood: Application of the new HEC-RAS version 5. *Ribagua* **2016**, *3*, 25–33. [CrossRef]

45. Maskong, H.; Jothityangkoon, C.; Hirunteeyakul, C. Flood hazard mapping using on-site surveyed flood map, HEC-RAS V. 5 and GIS tool: A case study of Nakhon Ratchasima municipality, Thailand. *Int. J. Geomate* **2019**, *16*, 1–8. [CrossRef]

46. Cook, A.; Merwade, V. Effect of topographic data, geometric configuration and modeling approach on flood inundation mapping. *J. Hydrol.* **2009**, *377*, 131–142. [CrossRef]

47. Di Baldassarre, G.; Schumann, G.; Bates, P.D.; Freer, J.E.; Beven, K.J. Flood-plain mapping: A critical discussion of deterministic and probabilistic approaches. *Hydrol. Sci.* **2010**, *55*, 364–376. [CrossRef]

48. Cavalli, M.; Tarolli, P. Application of LiDAR technology for rivers analysis. *Ital. J. Eng. Geol. Environ.* **2011**, *11*, 33–44. [CrossRef]

Article

Application of Python Scripting Techniques for Control and Automation of HEC-RAS Simulations

Tomasz Dysarz

Department of Hydraulic and Sanitary Engineering, Poznan University of Life Sciences, ul. Wojska Polskiego 28, 60-637 Poznan, Poland; dysarz@up.poznan.pl or tdysarz@gmail.com; Tel.: +48-061-846-6586; Fax: +48-061-848-7726

Received: 1 August 2018; Accepted: 30 September 2018; Published: 2 October 2018

Abstract: The purpose of the paper was to present selected techniques for the control of river flow and sediment transport computations with the programming language Python. The base software for modeling of river processes was the well-known and widely used HEC-RAS. The concepts were tested on two models created for a single reach of the Warta river located in the central part of Poland. The ideas described were illustrated with three examples. The first was a basic simulation of a steady flow run from the Python script. The second example presented automatic calibration of model roughness coefficients with Nelder-Mead simplex from the SciPy module. In the third example, the sediment transport was controlled by Python script. Sediment samples were accessed and changed in the sediment data file stored in XML format. The results of the sediment simulation were read from HDF5 files. The presented techniques showed good effectiveness of this approach. The paper compared the developed techniques with other, earlier approaches to control of HEC-RAS computations. Possible further developments were also discussed.

Keywords: river flow modeling; sediment transport simulation; automation of flow modeling; HEC-RAS controller; python scripting

1. Introduction

River flow modeling is a very popular approach in science and education, as well as in small, medium-size, and large technical projects in the areas of water management [1,2], river regulation [3,4], sediment transport [5], flood protection [6], and many others [7,8]. There are several professionally prepared and widely used commercial and non-commercial models for river flow, e.g., HEC-RAS [9], MIKE 11 [10], Delft Sobek [11], BASEMENT by ETH [12], and SRH1D [13]. The first of the mentioned models, HEC-RAS, is very specific. The acronym HEC means Hydrologic Engineering Center, and it is the name of the software developer. The second element, RAS, means River Analysis System. It defines the software application area which is focused on modeling of flow and transport processes in rivers, floodplains, and reservoirs. All of the modeled elements are solved with highly accurate numerical methods [9]. Additionally, HEC-RAS is freeware with a professionally developed graphical user interface. It makes application of the software easier, and it is considered to be the reason that HEC-RAS is so popular. Today a strong focus on modeling of uncertainty related to river processes is observed. It is a particularly crucial problem in the areas of model calibration [14,15], sediment transport [16,17], and flood hazard mapping [18,19]. The effective management of such computational tasks is not possible without automation of simulation and integration with other tools, e.g., GIS software. The examples analyzed in the paper present crucial automation techniques for the mentioned applications.

The main purpose of this paper is to present selected techniques for controlling HEC-RAS computations with the programming language Python. Control of river flow simulations with external scripts opens several new opportunities in the fields of flood hazard assessment, water management, and river hydraulics. It also enables application of more sophisticated methods in these fields, such as sensitivity analysis, automatic model calibration, and probabilistic flood mapping. The control of the HEC-RAS computations is possible due to the fact that this package is installed with the Component Object Module (COM). This COM library of methods is called HECRASController. The COM library enables outer access to the computational elements of HEC-RAS. The discussion presented here is focused on three specific examples. The ideas presented are not limited to the basic running of flow model simulation. Some concepts are aimed at integration of HEC-RAS with specific tools available in Python modules, e.g., SciPy, while others extend the capabilities of HECRASController and open access to more sophisticated data formats, e.g., XML (eXtensible Markup Language) and HDF5 (Hierarchical Data Format 5).

The HECRASController is a collection of Visual Basic subroutines and functions within the HEC-RAS package. It may be effectively used with VBA (Visual Basic for Applications) in Excel [20,21]. The main capabilities of this collection include running HEC-RAS and particular HEC-RAS editors, running computations plans, reading results of flow simulation, etc. One very interesting feature is the possibility of manipulation of roughness coefficients. It is a very important problem studied by many researchers, e.g., Yang et al. [22] and Lacasta et al. [23]. This capability is implemented in AHYDRA software [24]. The AHYDRA supports calibration of roughness coefficients and boundary conditions for steady flow conditions. The software cooperates with the initially configured HEC-RAS model, enabling automatic change of settings.

The capabilities of HECRASController are still limited. Even the main author promoting the use of this programming tool, Goodell [20,21], advises manipulation of the HEC-RAS data files in ASCII format. Another problem is the use of Visual Basic/VBA. Although it is a very handy programming language, its use is rather limited and still narrowing, e.g., VBA was replaced in ArcGIS by Python and its extension ArcPy. Hence, other languages are also implemented to handle HECRASController, e.g., MATLAB [25].

Python [26] is one of the most popular programming languages used today worldwide. It is easily applied in practical cases, as well as in highly scientific problems [27,28]. Its popularity follows from two factors: (1) relative simplicity and (2) a broad range of application areas. Considering the specific problem considered here, Python has several advantages and unique features. They include very easy access to text files in standard form, e.g., [26]; several specific libraries included, e.g., [29–33]; opportunities to handle the components of the Windows operation system, e.g., [34,35]; access to sophisticated data formats, e.g., [36,37]; and linking with a number of Windows application, e.g., [38,39].

It is no surprise that the first approach to control of HEC-RAS with Python was developed. The PyRAS module prepared by Peña-Castellanos [40] is one such approach. Unfortunately, PyRAS only transforms objects and functions of HECRASController available in VBA to Python. Hence, the limitations of HECRASController still apply. Another disadvantage is that the module was prepared for older versions of HEC-RAS, namely 4.1 and 5.0.0. It has not been developed further to date. Despite these facts, the codes of PyRAS could be an inspiration for independent development. Another open-source application linking Python and HEC-RAS is the example included in the package PyFloods [41]. Unfortunately, it is the only example for the previous version of HEC-RAS with some important elements hard-coded.

No general approach to linking Python and HEC-RAS has been proposed as yet, and it seems to us that it should be done on the basis of available Python modules for management of COM libraries. Another requirement is to develop the potential of HECRASController with additional Python modules, e.g., optimization of model parameters, access to specific data in XML files, and reading and handling of results in HDF5 files. These problems are addressed in the present paper.

The next section, Methods, includes general description of all important elements in the presented manuscript. Theoretical aspects of open channel flow, as well as Python scripting issues related to the applied codes, are discussed. In the section Results and Discussion, examples illustrating the invented concepts are described in detail. At the beginning of this section, the HEC-RAS models used are presented. In the final section, the summary, concluding remarks, and further developments are discussed.

2. Methods

2.1. HEC-RAS

HEC-RAS is a well-known hydrodynamic model for rivers and water reservoirs. The concepts applied in the package are well described by Brunner [9]. The HEC-RAS is applied for simulation of flow and transport processes in river networks, including floodplains and reservoirs. The modeled flow conditions include steady and unsteady longitudinal flow. The first is based on the simple equation of energy balance. The form of this equation implemented in HEC-RAS is shown below:

$$z_1 + h_1 + \alpha_1 \frac{u_1^2}{2g} = z_2 + h_2 + \alpha_2 \frac{u_2^2}{2g} + h_e, \tag{1}$$

where z is bottom elevation, h is the depth, and u is mean velocity in the channel cross-section. α is called the St. Venant coefficient and plays the role of correction factor, including the effect of velocity profile non-uniformity. g is well known acceleration of gravity. The Equation (1) is written for gradually-varying flow, when the assumption of hydrostatic pressure distribution may be suitable. The subscripts 1 and 2 denote two different cross-sections in the same channel reach. The basic assumption is that the cross-section number 1 is located upstream of the cross-section number 2. The first three terms of both sides represent the potential energy of the stream, work of pressure forces, and kinematic energy of the stream. The last element of the right side, h_e, describes friction losses due to bed and banks influence on flowing water. It also includes effects of channel contraction and extension. If the depth is known in one cross-section, on this basis, the depth may be also determined in the second cross-section. To calculate the distribution of the depth along the channel, its value must be known in one cross-section a priori. In practical cases, this cross-section is the inlet or outlet boundary of the channel reach. Hence, the condition is frequently called "boundary condition", but it is rather hydraulic jargon than strict mathematical terms. The discharge Q is a very important parameter of the Equation (1). The kinematic energy terms, as well as friction losses, depend on the magnitude of the flow. The influence of floodplains is included in the calculation of the St. Venant coefficients and calculation of weighted distance between cross-sections. The last element is important for the determination of the total losses h_e. More detailed description of this equation and its implementation may be found in Reference [9].

For description of the second model, the unsteady flow, the numerical solution of the St. Venant system of equations for 1D flow is implemented. This system consists of two partial differential equations shown below.

$$\frac{\partial A}{\partial t} + \frac{\partial (\phi Q)}{\partial x_c} + \frac{\partial [(1-\phi)Q]}{\partial x_f} = 0 \tag{2}$$

$$\frac{\partial Q}{\partial t} + \frac{\partial}{\partial x_c}\left(\phi^2 \frac{Q^2}{A_c}\right) + \frac{\partial}{\partial x_f}\left[(1-\phi)^2 \frac{Q^2}{A_f}\right] + gA_c\left(\frac{\partial H}{\partial x_c} + S_{fc}\right) + gA_f\left(\frac{\partial H}{\partial x_f} + S_{ff}\right) = 0 \tag{3}$$

The Equation (2) describes mass balance in the open channel flow. The second equation represents the momentum balance. It is written for gradually-varying flow conditions in the channel. There are two independent variables: t—time and x—distance. The distance is measured not only along the channel, but also along the floodplains. Hence, there are denotations x_c and x_f for channel and floodplain flow paths, respectively. It is important that the pair of distance (x_c, x_f) is unique for each cross-section. Q is still the total discharge in the cross-section and g is acceleration of gravity. H is the water surface elevation, which is the sum of the bottom elevation z and the depth h shown in the previously discussed Equation (1). It is assumed that the transversal slope of the water surface is zero in the single cross-section. Hence, H is constant in the cross-section, but varies along the channel. Denotation A without subscript is the total cross-section area in a single cross-section. However, A_c and A_f are the area of channel and floodplain parts of the cross-section, respectively. The hydraulic slopes describing the friction losses along the channel and along the floodplains are denoted as S_{fc} and S_{ff}. The HEC-RAS specific element is coefficient ϕ. It describes the distribution of the discharge between the parts of the cross-section. Hence, ϕQ is the flow along the channel and $(1 - \phi)Q$ is the flow along the floodplains. The coefficient ϕ depends on the hydraulic parameter called conveyance, which is the function of the water surface elevation H and other elements of the cross-section, such as shape and roughness. In the Equation (2), the first term represents the local storage of water, whilst two other terms describe the net inflow of water along the channel and floodplains. In Equation (3), the first three terms represent the inertia forces, the terms including gradient of H describe the gravity and pressure forces, and the terms with hydraulic slopes S_{fc} and S_{ff} represent the friction forces. The solution of the system (2)–(3) are two functions of time t and distance pair (x_c, x_f). These are discharge Q and water surface elevation H. To solve the system (2)–(3), the initial and boundary conditions have to be specified. The basic forms of the initial condition are known distributions of Q and H along the modeled reach. In each inlet and outlet cross-section, there should be one boundary condition imposed in gradually-varying flow. These conditions may have a basic form of discharge or water stage hydrographs. They can also be imposed as other more complex relationships of discharge and water stage, e.g., normal depth, rating curve, etc. The Preissmann scheme is used for the approximation. More detailed description of the St. Venant equations used in HEC-RAS and their implementation may be found in Reference [9].

The number of hydro-structures, such as bridges, culverts, etc., is also available for incorporation with the prepared water system. Both these mathematical models of open channel flow are well described in many books [42–45], as well as software manuals mentioned above [9–13]. In HEC-RAS, there is also a module enabling so-called quasi-unsteady flow simulation, i.e., simplified flow simulation in unsteady conditions on the basis of the energy equation. In the latest versions of the package, a 2D flow module is also available. The HEC-RAS modules for simulation of transport processes include different transport of solutes, heat, and sediments with deposition and erosion. Heat and solute transport is based on numerical solution of convection–dispersion equations with source terms. Models of this kind are described in some books on mathematical modeling of river processes, e.g., References [9,46,47]. The QUICKEST scheme with the ULTIMATE limiter is applied as an approximation method [48]. The sediment transport is modeled with the standard Exner equation, enabling simulation of different fractions of sediments. The Exner equation is shown below:

$$(1 - p)B\frac{\partial z}{\partial t} + \frac{\partial Q_s}{\partial x} = 0 \tag{4}$$

where x is the distance along the channel and t is time. z represents bottom elevation, B is the width of this bottom part, where the sediments are deposited or removed from. p is porosity of the bed layer. Q_s is volumetric intensity of sediment transport. This element is described by subjectively chosen empirical formulae, e.g., Meyer-Peter and Müller (MPM). It depends on the hydraulic parameters and the sediment characteristics. The first term represents the local change of the sediment deposited. The second term is the net inflow of sediments. The solution of Equation (4) is the change of bottom elevation z in time t, and along the channel x. The initial and one boundary condition have to be imposed to solve the problem. The initial condition is simple initial bottom elevation profile along the modeled channel. The boundary condition is imposed in the inlet cross-section. It may have the form of bottom elevation changes, sediment inflow intensity varying in time, or equilibrium sediment load in the stream. In HEC-RAS, the applied solution method is the basic upwind scheme [9,45]. More details may be found in Reference [9].

The HEC-RAS package also includes several useful tools for data preparation and results processing. These tools include the module for GIS data processing called RAS Mapper [49].

2.2. HECRASController

HECRASController is a part of the HEC-RAS application programming interface (API). It is a collection of programming tools: Classes, functions and subroutines. Access to the HEC-RAS elements is possible because this package is compiled as the Component Object Module (COM). Any program able to read COM DLL (Dynamic-Link Library) may be used to control HEC-RAS computations. The most convenient approach is to use Visual Basic in any form, i.e., as a separate suite for developers or linked with another application, e.g., MS Excel.

The most detailed description of the above-mentioned collection of programming tools was presented by Goodell [20,21]. Appendix A of his book [20] is a very good source of information about syntax, and usage of HECRASController functions and subroutines. The examples presented there, illustrate the great potential of these concepts. VBA in MS Excel is very powerful in computing, as well as in data analysis [50]. Linking HECRASController with VBA in Excel opens new opportunities for handling simulation data and results, as presented by Goodell [20,21]. The controller makes it possible to open HEC-RAS projects, run simulations, read results, and store them in other specific formats. It is also possible to open HEC-RAS editing tools for manual configuration of input data. The controller also enables direct manipulation of project data, but its capabilities in this area are rather limited. There are three procedures for changing roughness coefficients, and one for setting the area of the element called storage area.

Although this concept is not new and seems to be very useful, there are still areas of HEC-RAS computations in which controller functions are not able to help. The basic HEC-RAS data, such as geometry and flow boundary conditions, are stored in ASCII files. They may be accessed by manually written code. Goodell [20] has advised such an approach. However, some HEC-RAS data are stored in formats other than pure text, e.g., sediment data in XML format. Access to such data requires application of additional modules. The results of more complex simulations are also beyond HECRASController access, e.g., 1D sediment and 2D flow simulation results are stored only in HDF files.

In addition, the opportunities afforded by the controller depend on the application areas of VBA. This powerful language is widely used. However, there are also very important areas where VBA has been withdrawn, e.g., scripting in ArcGIS. There are also areas in which VBA has never been present, e.g., scripting in QGIS. Relying on VBA for automation of HEC-RAS simulation could limit possible applications.

2.3. Python Scripting and Applied Modules

Python is one of the most popular programming languages today, and it is still under development. In this paper, version 2.7.12 was used. Its usefulness and popularity in many areas have been reported in References [51–53]. This scripting language is relatively simple for the beginner, but it is also very powerful if applied by an experienced coder. A description of this language may be found in many books or on internet websites, e.g., Downey et al. [54], Python Software Foundation [26]. Considering the problem considered here, Python has several specific advantages:

(1) The basic language is extended by additional modules for numerical simulation and general scientific analysis, e.g., NumPy and SciPy [29–31].
(2) The usefulness of Python in numerical simulations may be extended by the user with modules enabling transformation of Fortran or C routines to Python code, e.g., F2Py [32] and ctypes [33] modules.
(3) It is possible to access COM libraries using additional modules, e.g., Pywin32 [34,35].
(4) Python enables access to more sophisticated data structures, e.g., XML [36] and HDF5 [37].

(5) There are Python modules integrating this language with geoprocessing software, such as ArcGIS [38] and QGIS [39].

VBA is a compiled language, which makes it faster in comparison with any scripting language in stand-alone applications. However, in the analyzed cases, the programming language is used to run an external application, HEC-RAS. Hence, the efficiency of the external application determines the efficiency of the whole concept, which reduces this advantage of VBA. The above list of Python features is the list of undoubted advantages of this scripting language over standard VBA. In some cases, the application of Python is as easy as VBA, e.g., access to COM libraries or XML data. Other elements, such as libraries for numerical computations, make Python implementation more effective. There are also features that are unique to Python, e.g., access to HDF5 data, and linking with Fortran and C. The undeniable advantage of Python is also clearly visible in the area of geoprocessing. This area is important for any hydrological or hydraulic modeling of flow and transport processes in natural water systems. As mentioned above, the former applications of VBA in older versions of ArcGIS are now replaced by modern scripting languages, such as Python [38], JavaScript [55], and R [56]. Python is the primary scripting language used in commercial Esri software, such as ArcGIS 10.x and ArcGIS Pro. This is also the only scripting language implemented in QGIS, which is the open source and cross-platform equivalent of Esri's packages.

Several Python modules are used in this research, including: (1) Pywin32, (2) NumPy, (3) SciPy with optimization, (4) ElementTree XML API, and (5) HDF 5 for Python. The first of them, the Pywin32 module, is used to access COM objects and COM servers in Python code [34,35]. Brief descriptions of available Pywin32 packages may be found in the PythonCOM Documentation Index [35]. The part of the module win32com.client is used in this paper. To access COM objects in the presented scripts, the Dispatch function is applied.

NumPy is a well-known module including objects and tools for scientific computing [20,57]. The functions and objects from the NumPy module work faster in numerical applications than standard Python functions and variables. The reason is that the main elements of NumPy were translated from Fortran and implemented in Python. This mechanism of method transfer may be continued by the user with the F2PY package [32] or ctypes module [33]. In the present research, only the array constructors available in NumPy were applied. These are array and empty methods. The string, integer, and float element arrays were used. However, extension of some presented examples, e.g., Examples 2 and 3, enables application of one's own methods and opens the way for more sophisticated techniques.

Another important package used here was SciPy [29,31]. The optimization part of SciPy was implemented in Example 2. The scipy.optimization module contains several optimization methods and root finding algorithms [31]. In this study, the most basic and well-known Nelder-Mead downhill simplex algorithm was used. A detailed description of it was provided by Press et al. [58].

The sediment simulation data in HEC-RAS are stored in XML format. It is a text-based markup language, derived from the Standard Generalized Markup Language. A detailed description of XML format and usage may be found in Reference [59]. To access data stored in an XML file, the ElementTree module is applied as described in References [35,60]. This package includes objects of the Element type. They are container objects designed to store hierarchical data structures, such as XML. The Element type is a cross between a list and a dictionary in Python. The module imported is xml.etree.ElementTree.

The results of the sediment simulation are stored in HDF5 format. Large sets of numerical data are easily accessed in HDF5 through the well-described hierarchy. A description of this format may be found in HDF Group [37]. To access these data the h5py module is applied [61].

2.4. General Remarks on Analyzed Examples

Three examples were tested. The first was a code of a basic simulation composed in a way that enabled comparisons with the examples presented by Goodell [20,21]. The comparison revealed the most important differences between VBA and Python, such as initialization of HECRASController, access to controller methods, handling of data and results.

The second example was the calibration of the steady flow model. Although there is a procedure for the calibration of roughness coefficients in HEC-RAS, it works only with the unsteady flow model. However, calibration of the steady flow model is useful in many cases. Additionally, the approach presented in this example could also be applied in sensitivity analysis of the model performance or probabilistic flood risk mapping. The Nelder-Mead simplex algorithm was applied to search for the proper roughness coefficients. This is routine from the optimization part of the SciPy module, namely scipy.optimize. It works for similar purposes as the AHYDRA application [24], but the optimization algorithm is applied instead of manual alteration of roughness coefficients. The code is prepared in such a way that the applied Nelder-Mead method may be replaced with any optimization algorithm, e.g., the Genetic Algorithm as proposed by Leon and Goodell [25] with MATLAB.

The objective of the third example was to automatically change sediment samples in sediment data and run a sediment simulation. Such an approach has not been found by the author in the literature to date. The sediment sample data describe the percentage of sediment fractions in a sample. In HEC-RAS, they are called gradations. This function plays the role of a model parameter. It is necessary for the calculation of the total sediment transport, as well as transport of particular fractions. However, in practical applications it should be a piece of material sampled from the river bed. Collection of sediment samples is not very common. The measurements are local and extrapolated to a region, e.g., the cross-section. Hence, such measurements are highly uncertain, which is an additional reason for presentation of Example 3. This example is very simple, but may be extended for more sophisticated cases. It requires handling of XML and HDF5 data with special Python modules. Although, the control of sediment simulation is a more complex problem, the example is focused on change of the sediment samples, because these data are accessible in the most difficult way. If the sediment fractions may be changed by the Python script, all other elements of the sediment simulation, e.g., sediment transport formula, parameters of chosen formula, method for calculation fall velocity, etc., may be changed too. Hence, the adjustment of sediment sample in the XML file is a good illustrative example of sediment data management in general.

There are two main application areas for the mechanism presented in Example 3. The first is obviously the uncertainty or sensitivity analysis of sediment transport simulation. Considering the complexity of such computations, the analysis of sensitivity is too difficult to perform in the standard manual way. However, it is necessary if the sediment transport models are going to be used for forecast purposes, which sometimes happens today. The sediment, as well as flow data, are too uncertain to neglect this problem. The second application area is related to historical data and their processing. Sometimes data collected in the past include the full longitudinal profiles, as well as cross-sections, for different moments in time. The present author had at his disposal such data for Ner river collected for 1983, 2003, and 2007 [62]. These data were collected by the company BIPROWODMEL, for the analysis of regulation requirements. They illustrate the effects of the sedimentation process along the analyzed channel. Unfortunately, the company taking measurements did not collect sediment samples at the same time. Hence, the modeling of sediment transport in the Ner river is possible only if the model is calibrated with respect to these data. There is one important aspect which should be carefully considered. The calibration should also consider other factors, such as the sediment transport formula. However, the differences in the role and structure of the factors influencing the sediment transport simulation lead to the design of separate computational processes for each factor. One such process should be searching for an optimal sediment sample or samples, with other factors kept as constants. The third example presents the powerful capabilities of Python in this area.

An example with change of data stored in ASCII files of HEC-RAS is not presented here. Such examples have been discussed by other authors [20,25] with VBA and MATLAB. In the author's personal opinion, the handling of the ASCII files with Python is easier than with the other languages mentioned above. Hence, such an example would not bring anything new. The third and more difficult example, with XML and HDF files, may be treated as a help or guide in dealing with ASCII files. If Python enables reading and writing of information stored in such complex formats as XML and

HDF5, and it is quite easy to access the data in the ASCII files, it means that the presented scripting mechanisms could be applied to manipulate all data and results of any HEC-RAS model. This is also a huge advantage of the presented approach.

2.5. Applied Rules of Coding

There are important differences between VBA and Python, which have to be discussed before the particular examples are presented. To explain them, the well-known examples given by Goodell [20,21] are quoted and compared with the devised Python code.

The first difference is declaration of variables. In Python, the variables are created dynamically during running of the program. Their type is recognized as the values that are assigned to them. The assignment operator may change this type, any time in the script. There are exceptions to these rules, e.g., special variables such as NumPy arrays. Their shape and type have to be declared before the variable is used. In VBA, dynamic creation is also available. However, it is more convenient to switch off this mechanism with the command Option Explicit at the beginning of the module. Then the variables have to be declared statically with the Dim statement.

To use the HECRASController object in the VBA code, it has to be declared as a variable [20,21]. In Python, the object is created in a similar way by calling the Dispatch function from the module win32com.client. Then all the functions and subroutines of the HEC-RAS controller are available in Python. The access is similar to the VBA calling of the HECRASController. The scheme is shown in Scripts 1–4.

In Script 1. HECobject is the HECRASController variable in the project, arguments is the list of input parameters and return_values is the list of the values we would like to get from the subroutine. The HECobject has to initialized with the Dispatch function of the win32com.client module. Although it seems to be quite simple, there is one technical problem. The subprograms in VBA and Python have different structures and different mechanisms of data exchange with the main body of the program. There are two types of arguments used by VBA subroutines. These are the "called-by-value" and "called-by-references" arguments. Exchange of information between the subprogram and the main program based on "by-value" and "by-reference" mechanisms are well known. The arguments of the first type are used only as inputs. The second calling mechanism may work as input and output. When the function construction is used, another method for output is available, i.e., the returned value of the function. Although these mechanisms are very common and are applied in many programming languages, e.g., VBA, Pascal, Fortran, C/C++, etc., the subprograms used in Python are constructed in a different way. In general, the Python subprograms are functions. The parameters set in the function head are only inputs. The output is exchanged with the outer code only by returned values of the function. It is important to indicate that the function may return a list of values and each of them can be of different type.

```python
import win32com.client
HECobject = win32com.client.Dispatch("RAS503.HECRASController")
# ...
return_values = HECobject.method(arguments)
```

Script 1. Code sample 1. Initialization of HEC-RAS object and schematic use of its methods.

To adopt HECRASController methods in Python, a change in the calling is necessary. Imagine such an example: One of the subroutines determines the value of the parameter "called-by-reference". Let this argument be x. In the case of arguments "called-by-reference", the variable storing the output value must be created before the subroutine is run. In Python, the variable storing the None value must be created, which means the variable without a value assigned. Such a statement is equivalent to the variable declaration in VBA and other programming languages. Then the x variable is set as the argument of the subroutine and repeated as the variable in the return list, as shown in Script 2.

```
import win32com.client
HECobject = win32com.client.Dispatch("RAS503.HECRASController")
# ...
x = None
x = HECobject.method(x)
```

Script 2. Code sample 2. Initialization of the HEC-RAS object and use of its methods requiring passing of the argument by reference.

In more complex cases it should be noted that the return list should follow the sequence of the argument list. An example is presented in Script 3.

```
import win32com.client
HECobject = win32com.client.Dispatch("RAS503.HECRASController")
# ...
x, y, z = None, None, None
x, y, z = HECobject.method( x, y, z)
```

Script 3. Code sample 3. Passing the arguments by reference to the method of the HEC-RAS object playing the role of a subroutine.

If the function is run, the first variable in the return list stores the original return value of the function. It is illustrated in Script 4. The variable RV used there is the return value of the function. This approach is used in all three presented examples.

```
import win32com.client
HECobject = win32com.client.Dispatch("RAS503.HECRASController")
# ...
x, y, z = None, None, None
RV, x, y, z = HECobject.method( x, y, z)
```

Script 4. Code sample 4. Passing the arguments by reference to the method of the HEC-RAS object playing the role of a function with one return value.

The main codes presented in Scripts 5–8 consist of the main elements necessary to access capabilities of the HEC-RAS application in Python code. There are also additional codes written in support.py. They are used for support of the main scripts. The support.py file contains the subroutines for data loading from ASCII files, for handling files and directory names, etc. All these additional subroutines are subject to the coder and may be written in several ways. Hence, they are not discussed here. The additional codes are loaded to the particular scripts, if there is such a need.

3. Results and Discussion

3.1. Case Study—Model of a River Reach

Two HEC-RAS models, called model A and model B, were used to test the presented concepts. Both of them represented a short reach of the Warta river in the central part of Poland. The reach is about 10 km long. The average width and depth of the main channel are 40 m and 1.5 m, respectively. There are two bridges along the reach. The estimated mean flow in the analyzed channel was about 47.45 m^3/s. This value was estimated on the basis of the hydrological data collected at the Sieradz gauge station during 1970–2013. The observed flows for the same period vary from the minimum of 15 m^3/s to the maximum of 408 m^3/s [63].

In both models, the number of cross-sections was 35. The average distance between cross-sections was about 300 m. The differences between the model A and model B include the geometry and configuration of flow conditions. Model A was based on the full geometry, including bridges. The flow conditions were steady with basic discharge of 120 m^3/s for the whole reach. The downstream boundary condition was the simple Normal Depth with bottom slope as a parameter. This model was used for basic simulation in Example 1. It was also applied in Example 2, for the test of calibration.

In model B, the geometry was simplified. The bridges were removed, because the model was used for simulation of sediment transport with different sediment samples. Hence, the configuration of flow conditions was based on a quasi-unsteady flow module. The upstream boundary condition was Flow Series with a constant discharge of 269 m^3/s. In the downstream boundary the Normal Depth was applied once again. The slope equals 0.2‰. The sediment transport intensity was calculated on the basis of the Meyer-Peter and Müller (MPM) formula. The simulation horizon was set to 4 months, with a 6-h time step. The time step guarantees necessary numerical stability and accuracy. The simulated period was rather short. In practice, the changes of the river bed due to sediment deposition and erosion are observed in longer periods, e.g., years. Such a configuration is chosen to avoid too long computational time in this illustrative example.

3.2. Example 1—Basic Simulation

The first example presented here was composed in a way similar to the introductory example discussed by Goodell [20,21] for the reach of Beaver Creek. In this paper, model A described above was used. The main scheme of computations is shown in Figure 1. The code of this example is presented as Script 5.

```
1.   import win32com.client
2.   import os, numpy
3.   from support import sc1_ShowNodes
4.
5.   # Initiate the RAS Controller class
6.   hec = win32com.client.Dispatch("RAS503.HECRASController")
7.   hec.ShowRas()                          # show HEC-RAS window
8.   # full filename of the RAS project
9.   RASProject = os.path.join(os.getcwd(),r'test01\test01.prj')
10.  hec.Project_Open(RASProject)           # opening HEC-RAS
11.  # to be populated: number and list of messages, blocking mode
12.  NMsg,TabMsg,block = None,None,True
13.  # computations of the current plan
14.  v1,NMsg,TabMsg,v2 = hec.Compute_CurrentPlan(NMsg,TabMsg,block)
15.  # ID numbers of the river and the reach
16.  RivID,RchID = 1,1
17.  # to be populated: number of nodes, list of RS and node types
18.  NNod,TabRS,TabNTyp = None,None,None
19.  # reading project nodes: cross-sections, bridges, culverts, etc.
20.  v1,v2,NNod,TabRS,TabNTyp =                               \
21.  hec.Geometry_GetNodes(RivID,RchID,NNod,TabRS,TabNTyp)
22.  # ID of output variables: WSE, ave velocity
23.  WSE_id,AvVel_id = 2,23
24.  TabWSE = numpy.empty([NNod],dtype=float)    # NumPy array for WSE
25.  TabVel = numpy.empty([NNod],dtype=float)    # NumPy array for velocities
26.  for i in range(0,NNod):             # reading over nodes
27.      if TabNTyp[i] == "":            # simple cross-section
28.          # reading single water surface elevation
29.          TabWSE[i],v1,v2,v3,v4,v5,v6 =                    \
30.          hec.Output_NodeOutput(RivID,RchID,i+1,0,1,WSE_id)
31.          # reading single velocity
32.          TabVel[i],v1,v2,v3,v4,v5,v6 =                    \
33.          hec.Output_NodeOutput(RivID,RchID,i+1,0,1,AvVel_id)
34.  hec.QuitRas()            # close HEC-RAS
35.  del hec                  # delete HEC-RAS controller
36.
37.  sc1_ShowNodes( NNod, TabRS, TabNTyp, TabWSE, TabVel )
```

Script 5. Example 1. Calculation of a single water profile.

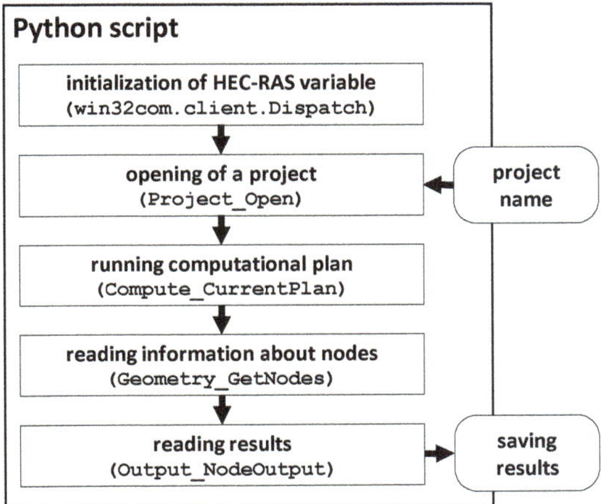

Figure 1. Main elements of computations in Example 1.

As can be seen in the first block of the diagram in Figure 1, the computations have to start with initialization of the variable, which is the HECRASController object. In the brackets, the method used for this purpose is indicated. In the second step, the HEC-RAS project is open. The name of the project is the necessary input. Then the computations are run and the results are processed in two steps. The first is reading information about computational nodes. It is done with one of the "geometric" routines of HECRASController. The second step is reading the output of computations. Then the results may be saved, displayed, or applied in another simulation.

The program presented in Script 5 includes four main steps reflecting the sequence presented in Figure 1. These are: (a) Opening HEC-RAS and loading the project (lines 9–10), (b) running the simulation (lines 12–14), (c) reading the output variables (lines 16–35), and (d) processing the output variables (line 37), e.g., displaying them on the screen. At the beginning, the necessary modules and subroutines are loaded. They are win32com.client for handling COM libraries [34,35], os for management of files and directories and numpy for numerical data. Additionally, the subroutine sc1_ShowNodes is loaded from the script support.py. It is used for displaying results.

The first step of the program is initialization of the HECRASController variable hec (line 6). The Dispatch function of the win32com.client module is applied. The HEC-RAS version used is 5.0.3. When the hec variable is created, the ShowRas and Project_Open functions may be used to open the HEC-RAS windows and load the project. Detailed descriptions of these and other functions used in the paper were provided by Goodell [20]. The next two methods of HECRASCcontroller used are Compute_CurrentPlan (line 14) and Geometry_GetNodes (line 21). The first function runs the current computational plan. The second one reads tables of River Stations, TabRS, and node types, TabNTyp. Both of them need initialization of proper arguments (lines 12 and 18). The approach presented in Code Samples 2-4 is applied. The river and reach ID numbers are also set in line 16. The variables v1, v2, and in general vX, where X is some digit, are used when the return value does not have to be stored.

The next step is reading of the output variables. It starts at line 23, with setting of the ID numbers for water stages and velocities, WSE_id and AvVel_id. Then the NumPy arrays, TabWSE, and TabVel, for storing the results are prepared (lines 24–25). The results are read in a loop over nodes (line 26–33), with calls to the function Output_NodeOutput. The values of water stages and velocities in regular cross-sections are accessed. In such nodes, the node type is an empty string (line 27).

Finally, the HEC-RAS windows is closed with QuitRas. Then the hec variable may be deleted. The last element of the code is the display of results. It is performed by sc1_ShowNode from the support module. A screenshot with the run of Example 1 and display of results is shown in Figure 2.

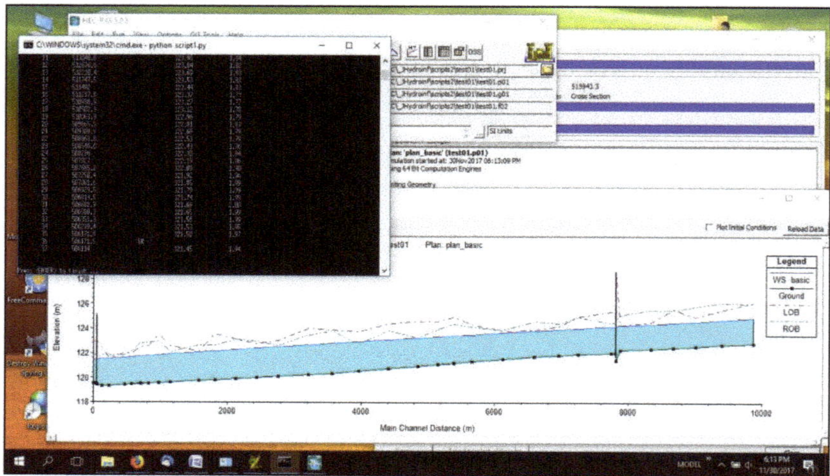

Figure 2. Run of Example 1 and display of results.

3.3. Example 2—Calibration of Roughness Coefficients

The problem of model calibration was considered. Model A described above was used here. The roughness coefficients were sought. The computations were performed in the steady mode. The calculated water surface profile should fit the imposed "observed" values. In fact, the "observed" values were calculated earlier for known roughness coefficients. Very simple distribution of their values was assumed. For the main channel along the whole reach, the value 0.025 was applied. For the left and right floodplains, one value of 0.04 was set in the same way. The purpose of the program was to find the set of optimal coefficients. It was assumed that only three values were necessary. Hence, the dimension of the problem was relatively small, but the generality of this solution is not lost.

The "observed" values were determined for each cross-section along the reach. However, they were applied with different frequencies in the search algorithm. It means that only some values were selected, and the simulated water surface was fitted only in the selected cross-sections. Hence, the objective function is as follows:

$$F(x) = \sqrt{\frac{1}{N} \sum_{i=1}^{N} \left(WSE_i^{(c)} - WSE_i^{(o)} \right)^2}$$

(5)

In the above formula, N is number of observations, $WSE_i^{(c)}$ is the computed water surface elevation, and $WSE_i^{(o)}$ is the "observed" one determined in the same i-th cross-section. This function should be minimized to find the optimal set of roughness coefficients x, where $x = [n_{LOB}, n_{Ch}, n_{ROB}]$ for the left floodplain, the main channel, and the right floodplain, respectively.

The algorithm for calculation of the objective function is presented in Figure 3. It starts with processing of the input parameters. At this step, the optimization variable is transformed into computational roughness coefficients. Then the coefficients are set in all model cross-sections. This step requires access to the global variable representing the object of the HECRASController. The next elements of the algorithm are similar to the operations presented in Figure 1. The computations start and the results are read. In both these steps, the HECRASController object is processed. To read the results properly, identification of the nodes where referenced or "observed" values of the water surfaces are located is necessary. This information is taken from another global variable. Then the calculated and referenced water surface elevations are used to determine the final value of the objective function according to Equation (5). The reference water surfaces are also stored in the global variable.

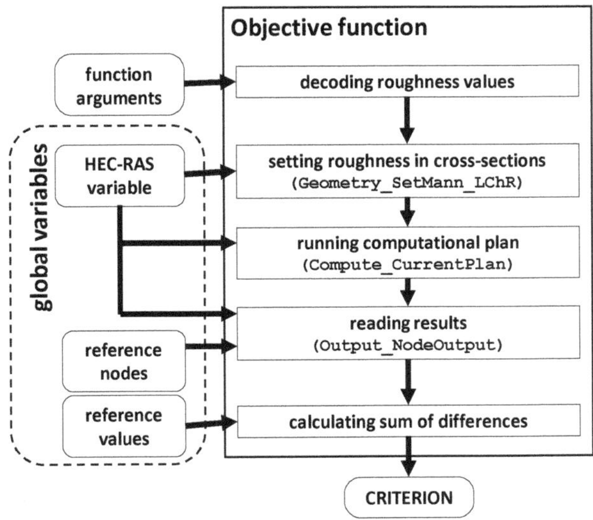

Figure 3. Algorithm for calculation of objective function values.

The main elements of the computations in this example are shown in Figure 4. The computational process starts with initialization of the HECRASController object. Then the HEC-RAS variable is used for opening of the HEC-RAS project, setting the current computational plan and running preliminary computations. These steps are necessary to compare the information about the reference nodes and values with the structure of the computational model. For this purpose, the project name and the reference values must be loaded. After these steps, the object of the HECRASController is ready and the numbers of computational nodes for comparisons with reference nodes are known. After loading of the initial estimate, which is the set of preliminary roughness coefficients, the optimization process starts. It uses the objective function presented in Figure 3.

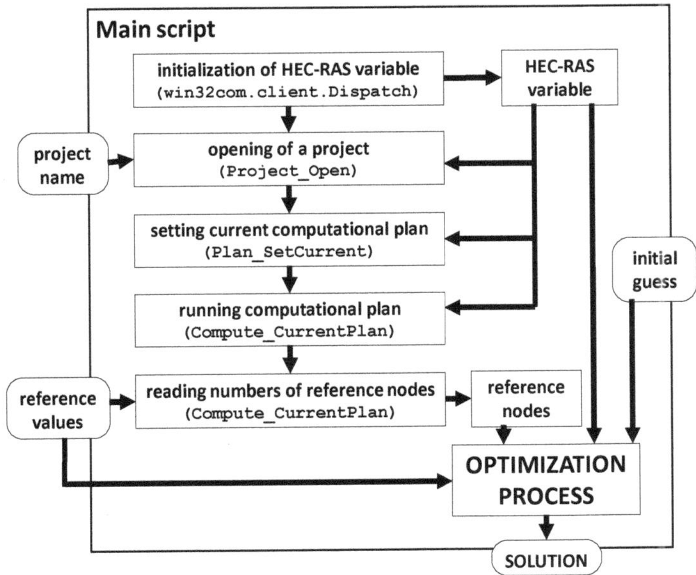

Figure 4. Main computational elements of Example 2.

The starting point is an arbitrarily chosen set of roughness coefficients. The values chosen were 0.08 for the left floodplain, 0.014 for the main channel, and 0.08 for the right floodplain.

In Script 6, such modules as win32com.client, os, math, numpy and scipy.optimize are necessary. They are loaded at the beginning of the script. The function for reading the "observed" values is also loaded from the support module.

```
1.    import win32com.client, os, math
2.    import numpy as np
3.    from scipy import optimize as opt
4.    from support import sc2_ReadObs
5.
6.    def ObjFun(x):                        # objective function
7.        global hec,RivID,RchID,WSE_ID,ProfID,RivName,RchName,nRS,RScmp,RStyp
8.        global Nobs,iRSobs,WSEobs
9.        nLOB,nCh,nROB = x[0],x[1],x[2]
10.       for i in range(0,nRS):
11.           ErrMsg    = None       # list of error messages
12.           v0,v1,v2,v3,v4,v5,v0,ErrMsg =                      \
13.             hec.Geometry_SetMann_LChR(RivName,RchName,RScmp[i],nLOB,nCh,nROB,ErrMsg)
14.       NMsg,ListMsg,block = None,None,True    # no. and list of messages, blocking
15.       v1,NMsg,ListMsg,v2 = hec.Compute_CurrentPlan(NMsg,ListMsg,block)
16.       TotSum = 0.0          # total sum of square errors
17.       for i in range(0,Nobs):
18.           wse,v1,v2,v3,v4,v5,v6 =                            \
19.             hec.Output_NodeOutput(RivID,RchID,iRSobs[i],100,ProfID,WSE_ID)
20.           TotSum += (wse - WSEobs[i])**2       # single square error added
21.       TotSum = math.sqrt( TotSum / Nobs )      # final value
22.       return TotSum
23.
24.   def PokazIter(xk):                    # iteration control function
25.       print 'nLOB=%15.6f, nCh=%15.6f, nROB=%15.6f, ' %          \
26.       (xk[0],xk[1],xk[2]),
27.       print '    funk=%15.6f' % (ObjFun(xk))
28.       return 0
29.
30.   # observed: number, RSes as strings, RSes as floats, WSEs
31.   Nobs,lRSobs,dRSobs,WSEobs = sc2_ReadObs('test01','ObsH','ObsH1x.txt')
32.   # init HEC-RAS Controller
33.   hec = win32com.client.Dispatch('RAS503.HECRASController')
34.   RivID,RchID = 1,1     # ID of river and reach
35.   ProfID,WSE_ID = 1,2  # ID of profile, ID of WSE variable
36.   projekt = os.path.join(os.getcwd(),r'test01\test01.prj')      # project file
37.   hec.Project_Open(projekt)               # opening HEC-RAS
38.   hec.ShowRas()                           # show HEC-RAS window
39.   # setting current computational plan
40.   test1 = hec.Plan_SetCurrent('plan_basic')
41.   # river and reach names
42.   RivName,RchName = 'Warta','Dop_Jeziorsko'
43.   # number and list of messages, blocking mode
44.   NMsg,TabMsg,block = None,None,True
45.   # computations of the current plan
46.   v1,NMsg,TabMsg,test2 = hec.Compute_CurrentPlan(NMsg,TabMsg,block)
47.   # numbers of nodes with observations
48.   iRSobs = np.empty([Nobs],dtype=int)
49.   for i in range(0,Nobs):
50.       iRSobs[i],v1,v2,v3 = hec.Output_GetNode(RivID,RchID,lRSobs[i])
51.   # computational: number, RS and types
52.   nRS,RScmp,RStyp = None,None,None
53.   v1,v2,nRS,RScmp,RStyp = hec.Output_GetNodes(RivID,RchID,nRS,RScmp,RStyp)
54.
55.   print "\nIteration process: Nelder-Mead simplex"
56.   x0 = np.array([0.08,0.014,0.08])                  # initial solution
57.   Xopt = opt.fmin(ObjFun,x0,callback=PokazIter)   # optimization
58.
59.   hec.QuitRas()          # close HEC-RAS
60.   del hec                # delete HEC-RAS controller
```

Script 6. Example 2. Calibration of HEC-RAS steady flow model.

Script 6 includes two functions. The first one plays the role of an objective function and is called ObjFun. It is an argument of the optimization algorithm. Its structure is discussed below. The second one, PokazIter, is used by the same algorithm to display results of the single iteration.

The main body of the script starts with the reading of "observed" values in line 31. They are stored as lists of River Stations, lRSobs and dRSobs, and a list of water surface elevations, WSEobs. Then the HEC-RAS variable, hec, is created (line 33). Before the project is opened (line 37), the HEC-RAS window is displayed (line 38) and the current plan is set (line 40), the project constants are set (line 34–35). They are the river and reach IDs, RivID and RchID, profile number, ProfID, and ID of the output variable storing water surfaces, WSE_ID. The names of the river and the reach, RivName and RchName (line 42), are also hard-coded in the script, though they could be read using the methods of the HEC-RAS controller.

The first computations are run in line 46. The necessary variables are prepared earlier (line 44). The only purpose of these preliminary computations is to check the numbers of nodes, at which the "observed" values are determined. The function Output_GetNode is used for this purpose. The important argument of the function is River Station, inserted as a string, lRSobs[i]. The numbers of the "observed" nodes are stored as NumPy array of integers, iRSobs.

The computational nodes, RScmp, and their types, RStyp, are also read from the results of the preliminary computations.

The optimization process with Nelder-Mead simplex starts at line 57. A detailed description of this function may be found at SciPy.org [28]. Before this the initial guess, x0, is prepared as a NumPy array. The main argument of the optimization algorithm is the objective function, ObjFun. It is defined as a separate function (lines 6–22).

The basic argument of the objective function is the set of searched parameters x. It is a 3-element NumPy array. At the very beginning, the parameters x are assigned to the variables, representing roughness coefficients nLOB, nCh, nROB. The roughness coefficients are set for any cross-section of the HEC-RAS project in a loop over nodes with the function Geometry_SetMann_LChR. The subsequent steps are running of the computations (line 15) and reading the results (line 19) in the loop over nodes (lines 17–20). These processes are performed with the HECRASController functions Compute_CurrentPlan and Output_NodeOutput. Both of them were used in the previous example. The difference between the computed, wse, and "observed", WSEobs[i], water surfaces in a single node is calculated in line 20. The final value of the objective function is determined beyond the loop, in line 21. When the optimization process stops, the HEC-RAS window is closed and the controller variable is deleted.

The results obtained in Example 2 are shown in Figure 5. Part (a) presents the convergence of the channel roughness during the run. Part (b) shows changes of the objective function (5) during the computations.

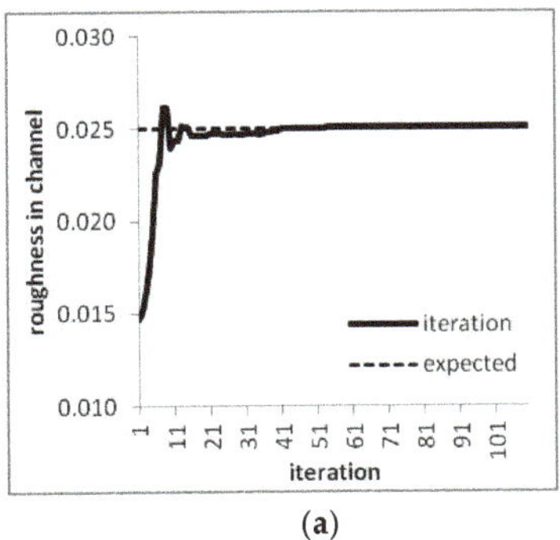

(a)

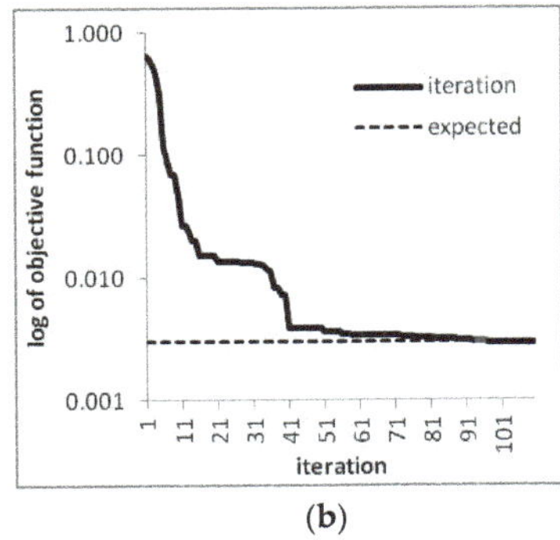

(b)

Figure 5. Results of Example 2: (a) convergence of channel roughness coefficient, (b) convergence of objective function.

3.4. Example 3—Control of Sediment Simulation

The purpose of this example is to show that sediment simulation in HEC-RAS may be run several times for different sediment samples. Model B described above is used in this case. The samples are changed automatically by the Python code. The HEC-RAS sediment data file has the XML format. Hence, the xml.etree.ElementTree module [36] is necessary to read and handle data. The results are then read from HDF files with the methods available in the h5py module [59]. The code in this example is very simple. However, the methods presented here may be extended and applied in a more sophisticated code, e.g., calibration of the sediment routing algorithm, analysis of sediment simulation uncertainty, or assessment of flow regime changes caused by sediments.

The third example is too long to present all lines of code in one script. Hence, it is split into Scripts 7 and 8. There are four files with data and results managed in this script. The first is the file storing the tested sediment samples. The format of this file is arbitrary and it will not be discussed here. The only requirement is that the loaded sediment samples should have a form similar to those stored in HEC-RAS. Three other files are more specific. They are the XML file with HEC-RAS sediment data, the HDF file with geometric data, and the HDF file with the results.

```
1.   import win32com.client
2.   import os, h5py, numpy
3.   import xml.etree.ElementTree as ET
4.   from support import sc3_LoadPrb,sc3_SaveRes
5.
6.   # loading samples
7.   pnames,sampA,sampB,sampC,sampD =            \
8.   sc3_LoadPrb('test_sedi','samples','samples.txt')
9.
10.  # init HEC-RAS Controller
11.  hec = win32com.client.Dispatch('RAS503.HECRASController')
12.  projekt = os.path.join(os.getcwd(),r'test_sedi\test_sedi.prj')   # project name
13.  hec.ShowRas()                                         # show HEC-RAS window
14.  # loading River Stations from geometry
15.  hdfname = os.path.join(os.getcwd(),r'test_sedi\test_sedi.g03.hdf')
16.  dane = h5py.File(hdfname,'r')                         # opening of HDF file
17.  # link to the River Stations in HDF file
18.  ListRS = dane.get('Geometry').get('Cross Sections').get('River Stations')
19.  GeomRS = numpy.empty([len(ListRS)],dtype='S15')       # NumPy array of RSes
20.  for i in range(0,len(ListRS)):
21.      GeomRS[i] = ListRS[i]
22.  dane.close()
23.  NXS = len(GeomRS)                    # number of cross-sections
```

Script 7. Example 3. Automatic control of sediment transport simulation—part 1.

An example of the sediment data file is shown in Figure 6a. It is presented in the XML Viewer. It has a structure similar to a tree. The groups include subgroups and so on. A description of the XML format can be found in the document prepared by the Python Software Foundation [36]. The group with bed gradation is opened in the XML Viewer. The name of the file and the full hierarchy to access these data are shown in Table 1. Particular gradations are stored as the group attribute. This attribute is a dictionary, where the class symbols 'GC1', 'GC2', etc. reflect the gradation in the format '%Finer'. Every XML file is an ASCII file, and it may be opened in Windows Notepad.

The HDF files may be opened in HDFViewer, as presented in Figure 6b. These are simulation results files in binary format. It is not possible to use Notepad to open them, but the HDF also has a structure similar to a tree with successive branches. Two elements are read from the opened HDF file. These are (a) the invert elevations and (b) the water surface elevations. The first is shown in Figure 6b. The hierarchy to access these data is also presented in Table 1.

```python
24.
25.  # test of loaded samples
26.  sname = os.path.join(os.getcwd(),r'test_sedi\test_sedi.s01')
27.  ii = 0
28.  for sample in [sampA, sampB, sampC, sampD]:
29.      plik = ET.parse(sname)    # opening XML file with sediment data
30.      dane = plik.getroot()     # access to bed gradations
31.      grad = dane.find('Bed_Gradation').find('Sample').find('Gradation')
32.      for elm in sample:        # assignment of tested sample
33.          grad.attrib[elm] = sample[elm]
34.      plik.write(os.path.join(os.getcwd(),r'test_sedi\test_sedi.s01'))
35.      del plik
36.      hec.Project_Open(projekt)                    # open HEC-RAS project
37.      test1 = hec.Plan_SetCurrent('plan sedi00')   # setting sediment plan
38.      NMsg,TabMsg,block = None,None,True   # no. and list of messages, blocking
39.      # computations of the current plan
40.      v1,NMsg,TabMsg,v2 = hec.Compute_CurrentPlan(NMsg,TabMsg,block)
41.      del NMsg, TabMsg, block
42.      hec.Project_Close()                 # close HEC-RAS project
43.      # full path to the HDF file results of the computations
44.      hdfname = os.path.join(os.getcwd(),r'test_sedi\test_sedi.p03.hdf')
45.      dane = h5py.File(hdfname,'r')    # opening of HDF file
46.      # accessing bed elevations
47.      XSbed = dane.get('Results').get('Sediment').get('Output Blocks')
48.      XSbed = XSbed.get('Sediment').get('Sediment Time Series')
49.      XSbed = XSbed.get('Cross Sections').get('Invert Elevation')
50.      NTS = len(XSbed)
51.      # accessing WSE
52.      XSwse = dane.get('Results').get('Sediment').get('Output Blocks')
53.      XSwse = XSwse.get('Sediment').get('Sediment Time Series')
54.      XSwse = XSwse.get('Cross Sections').get('Water Surface')
55.      # NumPy arrays for bed elevations and WSE
56.      InitBed = numpy.empty([NXS],dtype=float)     # initial bed and WSE
57.      InitWSE = numpy.empty([NXS],dtype=float)
58.      LastBed = numpy.empty([NXS],dtype=float)     # final bed and WSE
59.      LastWSE = numpy.empty([NXS],dtype=float)
60.      for i in range(0,NXS):              # assigment of bed and WSE form HDF
61.          InitBed[i] = XSbed[0][i]        # to NumPy array
62.          InitWSE[i] = XSwse[0][i]
63.          LastBed[i] = XSbed[NTS-1][i]
64.          LastWSE[i] = XSwse[NTS-1][i]
65.      dane.close()
66.      del dane
67.      sc3_SaveRes(pnames[ii],sample,GeomRS,InitBed,InitWSE,LastBed,LastWSE)
68.      ii += 1
69.
70.  hec.QuitRas()           # close HEC-RAS
71.  del hec                 # delete HEC-RAS controller
```

Script 8. Example 3. Automatic control of sediment transport simulation—part 2.

Table 1. Loading data and results from XML and HDF files: type of data, type of file, access hierarchy.

Data	File		Access Hierarchy
	Type	Name	
sediment sample	XML	test_sedi.s01	Data\Bed_Gradation\Sample\Gradation
River Stations	HDF	test_sedi.g03.hdf	Geometry\Cross Sections\River Stations
bed elevations	HDF	test_sedi.p03.hdf	Results\Sediment\Output Blocks\Sediment\Sediment Time Series\Cross Sections\Invert Elevation
water surface elevations	HDF	test_sedi.p03.hdf	Results\Sediment\Output Blocks\Sediment\Sediment Time Series\Cross Sections\Water Surface

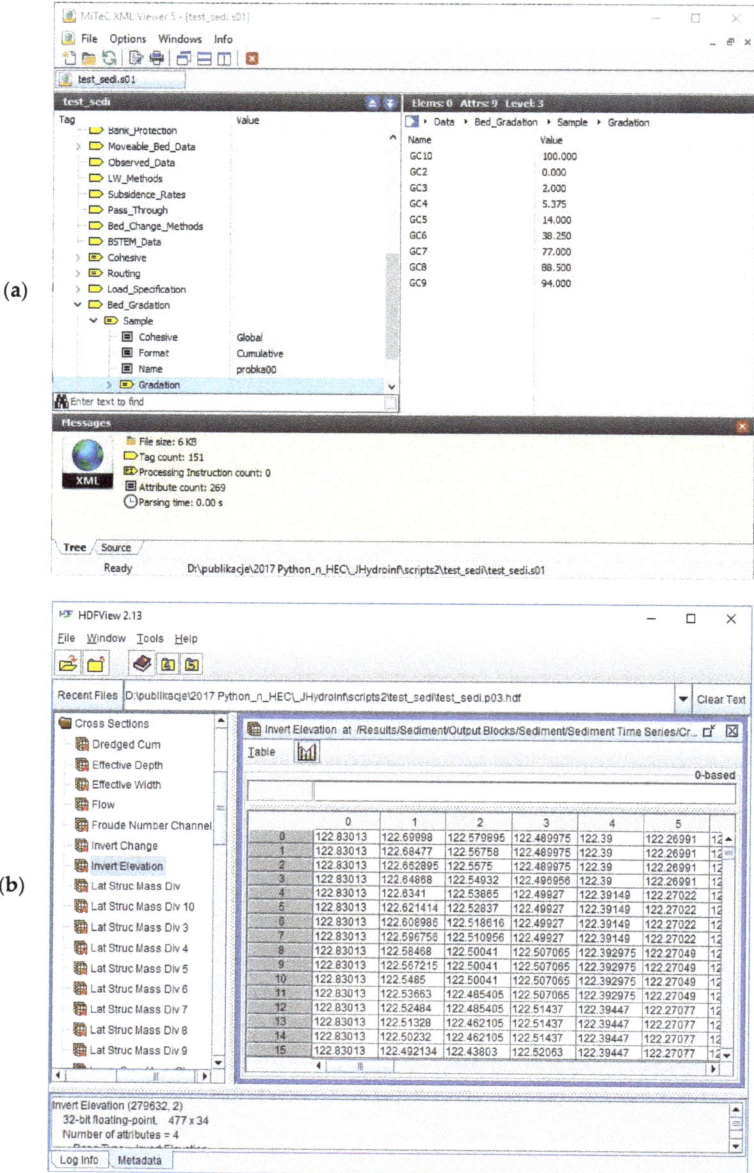

Figure 6. View of data and result files used in Example 3: (**a**) Sediment data file in XML Viewer, (**b**) sediment results in HDFViewer.

The flow chart of the main computations in this example is shown in Figure 7. The main element is a loop over sediment samples. Before the loop starts, the HECRASController object has to be initialized and the name of the project has to be loaded. If the HEC-RAS variable is ready, the information about computational nodes may be read from the HDF5 file storing geometry data. Before the loop starts, the sediment samples have to be loaded to the script. The first step in the loop is access to the XML file with sediment data. The grain fractions are written and the XML file is closed. Then the project can be opened, because any change of the sediment data in the XML file is noticed by the HEC-RAS project only in the phase of the data loading. Further changes of the data in files stored on the disk do not affect the project. Then the computations start and the results are read from the HDF5 file. At the end of each step, the bed elevations and positions of the computational nodes are written to the separate ASCII file, which enables preparation of the bed profiles.

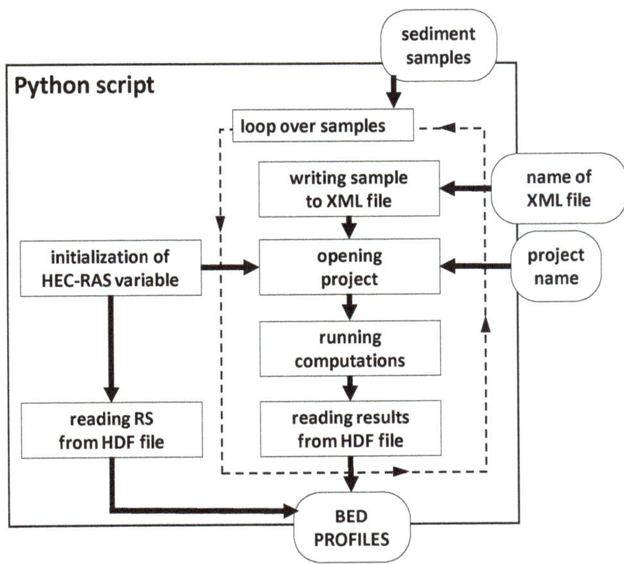

Figure 7. Main computational elements of Example 3.

The program presented in Scripts 7 and 8 starts with import of necessary modules (lines 1–4). The modules win32com.client, os, numpy are also applied in the previous examples. The new elements are xml.etree.ElementTree and h5py. The first one is loaded as ET, and it is used for handling the XML documents. The second one is used for reading of data and results from HDF files. Two functions are loaded from the support module. They are sc3_LoadPrb for loading sediment samples, and sc3_SaveRes for saving results in ASCII files. For the sake of simplicity, no local functions are defined.

The first command is reading of the sediment samples for tests (lines 7–8). The samples are stored as dictionaries sampA, sampB, sampC, and sampD. Their items are composed of sediment class denotation in the format 'GC#' and weighted gradation as '%Finer'. The keys of the dictionary should be the same as those used in the HEC-RAS and XML files with sediment data. The first element of the return list, pnames, is the table with samples' names.

The HECRASController starts at line 11. The HEC-RAS window is opened (line 13) and the full path of the project file is set, but the project is not loaded. Every change in the data requires reloading of the project. Hence, the project is loaded later.

The HDF file including geometry data is opened first (line 16). The link to the River Stations is created (line 18) with the get method of the h5py module. Then the NumPy array for storing these data is prepared (line 19). The data are assigned as value-by-value in a loop over nodes. Then the first HDF file is closed (line 22). The important element taken from the River Station array is the number of cross-sections NXS (line 23). The rest of the commands are shown in Script 8.

The longest part of the script is the loop over tested samples (lines 28–68). At the beginning of each step, the XML file with sediment data is opened (line 29). Then the link to bed gradation grad is created. In the internal loop (lines 32–33) over items of the sample, the bed gradations in the XML file are replaced with bed gradations of the current sample (line 33). Then the file is saved (line 34) and closed by deleting the variable representing this file. At this moment the XML file with sediment data is ready. The HEC-RAS project is opened (line 36) and the computational plan is set properly (line 37). After the current plan is computed (line 40), the project is closed (line 42). The last file of the current iteration is accessed (line 45). This is the HDF file with results of the computations (Table 1). Two elements are read from this file: Bed elevations and water surface elevations. These results are stored as list of successive time steps. Each time step is a list of values stored for each cross-section. The links to the bed and water surface elevations, XSbed and XSwse, are created in lines 47–49 and 52–54. The number of time steps is assigned to the variable NTS.

The NumPy arrays for storing the initial and final bed and surface elevations are created in lines 56–59. The results are assigned to them in another internal loop (lines 60–64) over cross-sections. Later, the HDF file is closed (lines 65–66). Finally, the results are saved in a file specific for each iteration, whose name is the name of the sample. At the end of the script, HEC-RAS is closed and the variable hec is deleted.

The sediment simulation running from the Python script is shown in Figure 8. The display of typical output for such simulation is also shown in the screenshot. The sieve curves of tested sediment samples are plotted in Figure 9a. The variability of assumed sieve curves is typical for a lowland river. The differences in the content of bed sediments are relatively small. However, more complex examples may also be tested in the same way. Figure 9b shows the results of the simulation as bed changes in the selected cross-sections. Considering the time horizon of the simulation, which is 4 months, the bed changes from −20 cm to +60 cm seem to be quite significant. The reason is the relatively huge discharge of 269 m^3/s, kept constant during the whole simulation. The differences between the changes of bed elevations vary from a few to 10 cm. If we compare this result with the range of changes, which is about 80 cm, it transpires that the differences between consecutive simulations are up to 12.5% for a very short simulation period. This stresses the importance of the sediment sample variability for the sensitivity of the sediment transport model.

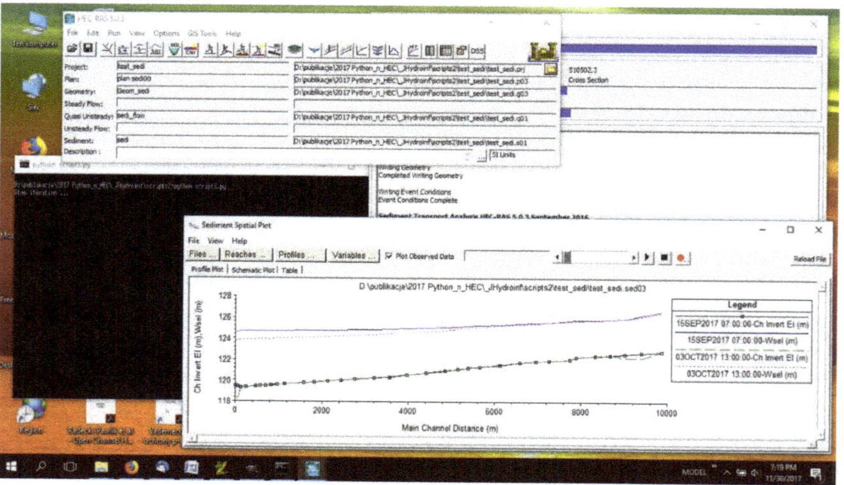

Figure 8. Run of sediment simulation and typical display of sediment output.

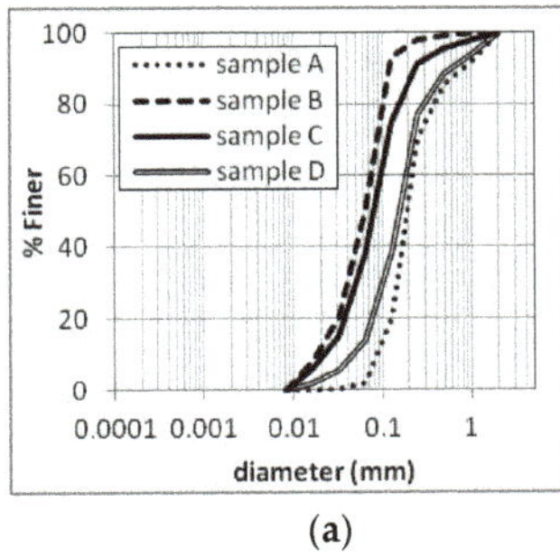
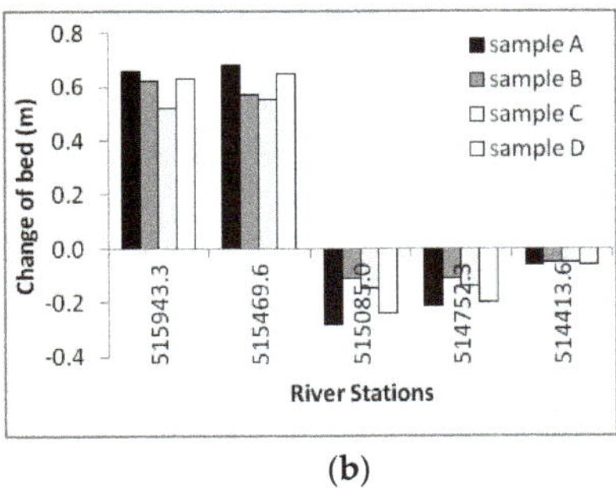

(a) (b)

Figure 9. Sediment simulations: (**a**) Sieve curves of tested samples, (**b**) bed changes in selected River Stations.

4. Conclusions

The examples presented in this paper explain the techniques necessary to control HEC-RAS computations with Python. The main element of the presented methods is the application of Python module Pywin32 for handling COM libraries. This module was used in all three examples. The comparison of the code developed for Example 1, with basic examples discussed by Goodell [20,21] and Leon and Goodell [25], illustrates the ease of Python application for handling HECRASController. The use of Python is not more difficult than VBA or MATLAB. Additionally, Example 1 presents a more general approach for application of Python for control of HEC-RAS computations than described in previous papers [40,41]. This example explains how to control any version of HEC-RAS, including the future developments. It also opens doors for further extension of HECRASController capabilities. Possible further extensions were shown in Examples 2 and 3.

The second example presented the application of HECRASController and Python specific modules, for the automatic calibration of the HEC-RAS model. The roughness coefficients were determined by the optimization method from the SciPy module. This approach was similar to that presented by Leon and Goodell [25], where the Genetic Algorithm from the MATLAB toolbox was applied for control of gate operation. Example 2 was also a more advanced approach for determination of roughness coefficients than that presented in the AHYDRA package [24]. Further development of the techniques presented in Example 2, could focus on application of a faster and/or more sophisticated optimization method than the Nelder-Mead simplex. It is also possible to develop a more general Python module for automatic calibration of HEC-RAS, on the basis of the concept presented here. Example 2 may also be improved by preparation of a user optimization algorithm. It could be implemented in Fortran, which is very fast and aimed at numerical computations, and then it could be translated to Python with the F2PY or ctypes modules.

Example 2 presented the linkage of HEC-RAS computations with numerical routines available in specific Python modules, NumPy and SciPy. The number of useful routines in SciPy is greater than those presented. Other extensions and HEC-RAS applications are also possible. The elements available in NumPy are also very useful for storing and handling of computational results. These features of Python are undoubted advantages of Python over standard VBA.

Example 3 illustrated an extension of HECRASController capabilities to control sediment simulation. To the best of the author's knowledge, such an approach or a similar one has not been tested yet. These examples presented a real advantage of Python, owing to the access to very complex data formats, XML and HDF5. The Python modules enabling such access are ElementTree and H5PY, respectively. There are several applications of the presented approach. The most basic is sensitivity analysis of sediment transport simulation. Another possible use is calibration of the sediment model on the basis of the historical data. These two application areas are discussed in the text. Other applications are also possible in the area of water management. Examples may be the impact of sediment deposition on flood hazard, prediction of morphodynamic changes in the river after regulation, etc. The uncertainty of sediment transport modeling may also be estimated with the techniques used in Example 3.

The illustrated capabilities of Python are not the only advantages of this scripting language over specific VBA applications. Another is access to geoprocessing tools, such as ArcGIS and QGIS. It makes integration of Python with HEC-RAS more necessary. Not only are VBA capabilities too poor to compete with Python in this area; more dedicated numerical software such as MATLAB does not have such broad capabilities for geoprocessing.

Application of Python for control of hydrodynamic simulations with HEC-RAS opens new opportunities in such areas as water management, river engineering, and flood risk management. Good opportunities are available by access to more different data formats from simple text files to more complex XML and HDF5. A huge advantage is the linking of HEC-RAS with other software, such as numerical routines or geoprocessing. More advanced future applications are also possible.

Software Availability: https://sourceforge.net/projects/hec-ras-and-python/.

Author Contributions: Everything, T. Dysarz.

Funding: This research received no external funding.

Conflicts of Interest: The author declares no conflict of interest.

References

1. Kachiashvili, K.J.; Melikdzhanian, D.I. Software realization problems of mathematical models of pollutants transport in rivers. *Adv. Eng. Softw.* **2009**, *40*, 1063–1073. [CrossRef]
2. Zischg, A.P.; Mosimann, M.; Bernet, D.B.; Röthlisberger, V. Validation of 2D flood models with insurance claims. *J. Hydrol.* **2018**, *557*, 350–361. [CrossRef]
3. Pinar, E.; Paydas, K.; Seckin, G.; Akilli, H.; Sahin, B.; Cobaner, M.; Kocaman, S.; Akar, M.A. Artificial neural network approaches for prediction of backwater through arched bridge constrictions. *Adv. Eng. Softw.* **2010**, *41*, 627–635. [CrossRef]
4. Shen, D.Y.; Jia, Y.; Altinakar, M.; Bingner, R.L. GIS-based channel flow and sediment transport simulation using CCHE1D coupled with AnnAGNPS. *J. Hydraul. Res.* **2016**, *54*, 567–574. [CrossRef]
5. Gibson, S.; Sánchez, A.; Piper, S.; Brunner, G. New One-Dimensional Sediment Features in HEC-RAS 5.0 and 5.1. In Proceedings of the World Environmental and Water Resources Congress 2017, Sacramento, CA, USA, 21–25 May 2017; pp. 192–206.
6. Horritta, M.S.; Bates, P.D. Evaluation of 1D and 2D numerical models for predicting river flood inundation. *J. Hydrol.* **2002**, *268*, 87–99. [CrossRef]
7. Rodriguez, L.B.; Cello, P.A.; Vionnet, C.A.; Goodrich, D. Fully conservative coupling of HEC-RAS with MODFLOW to simulate stream–aquifer interactions in a drainage basin. *J. Hydrol.* **2008**, *353*, 129–142. [CrossRef]
8. Drake, J.; Bradford, A.; Joy, D. Application of HEC-RAS 4.0 temperature model to estimate groundwater contributions to Swan Creek, Ontario, Canada. *J. Hydrol.* **2010**, *389*, 390–398. [CrossRef]
9. Brunner, G.W. *HEC-RAS River Analysis System Hydraulic Reference Manual*; US Army Corps of Engineers; Report No. CPD-69; Hydrologic Engineering Center (HEC): Davis, CA, USA, 2016.
10. DHI. MIKE 11—A Modeling System for Rivers and Channels User Guide, DHI Software. 2003. Available online: https://www.tu-braunschweig.de/Medien-DB/geooekologie/mike11usersmanual.pdf (accessed on 10 May 2018).
11. Deltares. SOBEK User Manual, Delft, The Netherlands, 2018. Available online: https://content.oss.deltares.nl/delft3d/manuals/SOBEK_User_Manual.pdf (accessed on 10 May 2018).
12. Caponi, F.; Ehrbar, D.; Facchini, M.; Kammerer, S.; Koch, A.; Peter, S.; Vonwiller, L. BASEMENT System Manuals–Reference Manual, VAW–ETH, Zurich, 2017. Available online: http://people.ee.ethz.ch/~basement/baseweb/download/documentation/BMdoc-Reference-Manual-v2-7.pdf (accessed on 10 May 2018).

13. Greimann, B.; Huang, J.V. *SRH-1D 4.0 User's Manual*; Bureau of Reclamation: Denver, CO, USA, 2018. Available online: https://www.usbr.gov/tsc/techreferences/computer%20software/models/srh1d/index. html (accessed on 10 May 2018).

14. Pappenberger, F.; Beven, K.; Horritt, M.; Blazkova, S. Uncertainty in the calibration of effective roughness parameters in HEC-RAS using inundation and downstream level observations. *J. Hydrol.* **2005**, *302*, 46–69. [CrossRef]

15. Vansteenkiste, T.; Tavakoli, M.; Ntegeka, V.; De Smedt, F.; Batelaan, O.; Pereira, F.; Willems, P. Intercomparison of hydrological model structures and calibration approaches in climate scenario impact projections. *J. Hydrol.* **2014**, *519*, 743–755. [CrossRef]

16. Shrestha, B.; Cochrane, T.A.; Caruso, B.S.; Arias, M.E.; Piman, T. Uncertainty in flow and sediment projections due to future climate scenarios for the 3S Rivers in the Mekong Basin. *J. Hydrol.* **2016**, *540*, 1088–1104. [CrossRef]

17. Guo, A.; Chang, J.; Wang, Y.; Huang, Q.; Zhou, S. Flood risk analysis for flood control and sediment transportation in sandy regions: A case study in the Loess Plateau, China. *J. Hydrol.* **2018**, *560*, 39–55. [CrossRef]

18. Dimitriadis, P.; Tegos, A.; Oikonomou, A.; Pagana, V.; Koukouvinos, A.; Mamassis, N.; Koutsoyiannis, D.; Efstratiadis, A. Comparative evaluation of 1D and quasi-2D hydraulic models based on benchmark and real-world applications for uncertainty assessment in flood mapping. *J. Hydrol.* **2016**, *534*, 478–492. [CrossRef]

19. Afshari, S.; Tavakoly, A.A.; Rajib, M.A.; Zheng, X.; Follum, M.L.; Omranian, E.; Fekete, B.M. Comparison of new generation low-complexity flood inundation mapping tools with a hydrodynamic model. *J. Hydrol.* **2018**, *556*, 539–556. [CrossRef]

20. Goodell, C. *Breaking HEC-RAS Code. A User's Guide to Automating HEC-RAS*; h2ls: Portland, OR, USA, 2014.

21. Goodell, C. Automating HEC-RAS, The RAS Solution. 2014. Available online: http://hecrasmodel.blogspot. com/2014/10/automating-hec-ras.html (accessed on 20 November 2017).

22. Yang, T.-H.; Wang, Y.-C.; Tsung, S.-C.; Guo, W.-D. Applying micro-genetic algorithm in the one-dimensional unsteady hydraulic model for parameter optimization. *J. Hydroinform.* **2014**, *16*, 772–783. [CrossRef]

23. Lacasta, A.; Morales-Hernández, M.; Burguete, J.; Brufau, P.; García-Navarro, P. Calibration of the 1D shallow water equations: A comparison of Monte Carlo and gradient-based optimization methods. *J. Hydroinform.* **2017**, *19*, 282–298. [CrossRef]

24. Vladyman. Automating Hydraulic Analysis (AHYDRA) v. 1.0. 2011. Available online: http://ahydra. yolasite.com/ (accessed on 7 November 2017).

25. Leon, A.S.; Goodell, C. Controlling HEC-RAS using MATLAB. *Environ. Model. Softw.* **2016**, *84*, 339–348. [CrossRef]

26. Python Software Foundation. Python 2.7.14 Documentation. 2017. Available online: https://docs.python. org/2/index.html (accessed on 8 November 2017).

27. Galiano, V.; Migallón, H.; Migallón, V.; Penadés, J. PyPnetCDF: A high level framework for parallel access to netCDF files. *Adv. Eng. Softw.* **2010**, *41*, 92–98. [CrossRef]

28. Patzák, B.; Rypl, D.; Kruis, J. MuPIF—A distributed multi-physics integration tool. *Adv. Eng. Softw.* **2013**, *60–61*, 89–97. [CrossRef]

29. SciPy Community SciPy Reference Guide. 2017. Available online: https://docs.scipy.org/doc/scipy/ reference/ (accessed on 24 November 2017).

30. SciPy.org NumPy. 2017. Available online: http://www.numpy.org/ (accessed on 8 November 2017).

31. SciPy.org Optimization and Root Finding (scipy.optimize). 2017. Available online: https://docs.scipy.org/ doc/scipy/reference/optimize.html (accessed on 8 November 2017).

32. SciPy.org F2PY Users Guide and Reference Manual. 2017. Available online: https://docs.scipy.org/doc/ numpy-dev/f2py/ (accessed on 24 November 2017).

33. Python 2.7.15 Documentation. Available online: https://docs.python.org/2/index.html (accessed on 11 September 2018).

34. Hammond, M.; Robinson, A. *Python Programming on Win32*; O'Reilly Media, Inc.: Sebastopol, CA, USA, 2000.

35. PythonCOM Documentation Index Python and COM. Blowing the Rest Away! 2017. Available online: http:// docs.activestate.com/activepython/2.4/pywin32/html/com/win32com/HTML/docindex.html (accessed on 8 November 2017).

36. Python Software Foundation. The ElementTree XML API, in Python Software Foundation. 2017. Available online: https://docs.python.org/2/library/xml.etree.elementtree.html (accessed on 8 November 2017).

37. HDF Group. High Level Introduction to HDF5. 2016. Available online: https://support.hdfgroup.org/HDF5/Tutor/HDF5Intro.pdf (accessed on 24 November 2017).

38. Zandbergen, P.A. *Python Scripting for ArcGIS*; Esri Press: Redlands, CA, USA, 2013.

39. QGIS Project PyQGIS Developer Cookbook Release 2.18. 2017. Available online: http://docs.qgis.org/2.18/pdf/en/QGIS-2.18-PyQGISDeveloperCookbook-en.pdf (accessed on 26 November 2017).

40. Peña-Castellanos, G. PyRAS—Python for River AnalysiS; MIT License. 2015. Available online: https://pypi.python.org/pypi/PyRAS/ (accessed on 7 November 2017).

41. Vimal, S. PyFloods. Python Module for Floods. 2015. Available online: https://github.com/solomonvimal/PyFloods (accessed on 7 November 2017).

42. Chow, V.T. *Open Channel Hydraulics*; McGraw-Hill: New York, NY, USA, 1959.

43. Henderson, F.M. *Open Channel Flow. Macmillan Series in Civil Engineering*; Macmillan Company: New York, NY, USA, 1966.

44. Abbott, M.B.; Minns, A.W. *Computational Hydraulics*; Ashgate Publishing: Farnham, UK, 1979.

45. Cunge, J.A.; Holly, F.M.; Verwey, A. *Practical Aspects of Computational River Hydraulics*; Pitman Advanced Publishing Program: Boston, MA, USA, 1980.

46. Wu, W. *Computational River Dynamics*; Taylor & Francis Group: London, UK, 2007.

47. Szymkiewicz, R. Numerical Modeling in Open Channel Hydraulics. In *Water Science and Technology Library*; Springer: Dordrecht, The Netherlands, 2010; Volume 83.

48. Leonard, B.P. A Stable and Accurate Convective Modelling Procedure Based on Quadratic Upstream Interpolation. In *Computer Methods in Applied Mechanics and Engineering*; North-Holland Publishing Company: Amsterdam, The Netherlands, 1979; Volume 19, pp. 59–98.

49. Brunner, G.W. *HEC-RAS River Analysis System User's Manual Version 5.0*; US Army Corps of Engineers; Report No. CPD-68; Hydrologic Engineering Center (HEC): Davis, CA, USA, 2016.

50. Green, J.; Bullen, S.; Bovey, R.; Alexander, M. *Excel 2007 VBA Programmer's Reference*; Wiley Publishing, Inc.: Indianapolis, IN, USA, 2007.

51. Sandler, R. The 14 Most Popular Programming Languages, According to a Study of 100,000 Developers. Buisness Insider, 2018. Available online: https://www.businessinsider.com/14-most-popular-programming-languages-stack-overflow-developer-survey-2018-4 (accessed on 8 September 2018).

52. Putano, B. Most Popular and Influential Programming Languages of 2018, Stackify. 2017. Available online: https://stackify.com/popular-programming-languages-2018/ (accessed on 8 September 2018).

53. Petkov, A. Here Are the Best Programming Languages to Learn in 2018. freeCodeCamp.org, 2018. Available online: https://medium.freecodecamp.org/best-programming-languages-to-learn-in-2018-ultimate-guide-bfc93e615b35 (accessed on 8 September 2018).

54. Downey, A.; Elkner, J.; Meyers, C. *How to Think Like a Computer Scientist. Learning with Python*; Green Tea Press: Wellesley, MA, USA, 2002; Available online: http://www.greenteapress.com/thinkpython/thinkCSpy.pdf (accessed on 8 November 2017).

55. Rubalcava, R. *Introducing ArcGIS API 4 for JavaScript: Turn Awesome Maps into Awesome Apps*; Apress: New York, NY, USA, 2017.

56. Pobuda, M. Using the R-ArcGIS Bridge: The Arcgisbinding Package. Available online: https://r-arcgis.github.io/assets/arcgisbinding-vignette.html (accessed on 11 September 2018).

57. Tutorials Point NumPy. Tutorials Point (I) Pvt. Ltd., 2016. Available online: https://www.tutorialspoint.com/numpy/ (accessed on 24 November 2017).

58. Press, W.H.; Teukolsky, S.A.; Vetterling, W.T.; Flannery, B.P. Numerical Recipes in Fortran 77: The Art of Scientific Computing, Volume 1 of Fortran Numerical Recipes. Second Edition, University of Cambridge. 1992. Available online: http://numerical.recipes/oldverswitcher.html (accessed on 24 November 2017).

59. Tutorials Point XML. Tutorials Point (I) Pvt. Ltd. 2017. Available online: https://www.tutorialspoint.com/xml/ (accessed on 24 November 2017).

60. Lundh, F. Elements and Element Trees. 2007. Available online: http://effbot.org/zone/element.htm (accessed on 24 November 2017).

61. Collette, A. h5py Documentation. 2017. Available online: http://docs.h5py.org/en/latest/ (accessed on 8 November 2017).

62. Dysarz, T.; Szałkiewicz, E.; Wicher-Dysarz, J. Long-term impact of sediment deposition and erosion on water surface profiles in the Ner river. *Water* **2017**, *9*, 168. [CrossRef]
63. Wicher-Dysarz, J.; Dysarz, T. Analiza procesu akumulacji rumowiska w górnej części zbiornika Jeziorsko. *Gospodarka Wodna* **2016**, *9*, 292–298. (In Polish)

Article

The Evaluation of Regional Water-Saving Irrigation Development Level in Humid Regions of Southern China

Lu Zhao [1,2], Lili Zhang [3], Ningbo Cui [1,4,*], Chuan Liang [1] and Yi Feng [1]

[1]   State Key Laboratory of Hydraulics and Mountain River Engineering & College of Water Resource and Hydropower, Sichuan University, Chengdu 610025, China; luya1121@163.com (L.Z.); lchester@sohu.com (C.L.); fengyscu@163.com (Y.F.)

[2]   Provincial Key Laboratory of Water-Saving Agriculture in Hill Areas of Southern China, Chengdu 610066, China

[3]   International School of Technical Education, Sichuan College of Architectural Technology, Deyang 618000, China; lindaz616@163.com

[4]   Key Laboratory of Agricultural Soil and Water Engineering in Arid and Semiarid Areas, Ministry of Education, Northwest A & F University, Yangling 712100, China

*   Correspondence: cuiningbo@126.com; Tel.: +86-136-8839-7865

Received: 20 December 2018; Accepted: 14 January 2019; Published: 18 January 2019

Abstract: Water-saving irrigation development level (WIDL) refers to reasonably and accurately judging a water-saving area based on the analysis of all factors affecting the water-saving irrigation development. The evaluation of regional WIDL is the premise of scientific planning guidance to irrigation work. How to select reasonable evaluation indexes and build a scientific and comprehensive model to evaluate WIDL is of great significance. In this study, the comprehensive evaluation index system of WIDL in 21 cities (states) of the Sichuan province in China (a typical humid region in southern China) was constructed, and the TOPSIS (Technique for Order Preference by Similarity to an Ideal Solution) method was improved to evaluate WIDL. Results showed that the overall development level of water-saving irrigation was "poor" in Sichuan province. The water-saving irrigation level turned out to be "good" in three regions with advantageous geographical conditions and developed economies, "general" in four regions with good economic levels where agronomy water saving has been popularized, and "poor" in fourteen regions of mountainous and hilly areas, especially Ganzi, Aba, and Liangshan, located in the Northwest plateau of Sichuan province, with poor natural resources and insufficient economies. The evaluation results were in good agreement with the actual situation, and in this area, there is enormous potential for the development of water-saving irrigation strategies. This study provides an important technical approach for the evaluation of water-saving irrigation development in humid regions of Southern China.

Keywords: water-saving irrigation development level; Sichuan province; TOPSIS method

---

1. Introduction

Water scarcity has become a major restraint factor for social and economic development among many regions, especially in semiarid and arid regions [1,2]. Agriculture consumes the largest amount of water resources among human activities: Irrigation water withdrawals represent 70% of the total human use of renewable water resources [3–6]. About 18% of croplands worldwide, or about 2% of the total land surface, are irrigated and meet 40% of the global food demand [7].

There are various available measures to solve global water scarcity, such as water re-pricing, water re-use, desalination, water diversion and distribution, improvements in water delivery systems, alternative plants, and water conservation through efficient irrigation, among which water-saving irrigation is one of the most feasible and effective measures attributing to significant water-savings [8–11]. The factors affecting water-saving irrigation mainly feature engineering (drip irrigation, micro-irrigation, etc.), agronomy (regulated deficit irrigation, water-saving varieties, etc.) and management (water price, government police, etc.) measures [12–18].

The natural conditions, such as hydrogeology, channel soil, irrigation soil, crop species, the management and maintenance level of irrigation districts, farmer's habits of water use, and water price policy vary in different regions, which will lead to different engineering technologies, economic benefits, environmental benefits, and development level of water-saving irrigation in various regions. Although many studies have focused on the evaluation of water-saving irrigation technology or comprehensive benefits [19–23], less research has been conducted on the comprehensive evaluation of the regional water-saving irrigation development level (WIDL). WIDL refers to reasonably and accurately judging a water-saving or high-efficiency water-saving area based on the analysis of all factors affecting water-saving irrigation development. It is an objective summary of past achievements and also a judgment of the current starting point. In most cases, we used the agricultural integrated gross irrigation quota to express water-saving irrigation development level. However, the regional WIDL is the organic embodiment of engineering, agronomy, and management the water-saving level under the conditions of optimal allocation of water resources. WIDL has never been reported in the humid regions of Southern China. Although water resources in these regions are relatively abundant, with annual precipitation greater than 1000 mm, seasonal drought often occurs, combined with serious engineering water shortages, leading to more serious droughts. Therefore, it is necessary to develop water-saving irrigation in the humid regions of Southern China. With the continuous enrichment of multi-index comprehensive evaluation methods such as the fuzzy comprehensive evaluation method [24], analytic hierarchy process method [25,26], set pair analysis [27], TOPSIS method [28] and so on, the comprehensive evaluation of WIDL is worthy of study and improvement.

Attaining the weights of indexes is a prerequisite for a comprehensive evaluation since the weight of the index represents the relative importance of the index in the evaluation system, and the accuracy of the weight directly affects the final evaluation results [29]. There are two kinds of methods to judge weight. One is the subjective method (such as analytic hierarchy process). Weights of indexes are obtained by the subjective judgment of experts. This method reflects the knowledge and experience of experts but easily leads to deviations due to personal subjective elements. The other is the objective weighting method (such as entropy weight method). Weights are judged according to the relationship among the original data with a strong mathematical theoretical basis. In order to make the evaluation results more convincing, in this study weights were obtained by combining the subjective and objective methods.

Sichuan province is a typical humid region of Southern China, with annual precipitation of 1000–1200 mm. The objective of this paper is to evaluate WIDL in 21 cities (states) of Sichuan province. How could we evaluate the development level of water-saving irrigation strategies? First, we established a comprehensive evaluation index system of WIDL from three aspects of engineering, agronomy, and management. Second, in order to simplify the evaluation indexes and enhance the accuracy of index weight, the principal component analysis (PCA) was used to extract the principal indexes, and the combined weight method was used to judge the relative importance of each index. Third, the TOPSIS method was improved to evaluate WIDL in 21 cities (states) of Sichuan province.

2. Materials and Methods

2.1. Study Area

Sichuan province, covering 21 cities (states), is the main producing area of grain in China (Figure 1). In this area, 500,000 small and medium-sized irrigation districts were built up until 2013. The effective irrigation area and water-saving irrigation area are 2647 thousand hm^2 and 1125 thousand hm^2, respectively. High-efficiency water-saving irrigation areas, with sprinkler irrigation, micro-irrigation, and pipe conveying irrigation of 37.9, 9.5, and 50 thousand hm^2 respectively, account for only 8% of the water-saving irrigation area. The demand for irrigation water in Sichuan province is 5595 m^3 per hm^2, the per capita income of the agriculture population is 5239 CNY, and the generalization area of agriculture water-saving technology is 667 thousand hm^2. A total of 3540 water user associations have been built to control 714.1 thousand hm^2 of irrigation croplands. A property rights system reform has been implemented in 380 thousand small water conservation projects. The current agriculture water price standard is 50% of the average water supply. The agricultural irrigation water fee in Sichuan province is supposed to be 320 million CNY, but the actual yield is less than 80%.

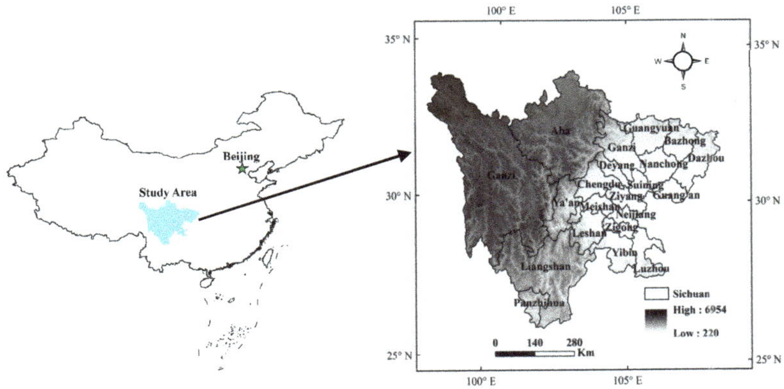

Figure 1. Twenty-one cities (states) of Sichuan province.

2.2. Methods

2.2.1. Principal Components Extracting of Evaluation Indexes

The principal component analysis (PCA) is one of the most widely applied tools which allows researchers to manipulate more variables [30–34]. The aim of the PCA algorithm is to reduce the dimensionalities of variables and meanwhile keep much information about the variables. When m principal components whose cumulative contribution rate ≥85% are selected in the real issue, this indicates that the first m principal components contain the total information of all indexes. In the paper, the indexes with the highest correlation with m principal components were chosen, which reduced the number of evaluation indexes and provided convenience for practical problems analysis.

2.2.2. Weight Assignments for the Indexes

The analytic hierarchy process (AHP) is a non-structure decision theory built by operational research experts Saaty [35]. It is applied perfectly to complex problems which are difficult to fully address with quantitative indexes for analysis. The basic idea is that the decision maker decomposes the complex problem into several levels and elements. A simple comparison, judgment, and calculation were carried out for the elements in order to get the weights of the different elements and the pending program. The entropy weight method (EWM) is found to be very useful in multi-attribute problems [36]. The smaller the value of the information entropy of the given parameter is, the larger the amount of contribution it will provide for the comprehensive evaluation.

We multiply the weights of the j-th index calculated by the above two methods, and normalize the product to obtain the combination weight ω_j:

$$\omega_j = \frac{\alpha_j \times \beta_j}{\sum\limits_{j=1}^{n} (\alpha_j \times \beta_j)}, \ 1 \leq j \leq n \tag{1}$$

where α_j, β_j are the weights of the j-th index determined by AHP and EWM, separately.

2.2.3. The Improved TOPSIS Method

TOPSIS (Technique for Order Preference by Similarity to an Ideal Solution), first developed by Hwang and Yoon (1981), is a simple ranking method which attempts to choose alternatives in the shortest distance from the positive ideal solution and the farthest distance from the negative ideal solution simultaneously [37,38]. The TOPSIS method can be summarized in a series of steps.

Step 1 involves the construction of the original performance rating matrix. A set of cases ($M = (M_1, M_2, \ldots, M_m)$) are compared with respect to a set of attributes ($C = (C_1, C_2, \ldots, C_n)$). The performance matrix can be obtained as follows:

$$
Z = \begin{bmatrix}
 & C_1 & C_2 & \cdots & C_n \\
M_1 & Z_{11} & Z_{12} & \cdots & Z_{1n} \\
M_2 & Z_{21} & Z_{12} & \cdots & Z_{2n} \\
\vdots & \vdots & \vdots & \vdots & \vdots \\
M_m & Z_{m1} & Z_{m2} & Z_{m3} & Z_{mn}
\end{bmatrix} \quad (i = 1, 2, \cdots, m; \quad j = 1, 2, \cdots, n) \tag{2}
$$

Step 2 involves the construction of the normalized performance rating matrix.

$$
v_{ij} = \frac{Z_{ij}}{\sqrt{\sum_{i=1}^{m} Z_{ij}^2}} \quad (i = 1, 2, \cdots m; \quad j = 1, 2, \cdots, n) \tag{3}
$$

Step 3 is the construction of the weighted normalized decision matrix. Attribute weights (w_j) have to be determined to indicate their relative importance and to calculate the weighted normalized values (r_{ij}) through:

$$
r_{ij} = w_{ij} v_{ij} \quad (i = 1, 2, \cdots, m; \quad j = 1, 2, \cdots, n) \tag{4}
$$

where w_j is the index weight determined by both AHP and EWM.

Step 4 indicates the determination of the positive ideal solution (PIS) and negative ideal solution (NIS).

$$
\begin{aligned}
PIS = A^+ &= \left\{ r_1^+(x), r_2^+(x), \cdots, r_j^+(x), \cdots, r_n^+(x) \right\} \\
&= \left\{ \max_i (r_{ij}(x) | j \in J_1), \min_i (r_{ij}(x) | j \in J_2), | i = 1, 2, \cdots, m \right\}
\end{aligned} \tag{5}
$$

$$
\begin{aligned}
NIS = A^- &= \left\{ r_1^-(x), r_2^-(x), \cdots, r_j^-(x), \cdots, r_n^-(x) \right\} \\
&= \left\{ \min_i (r_{ij}(x) | j \in J_1), \max_i (r_{ij}(x) | j \in J_2), | i = 1, 2, \cdots, m \right\}
\end{aligned} \tag{6}
$$

where J_1 and J_2 are the benefit and the cost attributes, and $r_j^+(x)$ and $r_j^-(x)$ are the maximum and minimum values for the j-th attribute.

Step 5 involves the calculation of the separation from the PIS and the NIS between alternatives. The determining distances S_i^+, S_i^- of each scheme away from the PIS and the NIS are as given as:

$$
S_i^+ = \sqrt{\sum_{j=1}^{n} \left[v_{ij}(x) - v_j^+(x) \right]^2}, \quad i = 1, \cdots, m \tag{7}
$$

$$
S_i^- = \sqrt{\sum_{j=1}^{n} \left[v_{ij}(x) - v_j^-(x) \right]^2}, \quad i = 1, \cdots, m \tag{8}
$$

Step 6 involves the calculation of the similarities to the ideal solution. The scheme decision and relative closeness between kinds of schemes and ideal solution are as follows:

$$\varepsilon_i = \frac{S_i^-}{S_i^+ + S_i^-} \qquad i = 1, 2, \cdots, m \tag{9}$$

Scheme M_i is sequenced according to ε_i. The larger the value of ε_i, the closer scheme M_i is to the ideal solution, which is better. In the opposite case, the scheme is worse.

The above TOPSIS method (traditional method) is only a ranking method. It cannot judge the level of the scheme M_i according to ε_i. In the study, the corresponding classification thresholds (the number of classification standard is k) of each index were seen as k schemes, which were evaluated together with the original program using the TOPSIS method. ε'_j ($j \leq k$) is the relative closeness between the k schemes and ideal solution. Then, the relative closeness was graded as $(\varepsilon'_1 \sim (\varepsilon'_1 + \varepsilon'_2)/2), \ldots, (\varepsilon'_{k-2} + \varepsilon'_{k-1})/2) \sim (\varepsilon'_{k-1} + \varepsilon'_k)/2, (\varepsilon'_{k-1} + \varepsilon'_k)/2)$, by which we can determine which level the scheme M_i is in.

3. Results

3.1. Construction of Evaluation System

The regional water-saving strategies were divided into three categories: Engineering, agronomy, and management water-saving. The indexes selected and the relationships among them were used as the foundation to establish the comprehensive evaluation index system of WIDL.

3.1.1. Engineering Water-Saving Evaluation Indexes

Engineering construction is the foundation of water-saving irrigation, and the contents are selected from the perspective of advanced and economic technology and reducing losses caused in the course of water transportation. Considering the present operation level and factors influenced by the engineering in the irrigation area, this research selected 12 indexes (Table 1) based on technical specifications for water-saving irrigation engineering (Chinese National Standard GB/T 50363-2006), and a detailed description of the following indexes was presented:

L_1 represents the ratio of the water-saving irrigation area to the effective irrigation area. This ratio indirectly reflects the water-saving level. The effective irrigation area refers to those areas of farmland, equipped with water sources and irrigation engineering subsidiaries, and their ability to conduct irrigation in normal years.

L_2 represents the ratio of the established high-efficiency water-saving irrigation area to the total water-saving irrigation area. High-efficiency water-saving engineering is also named as pressure irrigation, normally including some advanced irrigation methods such as sprinkler irrigation, micro-irrigation, drip irrigation, etc. Applying high-efficiency water-saving irrigation technology greatly increases water utilization efficiency.

L_3 represents the water efficiency of irrigation. This refers to the ratio of irrigated water available to crops in the field to the volume of water transported from the canal head. Since 2005, calculation of the water efficiency of irrigation has been performed all over China.

L_4 represents the economic benefits per hm^2 of water-saving irrigation. This is a comprehensive index used to evaluate the effects of water-saving irrigation. A complicated calculation method is required to figure out the value of it, and a lot of factors should be taken into consideration. On one hand, the comprehensive benefits of water-saving irrigation for economic crops vary a lot from food crops; on the other hand, the costs of economic crops and food crops vary greatly because of the use of different indexes such as operation costs, water-saving costs, and energy-saving costs, etc., used in the different engineering projects.

L_5 represents the water usage amount per hm^2 for irrigation. Water usage amount refers to the water consumed for the growth of crops, including loss amount in water delivery. It varies a great deal in different areas for different crops and in different weather conditions. The water usage amount equals the sum of irrigation water consumption in each growth periods of crops while considering the water loss caused by the irrigation system.

L_6 represents the ratio of water-saving irrigation area to cultivated area. Cultivated area refers to the field for growing crops and includes irrigation paddy fields, upland fields, and irrigated land. The cultivated area is larger than the irrigation area. The ratio of water-saving irrigation area to cultivated area, on the one hand, objectively reveals the local agricultural irrigation development level, and on the other hand, it reflects the present development level of water-saving agriculture.

L_7 represents the main crops yield per unit. This index significantly reflects the production benefit as well as the scientific development of the planting industry. The usual method used to increase the per unit yield of the main crops is the rational close planting method which provides relatively large space for sunlight and helps control organic consumption in respiration within a relatively small amount at the same time; an optimum density is selected by considering those two factors to increase the organic accumulation.

L_8 represents per capita GDP. Per capita GDP significantly reflects economic development and people's living standards. Water-saving irrigation commences with a rather late start, a large scale of engineering projects is under construction, and heavy investment for water-saving engineering construction is eagerly demanded. Investing capacity is closely related to regional economic development, so per capita GDP is considered as one index in engineering the water-saving evaluation.

L_9 represents the percentage of canal lining. This refers to the ratio of the calculated area of canal seepage proof to the maximum flow section area.

L_{10} represents the percentage of working lining channels Restrained by a depressed economy and poor technology during the period of construction, most channels were built below the construction standard in the irrigation area. Furthermore, after working for years, the channels have suffered some degree of damage. Therefore, maintenance for the channel system is needed. The percentage of working lining channels is significant to reflect the maintenance of damaged channels.

L_{11} represents the water usage amount for agriculture. This covers the water used for field irrigation, fishery, and forestry and fruit industries. Water used for forestry and the fruit industry belongs to the classification of agricultural irrigation water. Water used for agriculture is less likely to increase because of water shortage. Sichuan province measures the water usage amount for agriculture, and thus the amount can be directly obtained.

L_{12} represents agriculture investment. Agriculture investment reflects local economic development level as well as the degree of attention paid to agriculture. Funds invested in water-saving irrigation construction and agricultural investment come from various sources such as national finance, local finance, credit funds, collective economy, individual investment, or even foreign investment. With rapid economic development, the amount of funds invested in water-saving irrigation construction from local finance, collective agricultural economic organizations, and individual investments tends to increase year by year.

3.1.2. Agronomy Water-Saving Evaluation Indexes

Agronomy water-saving refers to a certain comprehensive agricultural technology which integrates the resources of water, soil, and crops by optimizing the cropping and farming system to effectively reduce soil evaporation and luxury transpiration of crops, and further improves water efficiency with regards to water-saving and productivity. The main features of the agronomy water-saving strategy are the rational layout of farm crops and the improvement of planting methods. Five indexes were selected from the contents to establish the evaluation system and a detailed description of the following indexes was presented as follows.

M_1 refers to the ratio of the rice dry nursery seedling area to the rice planting area. The rice dry nursery seedling ratio is distinctive with a short seedling stage and convenience of management, with the advantages including saving field planting resources, water conservation, high economic benefit, disease resistance, cold resistance, and so on. Compared with the water-raised seedling, the rice dry nursery seedling saves roughly 50–90% of water use.

M_2 represents the ratio of the area with the "Thin-Shallow-Wet-Dry" technique to the rice cultivated area. The "Thin-Shallow-Wet-Dry" irrigation technology is a scientific water-saving method used to plant rice. "Thin" means that a thin layer of water is needed for seedlings to take root when planting. "Shallow" means that rice seedlings turn green in shallow water. "Wet" refers to the practice of keeping the water-holding seedlings at an early tilling stage. "Dry" means stopping irrigation and adapting the seedlings to an external environment to train the seedlings at a late tilling stage. This technology is consistent with the water demand of the rice growth at different stages, which helps to save irrigation water as well as to invigorate its physiological activities to save water and increase production.

M_3 refers to the per capita income of the agriculture population. The development and popularization of water-saving irrigation technology has much to do with the economic development level, as applying a series of measures and technologies is concerned with the agronomy water-saving demand investment. So far, government finance-oriented investment has not been established in the application of agronomy technology. Therefore, the application of this technology is closely related to the income of the local population.

M_4 refers to the ratio of the drought-tolerant crops cultivated area to the cultivated area of all crops. The outstanding advantage of drought-tolerant crops is the water-saving advantage, hence one of the effective measures to solve water shortages in dry farming areas is to grow drought-tolerant crops. This practice has been popularized in Northwest of China; however, more efforts should be made to implement this in Sichuan province.

M_5 refers to water use efficiency. Water use efficiency (WUE) refers to the quantity of economic product yielded by water consumption and it is significant as a reflection of the water-saving irrigation efficiency. At present, WUE is the major economic index used to evaluate water-saving irrigation benefits.

3.1.3. Management Water-Saving Evaluation Indexes

Management water-saving includes all management and maintenance work after the completion of a water-saving construction project. Sixteen indexes, which were divided into 6 qualitative indexes and 10 quantitative indexes, were included in this section. Six indexes, i.e., the establishment of subsidiary policies and regulations (N_1), the degree of support from the government (N_4), propaganda and education level (N_5), water-saving engineering design level (N_7), the rationality of the irrigation system (N_{10}), and the sound level of the water-saving incentive mechanism (N_{14}), which are difficult to evaluate with data-measurement, were obtained using the investigation method. We described the 16 indexes as follows:

N_1 refers to the establishment of subsidiary policies and regulations. The perfection of legal systems and regulations is effective for the practice of irrigation water management. Thus, for the local government, promoting water-saving technology is a top priority in compiling an agricultural development plan, and much attention should be paid to develop, demonstrate, and extend key water-saving technologies.

N_2 refers to the reform execution situation of the property system. The reform execution situation of the property system refers to the proportion of the number of identified property rights and identified management and maintenance of main body projects to the number of small-scale agricultural water conservancy projects.

N_3 refers to the construction level of the technology popularizing system. This refers to the proportion of existing technical service institutions to the technical service institutions which are supposed to be established.

N_4 refers to the degree of support from the government. The attention paid to water saving strategies from governments is of benefit for the earlier execution of related water-saving regulations, measures, and funding.

N_5 refers to the propaganda and education level. Propaganda promoting water saving targets both government officials and citizens, especially farmers and other irrigation water users. Improving the public's water-saving consciousness is a long-term task. Water saving habits and consciousness should be cultivated among the public, and water saving knowledge should be popularized as well.

N_6 refers to the user participation level. This indicates the proportion of user-based management irrigation areas to total irrigation areas.

N_7 refers to the water-saving engineering design level. It is key to the promotion of water-saving irrigation to guarantee engineering quality, as well as ensure good operation of the system across long periods of time.

N_8 refers to the degree of perfection of engineering subsidiaries. This indicates the proportion of existing canal-attached facilities to attached facilities which are supposed to be constructed according to statistical data from the irrigation region.

N_9 refers to the management and maintain level of water-saving engineering. This indicates the proportion of the maintenance expense of irrigation engineering to the required maintenance cost.

N_{10} refers to the rationality of irrigation system. With the enforcement of the rationality of the irrigation system, the optimum design of water-saving irrigation systems has become increasingly important. The irrigation system design should be developed on the basis of comprehensive technologies of agricultural water-saving irrigation and experimental data, which underlies the basic research foundation of this method.

N_{11} refers to monitoring the coverage rate of soil moisture. The monitoring coverage rate of soil moisture refers to the proportion of existing monitoring stations to the total number of planned soil moisture monitoring stations.

N_{12} refers to the degree to which water measurement is popularized, indicating the proportion of practicing water measurement areas to total irrigation areas.

N_{13} refers to the ratio of planned water use irrigation areas to total irrigation areas.

N_{14} refers to the sound level of water-saving incentive mechanisms. Establishing water rights-based optimal water dispatching and allocation is the core of the water-saving incentive mechanism. It is urgent to accelerate the confirmation of agricultural irrigation water rights, to develop the water right trade market, and to establish agricultural water saving incentive mechanisms.

N_{15} refers to the popularity rate of the measurement charges of water, indicating the proportion of practicing water measurement charges areas to total irrigation areas.

N_{16} refers to the rationality of the water price. The rationality of the water price refers to the proportion of the collected price for agricultural irrigation water to the cost.

3.1.4. The Comprehensive Evaluation Index System of WIDL

The hierarchy of the comprehensive evaluation index system was classified into three layers: The target layer, system layer, and index layer. According to the actual situation in Sichuan province and available data, 12, 5, and 16 indexes were adopted for engineering, agronomy, and management WIDL, respectively. The target layer (first-class index) was the evaluation result of WIDL, the system layer (second-class index) contained engineering, agronomy and management WIDL, and the index layer (third-class index) included the 33 specific evaluation indexes (Table 1). The 33 indexes were divided into 2 categories according to the method of data acquisition. One included the 6 qualitative indexes using the questionnaire method, and the other included the 27 quantitative indexes from the 2014 Sichuan province Provincial Water Conservancy Statistical Yearbook, the 2014 Sichuan province Water Management Yearbook, and the 2014 Sichuan province Agricultural Yearbook.

Table 1. The index system of the comprehensive evaluation model for the water-saving irrigation development level (WIDL).

Target Layer	System Layer	Index Layer	Source
Water-saving irrigation developing level	Engineering water-saving	Ratio of water-saving irrigation area to effective irrigation area (%) (L_1)	B,D
		Ratio of established high-efficient water-saving irrigation area to the total water-saving irrigation area (%) (L_2)	C
		Water efficiency of irrigation (L_3)	C
		Economic benefits per hm^2 of water-saving irrigation (CNY) (L_4)	D
		Water usage amount per hm^2 for irrigation (m^3) (L_5)	B,D
		Ratio of water-saving irrigation area to cultivated area (%) (L_6)	B,C
		Main crops yield per unit (kg/hm^2) (L_7)	B
		Per capita GDP (CNY) (L_8)	D
		Percentage of canal lining (%) (L_9)	C
		Percentage of working lining channels (%) (L_{10})	C
		Water usage amount for agriculture (m^3/hm^2) (L_{11})	C,D
		Agriculture investment (CNY/hm^2) (L_{12})	B,C
	Agronomy water-saving	Ratio of rice dry nursery seedling area to rice planting area (%) (M_1)	B,D
		Ratio of area with the technique of "Thin-Shallow-Wet-Dry" to rice cultivated area (%) (M_2)	D
		Per capita income of agriculture population(CNY) (M_3)	B
		Ratio of drought tolerant crops cultivated area to all crops cultivated area (%) (M_4)	B,D
		Water use efficiency (Kg/m^3) (M_5)	C,D
	Management water-saving	Establishment of subsidiary policies and regulations (N_1)	A
		Reform execution situation of property system (%) (N_2)	B,C
		Construction level of technology popularizing system (N_3)	B,C
		Degree of support from the government (N_4)	A
		Propaganda and education level (N_5)	A
		User participation level (N_6)	B,C
		Water-saving engineering design level (N_7)	A
		Degree of engineering subsidiaries perfection(N_8)	B
		Management and maintained level of water-saving engineering (N_9)	C
		Rationality of irrigation system (N_{10})	A
		Monitoring coverage rate of soil moisture (%) (N_{11})	C,D
		Degree of popularizing water measurement (%) (N_{12})	B,C
		Ratio of planned water use (N_{13})	C
		Sound level of water-saving incentive mechanisms (N_{14})	A
		Popularity rate of measurement charges of water (%) (N_{15})	B,C
		Rationality of water price (N_{16})	C

Note: A, B, C, and D respectively indicate the qualitative indexes, the 2014 Sichuan province Provincial Water Conservancy Statistical Yearbook, the 2014 Sichuan province Water Management Yearbook, and the 2014 Sichuan province Agricultural Yearbook.

3.2. Index Quantification

We adopted a questionnaire method among the staff in the water administration department in Sichuan province and 21 cities (states) to quantify the 6 qualitative indexes. The staff had ample knowledge of water-saving irrigation management and were very familiar with the conditions of Sichuan province. Therefore, they could give a proper judgment or assessment for each index. Their work provided reliable information for our research. The current survey data of 6 qualitative indexes were quantified as follows:

$$Q = (f_1 \times M_1 + f_2 \times M_2 + f_3 \times M_3 + f_4 \times M_4 + f_5 \times M_5)/N \tag{10}$$

where $M_1 \sim M_5$ ($M_1 = 10$, $M_2 = 8$, $M_3 = 6$, $M_4 = 4$, $M_5 = 2$) were the quantitative index values corresponding to "better, good, general, poor, poorer". $f_1 \sim f_5$ were the sample numbers in the corresponding class and N is the total sample numbers.

Taking Sichuan province as an example, 20 qualified staff from the water administration department of Sichuan province were invited to the survey to quantify the 6 indexes. The staff had ample knowledge of water-saving irrigation and were very familiar with the conditions of Sichuan province. Therefore, they could give a proper judgment or assessment for each index. Their work provided reliable information for our research. The quantitative values of the 6 indexes in Sichuan province were obtained by Equation (10) as seen in Table 2.

Table 2. The quantification of the management WIDL indexes in Sichuan province.

No.	Qualitative Indexes	Quantitative Value
1	N_1	6.8
2	N_4	6.7
3	N_5	6.1
4	N_7	7.2
5	N_{10}	6.2
6	N_{14}	4.7

Similarly, 6 qualitative evaluation indexes of WIDL in 21 cities (states) of the Sichuan province were quantified, and 33 indexes of WIDL in the Sichuan province and 21 cities (states) were shown in Table 3.

3.3. Principal Component Extracting

Although the 33 evaluation indexes contained more comprehensive information on the development level of water-saving irrigation, they will increase the complexity of the evaluation process. We therefore chose PCA to simplify the engineering, agronomy, and management evaluation indexes, respectively. The principal components of engineering WIDL in Sichuan province and 21 cities (states) were analyzed using PCA in the DPS program. The eigenvalues and contribution rates of the principal components were shown in Table 4. From Table 4, the accumulative contribution rate of the first 5 principal components was up to 89.64% and above 85%. Therefore, the first 5 principal components can represent all the information of the 12 engineering WIDL.

The weights of coefficients between each index and the five principal components (F_1–F_5) were then shown in Table 5. The indexes with maximum weight coefficients were selected as the simplified principal indexes. Thus, the engineering WIDL evaluation indexes were simplified to 6 indexes as the water efficiency of irrigation (L_3), the economic benefits per hm^2 of water-saving irrigation (L_4), the ratio of water-saving irrigation area to cultivated area (L_6), the percentage of canal lining (L_9), and the agriculture investment (L_{12}).

Similarly, three indexes, including per capita income of agriculture population (M_3), water use efficiency (M_5), and the ratio of the area with the technique of "Thin-Shallow-Wet-Dry" to the rice cultivated area (M_2) were extracted from five indexes of agronomy WIDL using PCA with the accumulative contribution rate of 88.6%. The evaluation indexes of management WIDL could be simplified to six items as follows: the degree of support from the government (N_4), the degree of popularizing water measurement (N_{12}), the sound level of water-saving incentive mechanisms (N_{14}), the construction level of technology popularizing systems (N_3), the monitoring coverage rate of soil moisture (N_{11}), and the water-saving engineering design level (N_7). So, 33 evaluation indexes were simplified to 14 indexes, which greatly facilitated our evaluations (Table 6).

The grading standard of each index is the basis for determining the WIDL grade. Five grades of evaluation standards were established for each index of the regional WIDL according to the mean and standard deviation grading method and the actual situation of Sichuan province. These standards, (better, good, general, poor, and poorer) correspond to the I–V levels shown in Table 6 [39]. Although this standard will be further improved or updated in the future, at this stage we established a relatively objective evaluation system to ensure the accuracy of the evaluation results.

Table 3. The thirty-three indexes of WIDL in Sichuan province and 21 regions.

Region	L_1	L_2	L_3	L_4	L_5	L_6	L_7	L_8	L_9	L_{10}	L_{11}	L_{12}	M_1	M_2	M_3	M_4	M_5	N_1	N_2	N_3	N_4	N_5	N_6	N_7	N_8	N_9	N_{10}	N_{11}	N_{12}	N_{13}	N_{14}	N_{15}	N_{16}
Sichuan Province	42.00	1.55	0.40	975	5595	28	5254	21,369	45.16	36.80	3121	515	7.50	14.70	2802	22	1.20	6.80	0.67	0.67	6.70	6.10	0.31	7.20	0.73	0.45	6.20	0.43	0.10	0.66	4.70	0.08	0.33
Chengdu	67.50	3.00	0.43	1286	7365	63	6134	41,253	72.00	32.40	7550	746	25.60	17.30	4485	31	1.12	7.00	0.98	0.90	9.20	6.40	0.50	7.30	0.83	0.80	7.20	0.55	0.15	0.98	6.00	0.13	0.44
Zigong	29.10	0.04	0.38	630	5280	19	5856	23,613	41.15	38.20	1843	705	2.70	0.01	3188	23	1.33	6.00	0.8	0.80	5.80	5.70	0.37	5.50	0.74	0.40	5.80	0.34	0.08	0.50	3.00	0.07	0.40
Panzhihua	66.16	0.72	0.43	1388	9690	51	5459	43,959	33.29	46.60	4580	510	6.50	55.60	3463	19	1.28	7.60	0.68	0.65	8.00	6.80	0.39	5.80	0.96	0.36	7.20	0.60	0.08	0.72	5.80	0.08	0.60
Luzhou	35.04	0.23	0.40	837	3345	21	5441	16,698	45.41	31.00	1694	659	2.00	1.70	3165	25	1.30	6.70	0.77	0.90	6.90	6.00	0.49	8.00	0.78	0.16	6.90	0.28	0.05	0.53	3.90	0.05	0.35
Deyang	53.49	0.07	0.41	1082	9090	45	6435	25,335	58.53	46.00	6944	524	50.00	41.00	3585	33	1.13	7.70	0.98	0.80	7.50	7.00	0.40	6.80	0.9	0.59	5.60	0.51	0.13	0.82	4.40	0.13	0.62
Mianyang	41.53	0.80	0.40	1074	6150	33	5332	20,053	61.00	36.80	4430	478	20.20	0.80	3179	34	1.36	7.20	0.66	0.70	7.20	6.20	0.39	6.80	0.6	0.72	5.60	0.51	0.10	0.45	4.60	0.05	0.38
Guangyuan	36.95	1.27	0.38	788	4215	20	5306	12,313	40.45	48.00	1696	585	2.40	54.80	2000	11	1.31	5.60	0.38	0.60	6.60	5.90	0.13	6.90	0.35	0.32	6.60	0.30	0.10	0.50	4.00	0.05	0.46
Suining	28.40	0.51	0.39	767	3495	25	5257	14,498	46.05	48.80	1992	566	9.70	22.00	2828	18	1.11	6.30	0.72	0.75	5.00	5.80	0.25	5.80	0.93	0.13	5.80	0.51	0.10	0.50	3.40	0.02	0.50
Neijiang	39.81	4.14	0.39	844	2700	29	4923	18,022	43.85	37.60	2007	579	4.70	17.10	2986	25	1.21	7.20	0.79	0.70	6.60	5.40	0.24	6.00	0.78	0.32	6.00	0.40	0.02	0.49	5.00	0.12	0.55
Leshan	18.91	5.73	0.39	870	9825	13	4720	22,490	47.26	36.00	4269	464	0.60	8.90	3241	26	1.06	5.40	0.69	0.75	6.10	6.30	0.50	6.10	0.89	0.55	6.00	0.45	0.15	0.62	5.10	0.11	0.70
Nanchong	37.59	0.05	0.37	806	3060	28	5589	13,212	52.59	52.60	1738	479	1.40	0.80	2646	29	1.41	7.30	0.91	0.50	6.60	6.60	0.30	8.40	0.74	0.88	6.90	0.38	0.13	0.78	2.80	0.10	0.48
Meishan	39.14	6.24	0.39	900	7530	41	5617	18,586	50.83	30.00	2987	454	10.10	32.50	3284	20	1.23	7.70	0.65	0.60	7.30	6.70	0.15	2.80	0.72	0.45	5.50	0.38	0.13	0.72	4.90	0.06	0.45
Yibin	51.01	0.46	0.39	854	2520	26	5585	19,499	55.32	32.80	2036	690	6.20	6.70	3068	31	1.08	5.60	0.71	0.75	3.40	5.30	0.26	3.40	0.62	0.68	4.20	0.30	0.09	0.45	2.80	0.07	0.38
Guang'an	28.28	0.22	0.39	780	2490	18	5571	15,588	34.13	46.00	1790	509	2.70	7.90	2915	19	1.32	6.40	0.73	1.00	6.80	6.40	0.50	6.80	0.78	0.55	6.00	0.31	0.06	0.46	6.00	0.06	0.33
Dazhou	34.19	0.01	0.38	770	1710	19	5221	14,623	52.04	54.00	1073	812	10.30	27.60	2943	29	1.43	7.70	0.74	0.70	6.20	5.50	0.24	6.80	0.79	0.50	5.50	0.35	0.06	0.45	5.30	0.06	0.30
Ya'an	38.23	0.00	0.38	714	10,200	32	4243	18,881	32.03	44.40	2847	431	2.10	0.01	2829	10	1.14	6.10	0.8	0.80	6.30	5.60	0.25	7.80	0.77	0.42	5.50	0.35	0.06	0.45	2.80	0.07	0.45
Bazhong	27.44	0.18	0.37	819	1560	14	5372	8717	34.54	52.00	4479	800	3.00	19.10	2031	10	1.33	6.50	0.73	0.50	8.00	7.50	0.38	7.50	0.73	0.32	7.50	0.30	0.07	0.40	2.80	0.07	0.45
Ziyang	39.99	0.82	0.38	966	4065	26	4318	16,644	41.79	20.00	1957	500	16.10	0.01	2988	23	1.29	5.80	0.56	0.50	5.80	5.30	0.37	7.50	0.87	0.33	5.00	0.40	0.05	0.55	3.50	0.05	0.50
Aba	9.65	2.49	0.36	630	3540	3	3100	14,662	28.43	23.60	1421	439	0.01	0.01	1881	5	1.01	3.10	0.001	0.001	3.00	3.20	0.001	3.20	0.35	0.001	2.80	0.001	0.001	0.001	5.30	0.001	0.001
Ganzi	16.39	0.70	0.36	602	1905	6	2824	11,659	27.80	46.00	590	327	0.01	0.01	1309	4	1.01	7.00	0.001	0.001	6.50	7.80	0.001	6.20	0.45	0.001	6.90	0.001	0.001	0.02	5.10	0.001	0.001
Liangshan	58.16	0.05	0.37	713	9600	23	4358	17,560	49.82	31.00	4120	302	0.01	0.01	2438	28	1.25	6.10	0.41	0.30	7.80	7.20	0.12	7.20	0.94	0.13	7.00	0.001	0.02	0.24	2.90	0.001	0.25

Table 4. The eigenvalues and contribution rates of the principal components.

No.	Eigenvalues	Contribution Rate (%)	Cumulative Contribution Rate (%)
F_1	6.0462	50.3852	50.3852
F_2	1.7518	14.5984	64.9836
F_3	1.424	11.8667	76.8504
F_4	0.7881	6.5679	83.4182
F_5	0.7467	6.2224	89.6406

Table 5. The loading matrix of the principal components.

Evaluation Indexes	F_1	F_2	F_3	F_4	F_5
L_1	0.35	0.15	−0.04	0.15	−0.34
L_2	0.08	−0.55	−0.11	−0.13	0.41
L_3	0.57	0.09	−0.01	0.23	0.18
L_4	0.37	0.05	0.52	0.27	0.16
L_5	0.27	−0.33	0.29	−0.41	−0.33
L_6	0.58	0.05	−0.07	−0.01	−0.04
L_7	0.25	0.36	−0.32	−0.21	0.24
L_8	0.35	−0.07	0.14	0.28	0.01
L_9	−0.03	0.57	0.24	−0.52	0.27
L_{10}	0.25	0.06	−0.56	−0.18	−0.03
L_{11}	0.17	0.14	0.63	0.13	0.27
L_{12}	−0.32	0.26	−0.12	0.47	0.65

Table 6. The classification standard of the WIDL evaluation indexes.

Indexes	I	II	III	IV	V
L_3	>0.6	0.5~0.6	0.45~0.5	0.40~0.45	≤0.40
L_4	>2250	1650~2250	1050~1650	450~1050	≤450
L_6	>60	60~45	45~30	30~15	≤15
L_9	>75	60~75	50~60	35~50	≤35
L_{12}	≥1500	1200~1500	600~1200	300~600	<300
M_2	>50	35~50	15~35	5~15	≤5
M_3	>8000	6000~8000	4000~6000	2000~4000	≤2000
M_5	>1.75	1.35~1.75	1.0~1.35	0.8~1.0	≤0.8
N_3	>0.86	0.71~0.86	0.56~0.71	0.40~0.56	≤0.40
N_4	9~10	7~9	5~7	3~5	0~3
N_7	>7.94	7.02~7.94	6.10~7.02	5.18~6.10	≤5.18
N_{11}	>0.50	0.40~0.50	0.30~0.40	0.20~0.30	≤0.20
N_{12}	>0.12	0.10~0.12	0.07~0.10	0.04~0.07	≤0.04
N_{14}	>5.41	4.78~5.41	4.15~4.78	3.52~4.15	≤3.52

3.4. Weight Determination

Assigning weights to the 14 indexes was a key step in our evaluation. The weight vector of the second-class indexes determined by AHP was α = (0.395, 0.262, 0.344), which passed the consistency test. The weights of the third-class indexes were determined considering both subjective weights vector α by AHP and objective weights vector β by EWM, and the combined weight ω was acquired by Equation (9). The final weights of each index were shown in Table 7, and the distributions of several indexes with high weights in 21 regions were shown in Figure 2. From Table 7, the ratio of water-saving irrigation areas to cultivated land area (L_6), which was the direct reflection of the development level of water-saving irrigation, had the highest weight of 0.132, followed by the per capita income of the agriculture population (M_3), the water use efficiency (M_5), the degree of support from government (N_4), and the water-saving engineering design level (N_7) with higher weights of about 0.1. The construction level of technology popularizing system (N_3), the monitoring coverage rate of soil moisture (N_{11}), the degree of popularizing water measurement (N_{12}), and the sound level of water-saving incentive mechanisms (N_{14}) had the smallest weights (<0.05).

Table 7. The weights of each evaluation index.

Target Layer	System Layer	Second-Class Weights	Indexes Layer	Third-Class Weights	Final Weights
Regional water-saving irrigation developing level	Engineering water-saving	0.395	L_3	0.167	0.066
			L_4	0.163	0.064
			L_6	0.334	0.132
			L_9	0.167	0.066
			L_{12}	0.169	0.067
	Agronomy water-saving	0.262	M_2	0.246	0.064
			M_3	0.378	0.099
			M_5	0.376	0.098
	Management water-saving	0.344	N_3	0.119	0.041
			N_4	0.274	0.094
			N_7	0.266	0.092
			N_{11}	0.116	0.040
			N_{12}	0.114	0.039
			N_{14}	0.111	0.038

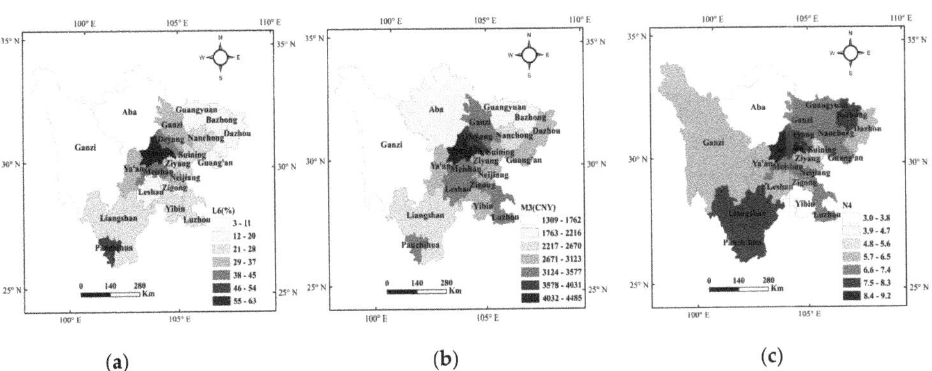

Figure 2. The key evaluation indexes between 21 regions of the Sichuan province. (**a**) The ratio of water-saving irrigation area to cultivated area (%) (L_6); (**b**) the per capita income of the agriculture population (CNY) (M_3); (**c**) the degree of support from the government (N_4).

3.5. The Calculation of Relative Closeness

In the study, we made some improvements when adopting TOPSIS method. Firstly, the corresponding classification thresholds of 14 indexes were seen as 5 schemes. Based on the classification thresholds in Table 6 and the principal indexes in the 21 cities (states) and in Sichuan province, decision matrix Z was constructed as follows. In matrix Z, the first 5 lines were the thresholds of the 14 indexes corresponding to the five categories (better, good, general, poor, and poorer), and the next 22 lines were the values of the 14 indexes in Sichuan province and 21 cities (states).

$$Z = \begin{vmatrix} 0.6 & 150 & 60 & 75 & 1500 & 1.75 & 50 & 8000 & 9 & 0.12 & 5.41 & 0.86 & 0.5 & 7.94 \\ 0.5 & 110 & 45 & 60 & 1200 & 1.35 & 35 & 6000 & 7 & 0.1 & 4.78 & 0.71 & 0.4 & 7.02 \\ 0.45 & 70 & 30 & 50 & 600 & 1 & 15 & 4000 & 5 & 0.07 & 4.15 & 0.56 & 0.3 & 6.1 \\ 0.4 & 30 & 15 & 35 & 300 & 0.8 & 5 & 2000 & 3 & 0.04 & 3.52 & 0.4 & 0.2 & 5.18 \\ 0 & 0 & 0 & 0 & 0 & 0 & 0 & 0 & 0 & 0 & 0 & 0 & 0 & 0 \\ 0.4 & 65 & 28 & 45.16 & 315.3 & 1.2 & 14.7 & 2802 & 6.7 & 0.1 & 4.7 & 0.67 & 0.43 & 7.2 \\ \vdots & \vdots & \vdots & \vdots & \vdots & \vdots & \vdots & \vdots & \vdots & \vdots & \vdots & \vdots & \vdots & \vdots \\ 0.37 & 47.5 & 23 & 49.82 & 302 & 1.25 & 0.01 & 2438 & 7.8 & 0.02 & 2.9 & 0.3 & 0.001 & 7.2 \end{vmatrix} \quad (11)$$

The standard decision matrix V was then obtained using the dimensionless method in Equation (3). The weight decision matrix R was established by multiplying the standard decision matrix V and weights w (Table 7) according to Equation (4). Finally, according to the 14 benefit indexes based on principal component analysis, the positive ideal solution (PIS) was obtained using Equation (5), and the negative ideal solution (NIS) was obtained using Equation (6). We calculated the distance S^+, S^- of 27 schemes to PIS and NIS using Equations (7) and (8), and further obtained the relative closeness ε between the 27 schemes and the ideal solution.

S^+ = (0.0055, 0.0267, 0.0523, 0.0737, 0.1023, 0.0571, 0.0432, 0.0627, 0.0437, 0.0601, 0.046, 0.0562, 0.0582, 0.057, 0.0554, 0.0661, 0.0593, 0.0523, 0.0598, 0.0622, 0.0561, 0.0626, 0.063, 0.0602, 0.0791, 0.0771, 0.0663);

S^- = (0.0991, 0.0763, 0.0525, 0.0334, 0, 0.0517, 0.0774, 0.0472, 0.0706, 0.0518, 0.0662, 0.0567, 0.0549, 0.0488, 0.0509, 0.0461, 0.0539, 0.0572, 0.0467, 0.0496, 0.053, 0.0506, 0.0493, 0.0499, 0.0339, 0.0385, 0.0474);

ε = (0.9125, 0.6301, 0.5498, 0.4124, 0, 0.4754, 0.6418, 0.4294, 0.6174, 0.4627, 0.5900, 0.5022, 0.4855, 0.4610, 0.4789, 0.4110, 0.4762, 0.5224, 0.4388, 0.4437, 0.4858, 0.4470, 0.4389, 0.4535, 0.2996, 0.3328, 0.4168);

The first 5 relative closeness (ε) values were 0.9125, 0.6301, 0.5498, 0.4124, and 0, respectively. So, the grading standard was (>(0.9125 + 0.6301)/2, (0.6301 + 0.5498)/2~(0.9125 + 0.6301)/2, (0.5498 + 0.4124)/2~(0.6301 + 0.5498)/2, (0.4124 + 0)/2~(0.5498 + 0.4124)/2, 0~(0.382 + 0)/2), that (>0.7713, 0.5899~0.7713, 0.4811~0.5899, 0.2062~0.4811, 0~0.2062), corresponding to the 5 categories (better, good, general, poor, and poorer). The WIDL in each region was judged using the grading standard, and the evaluation results are shown in Table 8 and Figure 3.

Table 8. The relative closeness (ε) of the water-saving irrigation development level in 21 regions of Sichuan province.

Region	ε	Level	Region	ε	Level
Sichuan province	0.4754	Poor	Nanchong	0.4762	Poor
Chengdu	0.6418	Good	Meishan	0.5224	General
Zigong	0.4294	Poor	Yibin	0.4388	Poor
Panzhihua	0.6174	Good	Guang'an	0.4437	Poor
Luzhou	0.4627	Poor	Dazhou	0.4858	General
Deyang	0.5900	Good	Ya'an	0.4470	Poor
Mianyang	0.5022	General	Bazhong	0.4389	Poor
Guangyuan	0.4855	General	Ziyang	0.4535	Poor
Suining	0.4610	Poor	Aba	0.2996	Poor
Neijiang	0.4689	Poor	Ganzi	0.3328	Poor
Leshan	0.4110	Poor	Liangshan	0.4168	Poor

From Table 8, the overall development level of water-saving irrigation was "poor" in Sichuan province with a ε value of 0.4754. Only 3 regions were in "good", 4 regions were "general", and the other 14 regions were "poor". By consulting experts or relevant managers from the Sichuan province

Water Conservancy Bureau, we determined that this result was consistent with the general situation of Sichuan province and was consistent with the previous reports of Lou [39].

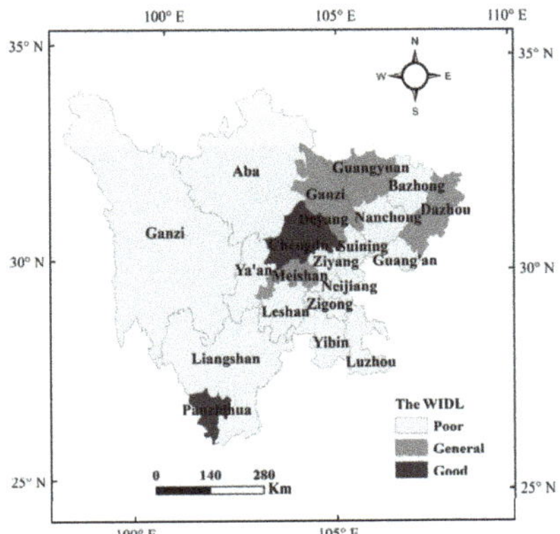

Figure 3. The water-saving irrigation development level (WIDL) in 21 regions of Sichuan province.

Only Chengdu, Panzhihua, and Deyang were at the "good" level, with ε values of about 0.6. The economic development levels in these regions were higher; the per capita GDP of Chengdu and Panzhihua was in the top 3 of Sichuan province, more than 60 thousand CNY higher than the average of 32.5 thousand CNY of Sichuan province. The per capita income of the agriculture population was relatively higher, so local people had a strong awareness of water conservation. The ratio of water-saving irrigation area to total cultivated land in those 3 regions ranks in the top 3. With better economic development level than other regions in Sichuan province and more funds invested to water-saving irrigation engineering maintenance in these regions, the existing lining canals were in better conditions than in the other 18 regions. The same was true with information management in irrigation areas. In 2004, Panzhihua was listed as the "national agricultural water-saving demonstration city", and it made a lot of exploration in engineering and management construction. The government attached great importance to water-saving irrigation development, leading to the improvement of water-saving incentive mechanisms.

The four regions of Meishan, Guangyuan, Dazhou, and Mianyang were at the "general" level, with a ε value of about 0.5. In these regions, the engineering and agronomy water-saving development level was higher than the average level of Sichuan province. The per capita GDP of these regions were up to or exceeding the average level of Sichuan province. With strong support by local government, the indexes of water use efficiency, the percentage of working lining canals, and the ratio of water-saving irrigation area to the total cultivated area were better in these 4 regions than in the other regions in Sichuan province. In GuangYuan, the agriculture planting structure was adjusted according to the differences of the water demand and drought resisting characteristics in different kinds of crops. The "Thin-Shallow-Wet-Dry" area of the rice was relatively high.

Fourteen regions were in "poor" condition, such as Neijiang, Bazhong, Ganzhi, and Aba, with ε values of less than 0.48. Most of these regions are located in mountainous and hilly areas of Sichuan province, especially Ganzi, Aba, and Liangshan, which are located on the Northwest plateau of Sichuan province, with poor natural resources and insufficient economies. This led to a lack of investment in country public utilities and a funds shortage for the construction and maintenance of water-saving engineering. With the water efficiency of irrigation below 0.4 in these regions, the ratio of the water-saving irrigation area to the total cultivated area is at the inadequate–average level in Sichuan province, only reaching around 5% in Ganzi and Aba. Agronomy water-saving and management water-saving levels were also relatively poor in these regions. Moreover, information management for irrigation (soil moisture content supervision and measurement technology) was hardly conducted in Ganzhi, Aba, and Liangshan.

4. Discussion

The shortage of water resources has become an important factor restricting the sustainable development of regional economies. The development of water-saving irrigation methods to improve the utilization efficiency of water resources is an important way to deal with the engineering water shortage and seasonal drought conditions in the humid regions of Southern China. A reasonable and accurate assessment of WIDL is not only a basis for further scientific planning and construction to promote the development of regional water-saving irrigation, but also a basis to promote and guarantee the sustainable development of agricultural production. Therefore, the objective evaluation of regional WIDL has great significance to promote the development of water-saving irrigation along a scientific, healthy, and positive track, and to ensure sustainable development of the national economy.

According to the evaluation results, great efforts should be made to increase water-saving irrigation levels in Sichuan province. Firstly, more focus should be placed on increasing the investment in agricultural water-saving irrigation and promoting water-saving engineering. The governmental authority should implement the policy that 10% of land transaction fees are extracted to invest into farmland water conservancy construction in order to solve problems such as the inadequate service of necessary facilities for key projects in the irrigation region, the low qualified canal (channel) number, the aging canal system structures, and so on. On the other hand, investment and financial reform should increase, and diversified investment should be established to play the role of the financial fund, and to fully utilize preferential policy, incentive measures, and credit aid to attract new major participants in agricultural operations such as leading enterprises, major planting farmers, and cooperatives to invest in water-saving efficiency.

Secondly, more focus should be placed on adjusting agricultural industry distribution and extending water-saving technologies. It is necessary to establish the agricultural industry layout so that it is appropriate for local water resources, land, and ecosystems, and to adjust the agricultural water utilization structure. Highly efficient water-saving irrigation strategies in accordance with water resources, planting structure, and geographic conditions, land management should be adopted. In hilly regions with water scarcity and poor water conservancy facilities, such as Guangyuan, Bazhong, and Nanchong, etc., the principle of exploring water sources combined with saving water should be conducted to plan water conservancy engineering in a scientific way. In plains and large-scale agricultural management areas, such as Chengdu and Deyang, high water-saving irrigation efficiency should be practiced with the assistance of a traceability system in agricultural production chains and a brand marketing system for agricultural products. Internet-accelerated speed must be introduced to advance the development of high-efficiency water-saving technologies. Efforts should be made to practice information management in agricultural water saving engineering, extend efficient water-saving technology in irrigation, and enhance the transformation of the achievements made in agricultural water-saving technology. Advanced water-saving technology from both at home and abroad should be assimilated and utilized.

Thirdly, more focus should be placed on amplifying the water price system and creating incentives for water-saving irrigation. Legislative, economic, and administrative incentives should be executed to reform the total agricultural water consumption and water quota. Firstly, an index system of water allocation and permits of water-withdrawal volume control in the administrative districts at the provincial, municipal, and county level ought to be established respectively, which would be convenient for decomposing the index throughout the three levels from the top-down. Secondly, water pricing for agricultural use should be reformed to encourage volumetric pricing, improve terminal water pricing, and enhance the ratio of agricultural water use. Meanwhile, compensation and incentive mechanism for saving water should be established to encourage local officials and individual farmers to actively practice water-saving irrigation.

Finally, more focus should be placed on enhancing management and maintenance capabilities. It is important to perfect the management and maintenance system by clearly defining the responsibilities of project property owners and water administrative supervisors. Meanwhile, technological guidance should be provided to constitute regulations and standards for highly-efficient water-saving irrigation and to compile standardized management manuals. Furthermore, integrated institutions of water conservancy services, professional teams, water users associations, and village-level water management should be established to guarantee long-term water use efficiency.

5. Conclusions

Regional WIDL evaluation is a task which involves multi-target and multi-attribution methods. In the paper, we choose 3 second-class indexes of engineering, agronomy, and management water-saving along with 33 third-class indexes to establish a comprehensive evaluation index system for WIDL. The principal components analysis method was adopted for simplifying the water-saving indexes of engineering, agronomy, and management, and the number of third-class indexes was simplified into 14. The combined weight which considered both objective weight and subjective weight was applied to judge the relative importance of each index. The combination weight offset the deficiency of the evaluation index weights determined only by expert subjective experience or indexes of sample data. Moreover, the TOPSIS method was improved to evaluate WIDL in 21 cities (states) of Sichuan province.

Results showed the overall development level of water-saving irrigation was "poor" in Sichuan province with a ε value of 0.4754. The water-saving irrigation level turned out to be "good" in the following three regions with advantageous geographical conditions and developed economies: Chengdu, Panzhihua and Deyang. The water-saving irrigation level was "general" in the following four regions with good economic levels where agronomic water saving has been popularized: Meishan, Guangyuan, Dazhou, and Mianyang. Water-saving irrigation level was and "poor" in fourteen regions of mountainous and hilly areas in Sichuan province, especially in Ganzi, Aba, and Liangshan. Located on the Northwest plateau of Sichuan province, these areas feature poor natural resources and insufficient economies. According to the evaluation results, efforts should be made to increase the water-saving irrigation level in Sichuan province. We should increase the investment in agricultural water-saving irrigation and enhance management and maintenance, especially in "poor" regions. Besides, we should adjust the agricultural industry distribution and extend water-saving technology to be more appropriate for local water resource, land, and ecosystem conditions. Furthermore, we should amplify the water price system and create new incentives for the promotion of water-saving irrigation strategies.

This paper also has limitations and shortcomings. The evaluation of the data collection needs to be broader to fully represent the actual conditions of the studied areas. In the meantime, the evaluation standard can only be adapted in Sichuan province, and more evaluation methods should be used to verify the results. Although water-saving irrigation policies, technologies, agronomy measures, and evaluation criterion vary in other areas, the evaluation of WIDL in Sichuan province established a demonstration of the potential effect of these strategies in other humid regions of southern China.

Author Contributions: All authors were involved in designing and discussing the study. L.Z. (Lu Zhao) performed the calculation, result analysis, and wrote the article; C.L. and Y.F. contributed data collection and result calculation and analysis; N.C. and L.Z. (Lili Zhang) contributed to the article's discussion and edition.

Funding: This work was also supported by the National Key Research and Development Program of China (No. 2016YFC0400206), the National Natural Science Foundation of China (51779161), the Central University special fund basic research and operating expenses (2018CDPZH-10, 2016CDDY-S04-SCU, 2017CDLZ-N22) and the National Key Technologies R&D Program of China (No. 2015BAD24B01).

Acknowledgments: We would like to thank the National Climatic Centre of the China Meteorological Administration for providing the climate database used in this study. Thanks are extended to the editors and anonymous reviewers for their valuable comments.

Conflicts of Interest: The authors declare no conflict of interest.

References

1. Salinas, J.G.; Pinto, M.; Pinto, M. Water security as a challenge for the sustainability of La Serena-Coquimbo conurbation in northern Chile: Global perspectives and adaptation. *Mitig. Adapt. Strateg. Glob. Chang.* **2016**, *21*, 235–1246. [CrossRef]
2. Niu, G.; Li, Y.P.; Huang, G.H.; Liu, J. Interactive Fuzzy-Boundary Interval Programming for Water Resources Management of the Hetao Basin, China. *J. Irrig. Drain. Eng.* **2016**, *142*, 04016056. [CrossRef]
3. Piedad, M.; Lin, H.; David, D.; Chen, C.D. The impact of on-farm water saving irrigation techniques on rice productivity and profitability in Zhanghe Irrigation System, Hubei, China. *Paddy Water Environ.* **2004**, *2*, 207–215.
4. Fisher, G.; Tubiello, F.N.; van Velthuizen, H. Climate change impacts on irrigation water requirements: Effects of mitigation, 1990–2080. *Technol. Forecast. Soc.* **2007**, *74*, 1083–1107. [CrossRef]
5. Sun, S.K.; Wang, Y.B.; Liu, J.; Wu, P.T.; Cai, H.J. Sustainability Assessment of Regional Water Resources Under the DPSIR Framework. *J. Hydrol.* **2016**, *532*, 140–148. [CrossRef]
6. Lee, S.H.; Yoo, S.H.; Choi, J.Y.; Engel, B.A. Effects of climate change on paddy water use efficiency with temporal change in the transplanting and growing season in South Korea. *Irrig. Sci.* **2016**, *34*, 443–463. [CrossRef]
7. UNCSD. Comprehensive Assessment of the Freshwater Resources of the World. United Nations Commission on Sustainable Development. 1997. Available online: http://www.un.org/esa/documents/ecosoc/cn17/1997/ecn171997-9.htm/ (accessed on 25 February 2016).
8. Ket, P.; Garré, S.; Oeurng, C.; Hok, L.; Degré, A. Simulation of Crop Growth and Water-Saving Irrigation Scenarios for Lettuce: A Monsoon-Climate Case Study in Kampong Chhnang, Cambodia. *Water* **2018**, *10*, 666. [CrossRef]
9. Cai, X.M.; Rosegrant, M.W. Irrigation technology choices under hydrologic uncertainty: A case study from Maipo River Basin. *Water Resour. Res.* **2004**, *4*, 1–10. [CrossRef]
10. Tal, A. Rethinking the sustainability of Israel's irrigation practices in the Drylands. *Water Res.* **2016**, *90*, 387–394. [CrossRef]
11. Tal, A. Seeking sustainability: Israel's evolving water management strategy. *Science* **2006**, *313*, 1081–1084. [CrossRef]
12. Horst, M.G.; Shamutalov, S.S.; Pereira, L.S.; Gonçalves, J.M. Field assessment of the water saving potential with furrow irrigation in Fergana, Aral Sea basin. *Agric. Water Manag.* **2005**, *77*, 210–231. [CrossRef]
13. Mushtaq, S.; Dawe, D.; Lin, H.; Moya, P. An assessment of the role of ponds in the adoption of water-saving irrigation practices in the Zhanghe Irrigation System, China. *Agric. Water Manag.* **2006**, *83*, 100–110. [CrossRef]
14. Ørum, J.E.; Boesen, M.V.; Jovanovic, Z.; Pedersen, S.M. Farmers' incentives to save water with new irrigation systems and water taxation-A case study of Serbian potato production. *Agric. Water Manag.* **2010**, *98*, 465–471. [CrossRef]
15. Nyakudya, I.W.; Stroosnijder, L. Effect of rooting depth, plant density and planting date on maize (*Zea mays* L.) yield and water use efficiency in semi-arid Zimbabwe: Modelling with Aqua Crop. *Agric. Water Manag.* **2014**, *146*, 280–296. [CrossRef]
16. Li, X.L.; Zhang, X.T.; Niu, J.; Tong, L.; Kang, S.Z.; Du, T.S.; Li, S. Irrigation water productivity is more influenced by agronomic practice factors than by climatic factors in Hexi Corridor, Northwest China. *Sci. Rep.* **2016**, *6*, 37971. [CrossRef]
17. Pérez-Sarmiento, F.; Mirás-Avalos, J.M.; Alcobendas, R.; Alarcón, J.J.; Mounzer, O.; Nicolas, E. Effects of regulated deficit irrigation on physiology, yield and fruit quality in apricot trees under Mediterranean conditions. *Span. J. Agric. Res.* **2016**, *14*, 1–12. [CrossRef]
18. Peng, Y.; Zhang, J.M.; Meng, J.P. Second order force scheme for lattice Boltzmann model of shallow water flows. *J. Hydraul. Res.* **2017**, *55*, 592–597. [CrossRef]
19. Jerry, R.W.; Orlan, H.B.; Gary, J.D. A microcomputer model for irrigation system evaluation. *South. J. Agric. Econ.* **1988**, *7*, 56–69.
20. Anagnostopoulosa, K.P.; Petalasb, C. A fuzzy multicriteria benefit-cost approach for irrigation projects evaluation. *Agric. Water Manag.* **2011**, *98*, 1409–1416. [CrossRef]

21. Adusumilli, N.; Davis, S.; Fromme, D. Economic evaluation of using surge valves in furrow irrigation of row crops in Louisiana: A net present value approach. *Agric. Water Manag.* **2016**, *174*, 61–65. [CrossRef]
22. Xue, J.; Ren, L. Evaluation of crop water productivity under sprinkler irrigation regime using a distributed agro-hydrological model in an irrigation district of China. *Agric. Water Manag.* **2016**, *178*, 350–365. [CrossRef]
23. Zhang, Q.W.; Cui, N.B.; Feng, Y.; Gong, D.Z.; Hu, X.T. Improvement of Makkink model for reference evapotranspiration estimation using temperature data in Northwest China. *J. Hydrol.* **2018**, *566*, 264–273. [CrossRef]
24. Luo, Y.F.; Fu, H.L.; Xiong, Y.J.; Xiang, Z.; Wang, F.; Bugingo, Y.C.; Khan, S.; Cui, Y.L. Effects of water-saving irrigation on weed infestation and diversity in paddy fields in East China. *Paddy Water Environ.* **2017**, *15*, 593–604. [CrossRef]
25. Okada, H.; Styles, S.W.; Grismer, M.E. Application of the analytic hierarchy process to irrigation project improvement Part I. Impacts of irrigation project internal processes on crop yields. *Agric. Water Manag.* **2008**, *95*, 199–204. [CrossRef]
26. Okada, H.; Styles, S.W.; Grismer, M.E. Application of the analytic hierarchy process to irrigation project improvement Part II. How professionals evaluate an irrigation project for its improvement. *Agric. Water Manag.* **2008**, *95*, 205–210. [CrossRef]
27. Tan, C.; Song, Y.; Che, H. Application of set pair analysis method on occupational hazard of coal mining. *Saf. Sci.* **2017**, *92*, 10–16.
28. Mao, N.; Song, M.J.; Deng, S.M. Application of TOPSIS method in evaluating the effects of supply vaneangle of a task/ambient air conditioning system on energy utilization and thermal comfort. *Appl. Energy* **2016**, *180*, 536–545. [CrossRef]
29. Sun, H.Y.; Wang, S.F.; Hao, X.M. An Improved analytic hierarchy process method for the evaluation of agricultural water management in irrigation districts of north China. *Agric. Water Manag.* **2017**, *179*, 324–337. [CrossRef]
30. Abazar, S.; Amir, P.; Ramin, B.; Heidar, Z. Pre-processing data using wavelet transform and PCA based on support vector regression and gene expression programming for river flow simulation. *J. Earth Syst. Sci.* **2017**, *126*, 65. [CrossRef]
31. Loucif, B.; Larbi, H. The effect of simple imputations based on four variants of PCA methods on the quantiles of annual rainfall data. *Environ. Monit. Assess.* **2018**, *190*, 569. [CrossRef]
32. Shieh, M.Y.; Chiou, J.S.; Hu, Y.C.; Wang, K.Y. Applications of PCA and SVM-PSO based real-time face recognition system. *Math. Probl. Eng.* **2014**, *2014*, 1–12. [CrossRef]
33. Zhang, J.; Song, W.; Jiang, B.; Li, M. Measurement of lumber moisture content based on PCA and GS-SVM. *J. For. Res.* **2018**, *29*, 557–564. [CrossRef]
34. Lin, T.K. PCA/SVM-based method for pattern detection in a multisensor system. *Math. Probl. Eng.* **2018**, *2018*, 1–11. [CrossRef]
35. Saaty, T.L. How to make a decision: The analytic hierarchy process. *Eur. J. Oper. Res.* **1990**, *48*, 9–26. [CrossRef]
36. Zhang, H.; Polytechnic, W. Application on the entropy method for determination of weight of evaluating index in fuzzy mathematics for wine quality assessment. *Adv. J. Food Sci. Technol.* **2015**, *7*, 195–198. [CrossRef]
37. Hwang, C.L.; Yoon, K. *Multiple Attribute Decision Making-Method and Applications, A State-of-the-Art Survey*; Springer: New York, NY, USA, 1981.
38. Aghajani, M.; Mostafazadeh-Fard, B.; Navabian, M. Assessing Criteria Affecting Performance of the Sefidroud Irrigation and Drainage Network Using TOPSIS-Entropy Theory. *Irrig. Drain.* **2017**, *66*, 626–635. [CrossRef]
39. Lou, Y.H.; Kang, S.Z.; Cui, N.B. Application of set pair analysis in the comprehensive water-saving irrigation development level evaluation. *J. Sichuan Prov. Univ. (Eng. Sci. Ed.)* **2014**, *46*, 20–27.

Article

Estimation of Precipitation Evolution from Desert to Oasis Using Information Entropy Theory: A Case Study in Tarim Basin of Northwestern China

Xiangyang Zhou [1,2], Zhipan Niu [2,3,*] and Wenjuan Lei [4]

[1]   College of Resource and Environmental Engineering, Guizhou University, Guiyang 550025, China; xyzhou6@gzu.edu.cn

[2]   State Key Laboratory of Hydraulics and Mountain River Engineering, Sichuan University, Chengdu 610065, China

[3]   Institute for Disaster Management and Reconstruction, Sichuan University, Chengdu 610065, China

[4]   College of Architecture & Environment, Sichuan University, Chengdu 610065, China; leixiaojuan333@126.com

*   Correspondence: niuzhipan@sina.com; Tel.: +86-134-3888-2922

Received: 25 August 2018; Accepted: 10 September 2018; Published: 15 September 2018

Abstract: The cold-wet effect of oasis improves the extreme natural conditions of the desert areas significantly. However, the relationship between precipitation and the width of oasis is challenged by the shortage of observed data. In this study, the evolution of annual precipitation from desert to oasis was explored by the model establishment and simulation in Tarim Basin of northwestern China. The model was developed from the principle of maximum information entropy, and was calibrated by the China Meteorological Forcing Dataset with a high spatial resolution of 0.1° from 1990 to 2010. The model performs well in describing the evolution of annual precipitation from the desert to oasis when the oasis is wide enough, and the R^2 is generally more than 0.90 and can be up to 0.99. However, it fails to simulate the seasonal precipitation evolution because of the non-convergence solved by nonlinear fitting and the unfixed upper boundary condition solved by the least square method. Through the simulation with the parameters obtained from the nonlinear fitting, the basic patterns, four stages of precipitation evolution with the oasis width increasing, are revealed at annual scale, and the current stages of these oases are also uncovered. Therefore, the establishment of the model and the simulated results provide a deeper insight from the perspective of informatics to understand the regional precipitation evolution of the desert–oasis system. These results are not only helpful in desertification prevention, but also helpful in fusing multisource data, especially in extreme drought desert areas.

Keywords: evolution of precipitation; model simulation; information entropy theory; desert–oasis areas; Tarim Basin

---

1. Introduction

Oasis serves to improve the extreme natural conditions of the arid regions by affecting the regional hydrometeorological factors [1–7], and the oasis–desert interactions are important for the stable co-existence of oasis and desert ecosystems [8] and water resources management [9]. Hence, quantifying the spatial evolution of precipitation with the oasis width increase is crucial to the regional eco-environmental security in the arid areas.

In fact, the spatial evolutions of precipitation are affected by the cold-wet effect of oasis. The cold effect is mainly resulted from the greater absorption of latent heat through evapotranspiration [10,11] and the higher surface albedo of the vegetation than the desert surface [3]. The wet effect is because of more water vapor source from the evapotranspiration of oasis [3,4]. As a result, the local water vapor can account for up to 20% of the precipitation in arid and semiarid regions [12,13] and 20–50% in humid regions [14–17]. In fact, the extreme arid environment leads to the much larger difficulties of data observation. Many studies are based on the field observation at a low spatial resolution [1–6,12–17], although some multiscale datasets are available in specific areas such as Heihe Basin located in the northwestern China [18]. Thus, generally, exploring the relationship between the precipitation and oasis width quantitatively is still challenged by the resolution of data.

Merging multisource data is a useful tool to solve the problem of data shortage and low resolution, especially fusing the calibrated remote sensing data with the land surface observed data. After the early attempts in the 1980s [19,20], the resolution of the assimilation data has improved substantially. For example, as for the ocean system, the current resolution is about 1/4–1/6 degrees [21,22]. In China, the current resolution is 0.1 degrees spatially [23,24], and its good spatial continuity has been demonstrated by several studies. These studies have uncovered the spatial distribution of precipitation and evaporation in the Nam Co basin in Tibetan Plateau [25,26], and the spatial patterns of permafrost based on the climatic factors of this dataset [27]. The higher resolution of fusion dataset provides an alternative source to analyze the spatial evolution of precipitation.

Entropy, combining the micro and macro status, has been widely used to simulate the evolution of a system. Information entropy is a better measure of variability than the variance and coefficient of variation when the probability distribution is not symmetric [28,29], because entropy may be related to higher order moments of a distribution [30]. Hence, it has been employed to evaluate the complexity of typical chaos [31] and uncertainty of precipitation and the potential water resources ability at different scales [32–35]. Spatially, entropy has been widely employed to evaluate the spatial distribution of rainfall gauge net [36,37], quantify the soil water dynamic processes [38–40], explore the multifractal generation [41,42] and simulate the shallow water and solution transportation based on the molecule collision [43,44]. These applications above indicate that entropy is not only a good index to describe the uncertainty of a time series, but also a good method to describe the spatial continuity.

Therefore, the objective of this study was to quantitatively explore the evolution of precipitation from desert to inner oasis based on the Tarim Basin, northwestern China. A model describing precipitation evolution was developed from the information entropy theories, and then the parameters of model were fitted by the China Meteorological Forcing Dataset with the spatial resolution of 0.1 degrees. Through the simulation, the typical precipitation evolution patterns weare revealed, and the current stage of these oases were also identified. The establishment of the model and the simulated results provide a deeper insight from the perspective of informatics to understand the regional precipitation evolution of the desert–oasis system. These results are useful for the desertification prevention and multisource data fusion, especially in extreme drought desert areas.

2. Study Area, Data and Methodology

2.1. Study Area

The Tarim Basin, covering an area of approximately 560,000 km^2, is located in the south of Xinjiang Province in northwestern China [45,46], as shown in Figure 1. The Taklamakan Desert, which formed at least 5.3 Ma years ago, is in the center of the Tarim Basin [47]. The total area of the oasis is approximately 103,900 km^2 [11,48]. In the desert region, the annual precipitation is from less than 15 to 60 mm, increasing from the east to the west [49], and the increase rate is 10.15, 6.29, and 0.87 mm per decade in the mountain, oasis and desert areas, respectively, based on the observed data from 1960 to 2010 [50]. However, the annual potential evaporation is over 3200 mm based on the evaporation from water surface, and most of the basin is a generally unsuitable or extremely unsuitable area for human settlement [51]. Currently, the irrational reclamation of land and overuse of natural resources have led to the destruction of vegetation and shrinkage of the water area [52–54] and have threaten the security and sustainable development of oases [48].

Figure 1. Spatial distribution of oasis in Tarim Basin, obtained from Google Earth. The linkage for Google's permissions is: https://www.google.ca/permissions/geoguidelines.

2.2. Data

In this study, the China Meteorological Forcing Dataset, developed by the Data Assimilation and Modeling Center for Tibetan Multispheres, Institute of Tibetan Plateau Research, Chinese Academy of Sciences [23,24], was used to test the performance of the model and to calibrate the parameters of the precipitation evolution form the desert toward oasis. The dataset was produced by merging a variety of data sources. More details on the dataset are given in the user's guide of the Dataset [55]. The spatial and temporal resolutions of the dataset are 0.1 degrees and 3 h, respectively, and the data length is 21 years (1990–2010). Based on the dataset, the spatial distributions of annual precipitation in Tarim basin are shown in Figure 2.

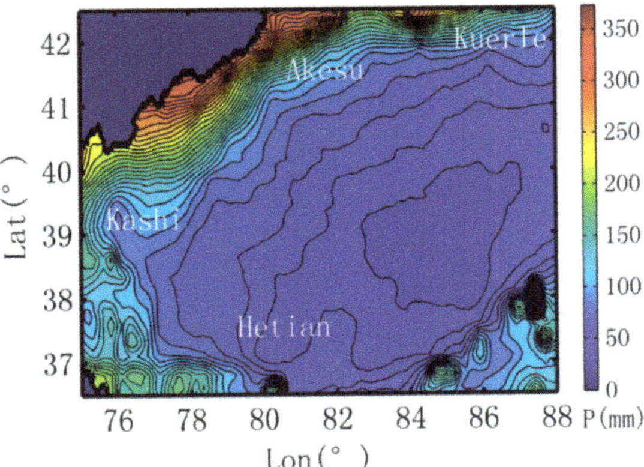

Figure 2. Spatial distribution of annual precipitation in Tarim Basin based on The China Meteorological Forcing Dataset.

2.3. Information Entropy and Principle of Maximum Entropy

Assuming the probability density function of a continuous variable x is expressed by $f(x)$, and the information entropy is calculated as [56,57]:

$$H(x) = -\int_a^b f(x) \ln(x) dx \tag{1}$$

$$\int_a^b f(x) dx = 1 \tag{2}$$

where a and b and the lower and upper boundaries of the variable x, respectively.

In addition, the known information of the variable represented by $g_i(x)$, such as mean and standard deviation, can be given by the following formula:

$$\int_a^b g_i(x) f(x) = N_i \tag{3}$$

Hence, the general solution of $f(x)$ can be obtained by solving the conditional extreme values using the method of Lagrange multipliers: the objective function is the maximum information entropy, and the constraint is the known information in Equation (3). More details about the mathematical derivation is presented in the literature [40,58]. The general form of $f(x)$ is given as:

$$f(x) = e^{\sum_{i=1}^{n} g_i(x)} \tag{4}$$

3. Establishment of the Model

Assuming the precipitation at the boundary area between oasis and desert is P_0, the precipitation evolution from desert toward the oasis is represented by the integration of a function $p(z)$, as given in Equation (5):

$$P(z) = P_0 + \int_0^Z p(z) dz \tag{5}$$

where z represents the distance from the desert–oasis boundary to a certain point of the oasis. $P(z)$ represents the precipitation at the location with distance z. Equation (5) can also be transformed into the following form:

$$P(z) - P_0 = \int_0^Z p(z) dz \tag{6}$$

Usually, $P(z)$ reaches a constant approximately when the oasis is wide enough. Here, Equation (6) can be written as:

$$\int_0^{+\infty} p(z) dz = C \tag{7}$$

where the constant C represents the maximum increment of precipitation with the increase of oasis width. Dividing both sides of the equation by the constant C, Equation (7) becomes:

$$\frac{1}{C} \int_0^{+\infty} p(z) dz = 1 \tag{8}$$

Hence, the item of $p(z)/C$ can be considered as a probability density function, and its information entropy is calculated by Equations (9) and (10):

$$H = -\int_0^Z \frac{1}{C} p(z) \ln \frac{p(z)}{C} dz \tag{9}$$

$$H = -\frac{1}{C}\int_0^Z p(z)\ln p(z)dz + \frac{1}{C}\ln C \tag{10}$$

As for a specific oasis, the spatial distribution can be considered as stable approximately in a mid-long term. Correspondingly, the arithmetic mean and geometric mean of precipitation evolution can be considered as constant, as given by Equations (11) and (12), respectively.

$$\int_0^\infty \frac{z}{C}p(z)dz = \mu_z \tag{11}$$

$$\int_0^\infty \frac{\ln(z)}{C}p(z)dz = \nu_z \tag{12}$$

where μ_z and ν_z represent the arithmetic mean and geometric mean. Here, the Lagrange function is obtained based on the information entropy of precipitation evolution with the constraints of Equations (8), (11) and (12), as given by Equation (13).

$$\begin{aligned}
L = {} & -\frac{1}{C}\int_0^Z p(z)\ln p(z)dz + \frac{1}{C}\ln C + \lambda_0\left(\int_0^Z \frac{1}{C}p(z)dz - 1\right) \\
& + \lambda_1\left(\int_0^Z \frac{z}{C}p(z)dz - \mu_z\right) + \lambda_2\left(\int_0^Z \frac{\ln(z)}{C}p(z)dz - \nu_z\right)
\end{aligned} \tag{13}$$

Let $\frac{\partial L}{\partial p(z)} = 0$, then the general solution of $p(z)/C$ is obtained by solving the conditional extreme values, as given in Equation (14):

$$p(z)/C = e^{\lambda_0 - 1 + \lambda_1 z + \lambda_2 z \ln z} \tag{14}$$

Equation (14) can also be written as:

$$p(z)/C = e^{\lambda_0 - 1}e^{\lambda_1 z}z^{\lambda_2} \tag{15}$$

Let $\lambda_2 = \alpha - 1$, $\lambda_1 = -1/\beta$. It is easy to obtain that $e^{\lambda_0 - 1} = -^\alpha/\Gamma(\alpha)$. Thus, the probability density function $p(z)/C$ can also be expressed as:

$$p(z)/C = \frac{1}{\Gamma(\alpha)\beta^\alpha}z^{\alpha-1}e^{-z/\beta} \tag{16}$$

Substituting Equation (16) into Equation (5), the evolution of precipitation with the oasis width increase is:

$$P(z) = P_0 + C\int_0^Z \frac{1}{\Gamma(\alpha)\beta^\alpha}z^{\alpha-1}e^{-\frac{z}{\beta}}dz \tag{17}$$

In fact, the right-hand term of Equation (16) is the probability density function of a gamma distribution. Therefore, Equation (17) shows that the precipitation evolution model is based on the scaled cumulative distribution function of a gamma distribution with an initial precipitation. In other words, the model is based on the linear transformation of the cumulative distribution function of a gamma distribution. The parameters of the model include the shape factor α, scale factor β, zooming constant C and initial precipitation P_0 at the boundary between desert and oasis.

Assuming the initial precipitation P_0 at the boundary between desert and oasis is 50 mm/year and the zooming constant C is 50, the illustration of the precipitation evolution are shown in Figure 3 with different shape and scale factors.

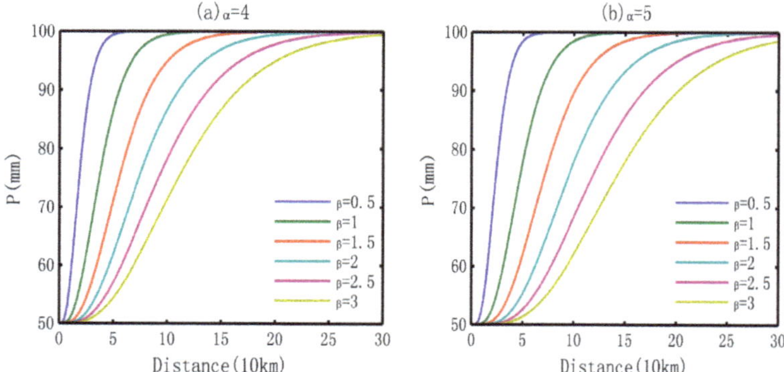

Figure 3. Illustration of precipitation evolution from desert to inner oasis. Assuming the initial precipitation P_0 at the boundary between desert and oasis is 50 mm/year and the zooming constant C is 50, the precipitation evolution is shown with different shape factor and different scale factor.

4. Calibration of the Model and the Simulated Results

4.1. Calibration of the Model

4.1.1. Performance of the Model

In this study, four main oases were used to test the performance of the model. These oases are Kashi Oasis located in the western basin, Akesu Oasis in the northern basin, Kuerle Oasis in the northeastern basin and Hetian Oasis in the southwestern basin. The evolution routes of precipitation from desert to oasis are shown in Figure 4. Generally, these routes are located in the middle of these oases.

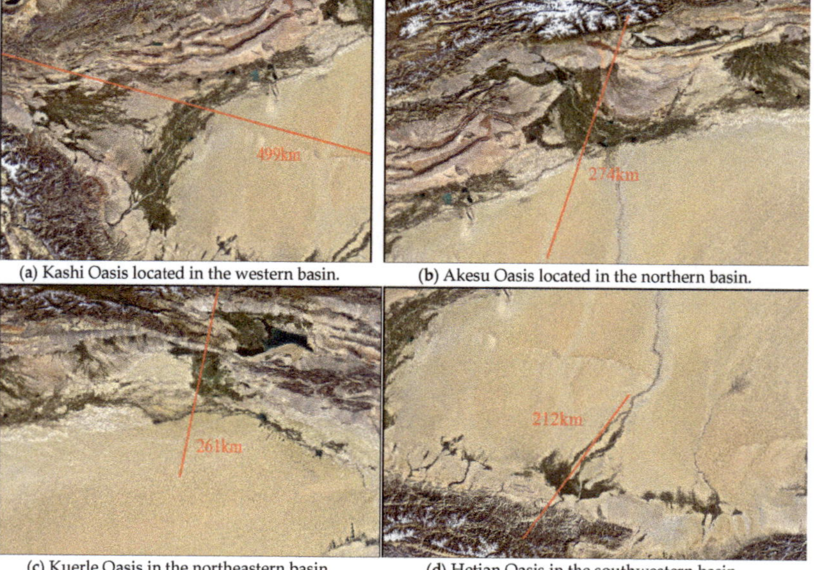

(a) Kashi Oasis located in the western basin.

(b) Akesu Oasis located in the northern basin.

(c) Kuerle Oasis in the northeastern basin

(d) Hetian Oasis in the southwestern basin.

Figure 4. Selected routes of precipitation transition from the desert to oasis in different location of the Tarim Basin, obtained from Google Earth. The permission of Google to use Google Earth Periodicals is given at: https://www.google.ca/permissions/geoguidelines.

Following the routes shown in Figure 4, the evolution of precipitation from desert to oasis is obtained, and the results are scattered in Figure 5. Here, the parameters of the model are obtained by nonlinear fitting. The correlation coefficient is used to evaluate the performance of the model given by Equation (17). The fitted results and the square of correlation coefficient, R^2, of the four selected routes are also shown in Figure 5.

The model performs well in three oases: Kashi Oasis, Akesu Oasis and Kuerle Oasis, corresponding to R^2 of 0.99, 0.93 and 0.97, respectively. However, the model does not perform well in Hetian Oasis and the R^2 is 0.47. To improve the accuracy trend of precipitation evolution, the model should be calibrated further for the Hetian Oasis.

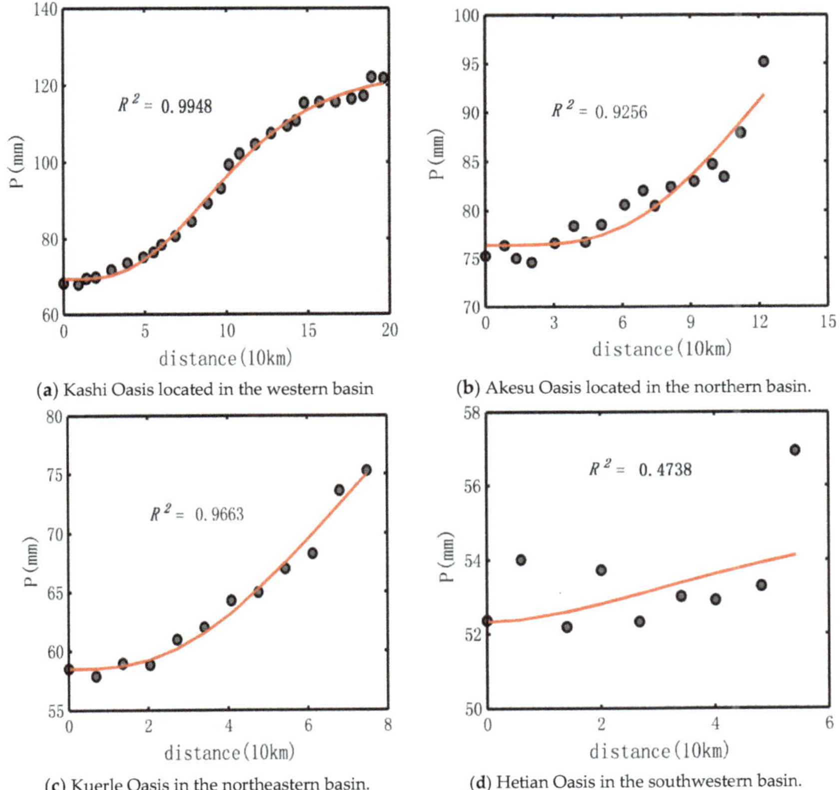

(**a**) Kashi Oasis located in the western basin

(**b**) Akesu Oasis located in the northern basin.

(**c**) Kuerle Oasis in the northeastern basin.

(**d**) Hetian Oasis in the southwestern basin.

Figure 5. Performance of the model simulating precipitation from desert to oasis in different areas of Tarim Basin.

4.1.2. Calibration of Model for Hetian Oasis

Through combining the spatial distribution of oasis and the local hydrological cycle, two main reasons were found to lead to the poor performance of the model. The first is the much smaller width of Hetian Oasis. As shown in Figure 5, the Hetian Oasis is about 55 km wide, while Kashi Oasis is 200 km, Akesu Oasis is 120 km and Kuerle Oasis is 80 km. The second reason is that the prevailing wind direction of the Hetian Oasis is opposite to the cold effect of oasis [59–61]. Our previous study revealed that the opposite direction leads to water vapor from the local evapotranspiration accumulating over the buffer zone located in the down wind direction of the oasis [49], as shown in Figure 6a. Hence, the evolution of precipitation in the Hetian Oasis should be divided into two phases: the buffer zone with abnormal high precipitation and the regular pattern of the precipitation evolution after the buffer zone, as shown in Figure 6b.

By overlapping the spatial distribution of water vapor content and oasis layout, it was found that the first four grids should be removed to calibrate the model. After their removal, the calibrated model performs much better, and the R^2 increased from 0.47 to 0.87, as shown in Figure 6b.

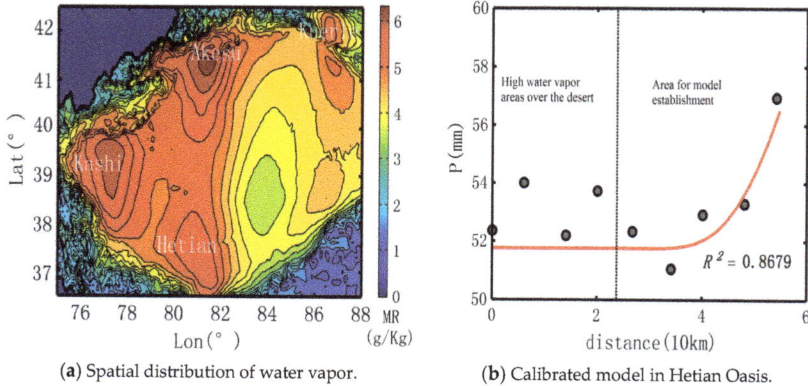

(a) Spatial distribution of water vapor.

(b) Calibrated model in Hetian Oasis.

Figure 6. Spatial distributions of water vapor in Tarim Basin and the calibrated model in Hetian.

4.1.3. Parameter Inversion

Based on the results obtained by nonlinear fitting, the parameters of the model for all four oases are obtained, as shown in Table 1. According to Equation (17) and Figure 3, the meanings of these parameters are clear. P_0 is the initial value that represents the precipitation at the boundary area between oasis and desert. Here, AKesu Oasis exhibits largest value, followed by Kashi, Kuerle and Hetian respectively. The zooming constant, C, represents the maximum potential increment of precipitation with the oasis width increase. The results of zooming constant C suggest the maximum promotion on local precipitation in Kashi, middle level in AKesu and Kuerle, and the minimum promotion in Hetian Oasis. Shape and scale factors represent the increase rate of the precipitation. A larger value is usually accompanied with a smaller increase rate, and vice versa (Figure 3).

Table 1. Performance of the model and the parameters inversion for precipitation evolution from desert to oasis in different locations of the Tarim Basin.

Main Oases	Initial Precipitation P_0 (mm)	Zooming Constant C (-)	Shape Factor α (-)	Scale Factor β (-)	Correlation R^2
Kashi Oasis (Western Basin)	69.31	54.24	4.64	2.32	0.99
Akesu Oasis (Northern Basin)	76.46	43.04	5.16	3.00	0.93
Kuerle Oasis (Northeastern Basis)	58.54	44.60	3.36	3.00	0.97
Hetian Oasis without calibration (Southwestern Basin)	52.34	3.34	2.00	3.00	0.47
Hetian Oasis after calibration (Southwestern Basin)	51.76	32.57	1.02	5.41	0.87

4.1.4. Influence of the Local Terrain

Apart from the cold-wet effect of oasis, the significant uplift of the terrain would lead to a sharp decrease in temperature and result in a substantial increase in precipitation at the local scale. As given in our previous study [49], the increment of the elevation is about 80–100 m. Meanwhile, the widths of Kashi, Akesu, Kuerle and Hetian Oases are approximately 210, 120, 80 and 55 km, respectively, as shown in Figure 5. This indicates that the increase rate is ordered by: Hetian > Kuerle > Akesu > Kashi.

However, this trend is opposite to the increment and increase rate of precipitation from desert to oasis. Hence, the increase of precipitation from desert to oasis is not dominated by the local terrain.

4.1.5. Performance of the Model at Seasonal Scale

The performance of the model at seasonal scale was also analyzed. The first problem confronted is that the result does not converge when solved by nonlinear fitting using MATLAB (MathWorks, Natick, MA, USA). Here, the results are based on least square method and the results are shown by Figure 7. Generally, the model seems to perform well with several exceptions, including the winter of Kuerle and Hetian Oases, and the summer of Hetian Oasis. The reason can be attributed to the transportation of water vapor. As discussed above, the wind blowing from the oasis to the desert would decrease the cold-wet effect of the oasis substantially [49].

However, the second problem is that the solutions of the model based on the least square method are substantially affected by the upper boundary conditions. Two examples based on different upper boundary conditions are given to illustrate this problem. As shown in Table 2, the initial precipitation P_0 is not affected by the upper boundary conditions. The other three parameters, however, are influenced significantly. For example, the difference of the parameter C estimated for the summer of the Kuerle Oasis can be up to 140 mm, although the R^2 is 0.99 for both solutions. Here, the failure of the model may be resulted from the resolution of the data or the more complex water vapor transportation. Hence, further studies should be carried out to explore the simulation of the season precipitation evolution from desert to the oasis.

Table 2. Comparisons on the sensitivity of parameter with different upper boundary conditions (UBC) at seasonal scale. R^2 are indicated in italics and Bold.

| Oasis | Season | UBC: $P_{0u}=10$; $C_u=60$, $\alpha_u=8$, $\beta_u=8$ | | | | | UBC: $P_{0u}=10$; $C_u=200$, $\alpha_u=20$, $\beta_u=20$ | | | | |
		P_0 (mm)	C (-)	α (-)	β (-)	R^2	P_0 (mm)	C (-)	α (-)	β (-)	R^2
(a) Kashi	Spring	16.87	30.80	2.47	8.00	*0.99*	16.53	57.55	1.78	20.00	*0.99*
	Summer	31.81	21.93	8.00	1.12	*0.98*	31.81	21.93	9.11	0.99	*0.98*
	Autumn	11.20	11.59	1.91	8.00	*0.99*	10.94	18.18	1.33	20.00	*0.99*
	Winter	8.44	7.39	5.49	2.55	*0.99*	8.44	7.39	5.49	2.55	*0.99*
(b) Akesu	Spring	15.14	60.00	8.00	2.37	*0.83*	15.24	32.22	20.00	0.65	*0.89*
	Summer	35.03	14.36	0.74	8.00	*0.97*	34.94	19.32	0.60	20.00	*0.98*
	Autumn	13.59	60.00	8.00	2.38	*0.87*	13.66	200.00	10.08	2.12	*0.88*
	Winter	9.19	2.00	1.42	2.80	*0.86*	9.19	2.00	1.42	2.80	*0.86*
(c) Kuerle	Spring	12.35	11.05	1.91	8.00	*0.98*	12.33	28.21	1.65	20.00	*0.98*
	Summer	27.10	60.00	3.32	4.11	*0.99*	27.03	199.94	2.71	9.94	*0.99*
	Autumn	11.65	4.41	1.47	8.00	*0.94*	11.64	5.87	1.36	11.83	*0.94*
	Winter	7.01	0.65	8.00	8.00	*0.02*	7.01	2.91	17.15	16.99	*0.01*
(d) Hetian	Spring	10.74	60.00	8.00	1.41	*0.96*	10.76	199.35	8.15	1.73	*0.96*
	Summer	23.76	60.00	8.00	1.92	*0.09*	23.76	200.00	20.00	0.54	*0.16*
	Autumn	8.11	60.00	6.02	3.03	*0.78*	8.11	200.00	5.60	4.52	*0.78*
	Winter	9.76	60.00	8.00	2.33	*0.01*	9.73	200.00	20.00	0.57	*0.05*

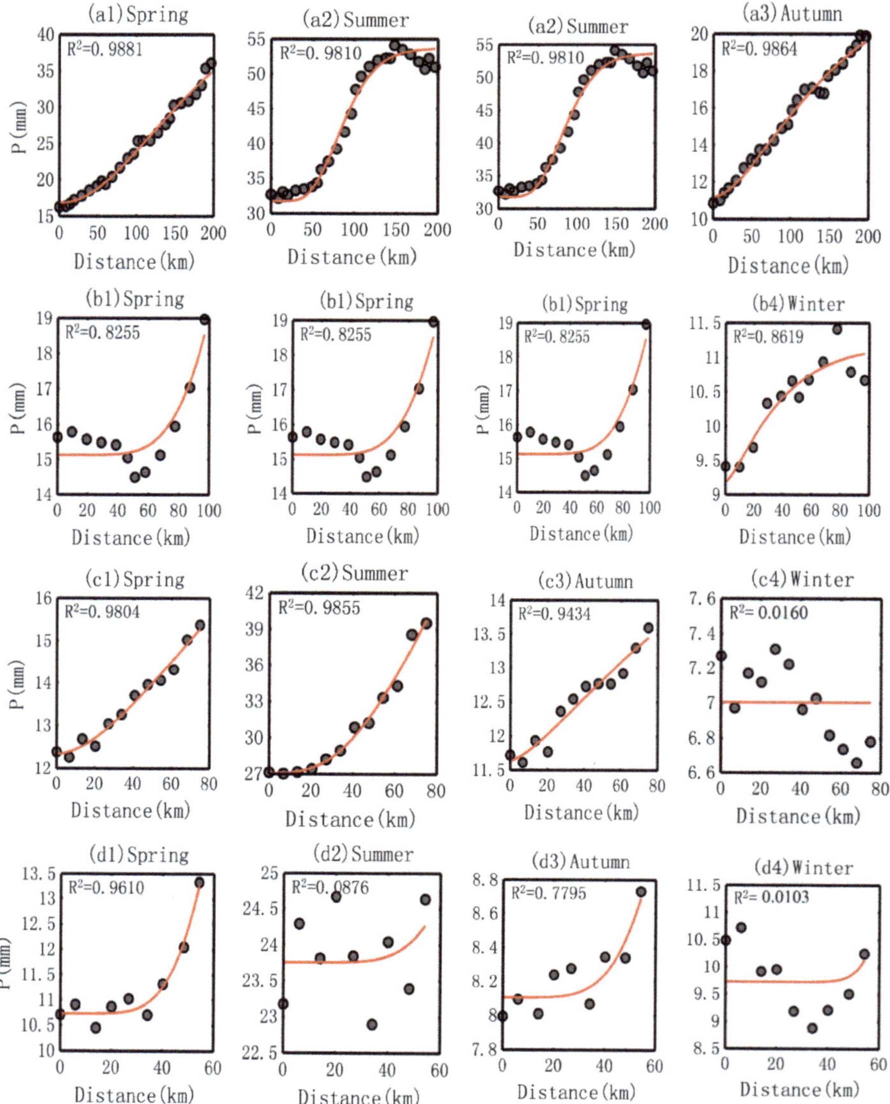

Figure 7. Performance of the information entropy based model at seasonal scale: (**a**) Kashi Oasis located in the western basin; (**b**) Akesu Oasis located in the northern basin; (**c**) Kuerle Oasis in the northeast basin; and (**d**) Hetian Oasis in the southwestern basin.

4.2. Simulated Results

Based on the performance of the model at different scales, the parameters are reasonable at annual scale. The evolution of precipitation with the increase of oasis width was calculated by Equation (17) using the parameters given in Table 1. The results are shown in Figure 8, from which the basic patterns of precipitation evolution are revealed and the current stage of these oases are also obtained. More details on the two aspects are as follows.

4.2.1. Basic Patterns of Precipitation Evolution

According to the simulated results (Figure 8), it could be found that the basic patterns of precipitation evolution exhibit four stages: a small increase at first, then a much faster increase after a threshold of oasis width, followed by a small increase rate again when the oasis width reaches a relatively large width, and finally the increase rate decreases to close to zero when the oasis is wide enough. The threshold values of the four stages are also obtained (Table 3).

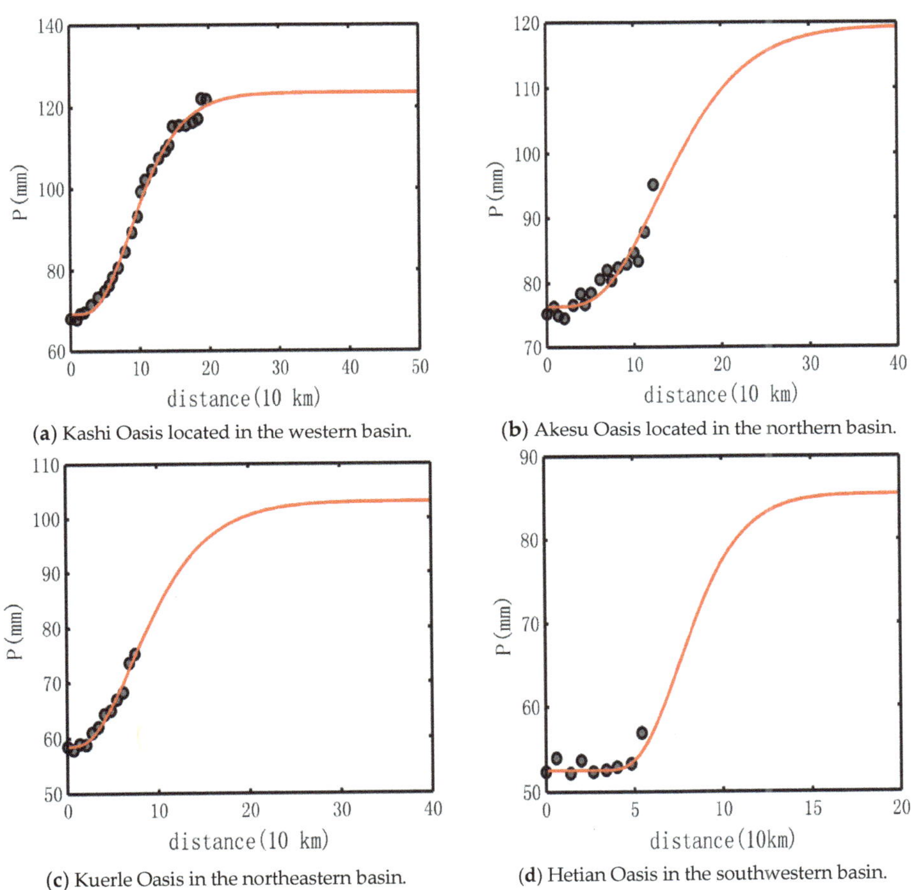

(**a**) Kashi Oasis located in the western basin.

(**b**) Akesu Oasis located in the northern basin.

(**c**) Kuerle Oasis in the northeastern basin.

(**d**) Hetian Oasis in the southwestern basin.

Figure 8. The simulated precipitation evolution from desert to oasis with the oasis width increase.

1. The first stage: Small increase rate. The oasis width of the first stage (small increase rate) is about 30–50 km. This means that, if the oasis width is less than the threshold value, the promotion of oasis' cold-wet effect on local precipitation is not substantial. More specifically, this width is about 40, 50, 40 and 55 km for Kashi, Akesu, Kuerle and Hetian Oases, respectively.

2. The second stage: Large increase rate. The second threshold value is about 150–185 km. In this period, the increase rate of precipitation is much larger than before. Generally, precipitation exhibits a linear trend with the oasis width increase. Hence, the increase rate of precipitation is approximately constant. As shown in Table 3, the increase rate varies substantially in the four oases, with the value of 3.32, 2.86, 1.76 and 0.65 mm/10 km corresponding to Kashi Oasis, Akesu Oasis, Kuerle Oasis and Hetian Oasis, respectively.

3. The third stage: Small increase rate again. The threshold value for the third stage is about 200–300 km. Two oases, Kashi and Hetian, display smaller value of about 200 km. The other two oases, Akesu and Kuerle, display larger values of 265 and 300, respectively. This means that Kashi and Hetian Oases would reach the maximum promotion on local precipitation easier than the other oases because of the smaller threshold value. However, the opposite direction between the cold effect of oasis and water vapor transportation lead to Hetian Oasis experiencing a longer first stage than the other three oases, as well as smaller precipitation locally.

4. The fourth stage: Stable status. When the oasis width is larger than the threshold value of the third stage, the precipitation evolution reaches the fourth stage, which is almost constant.

Table 3. Typical patterns of precipitation evolution with the increase of oasis width in different oases of the Tarim Basin based on the simulated results.

Stages	Threshold Value and Average Slope	Kashi Oasis	Akesu Oasis	Kuerle Oasis	Hetian Oasis
(1) Small Increase Rate	Threshold value	0–40 km	0–50 km	0–40 km	0–55
	Average slope	Not substantial	Not substantial	Not substantial	Not substantial
(2) Large Increase Rate	Threshold value	40–185 km	50–185 km	40–180 km	55–155 km
	Average slope	3.32 mm/10 km	1.76 mm/10 km	2.86 mm/10 km	0.65 mm/10 km
(3) Small Increase Rate Again	Threshold value	185–220 km	185–300 km	180–265 km	155–200 km
	Average slope	Not substantial	Not substantial	Not substantial	Not substantial
(4) Stable Status	Threshold value	>220 km	>300 km	>265 km	>200 km
	Average slope	Close to 0	Close to 0	Close to 0	Close to 0

4.2.2. Current Stages of Different Oases

The widths of Kashi, Akesu, Kuerle and HetianOases are 200 km, 135 km, 80 km and 60 km, respectively, as shown in Figure 5. By contrast with Figure 8, it can be obtained that the four oases belong to different stages that are manifested by the following aspects.

1. Kashi Oasis. Kashi Oasis is at the beginning of the fourth stage, in which the precipitation is stable with the oasis width increase. This implies that the annual precipitation will be the same as the maximum value (about 120 mm/year) if the oasis expands towards the desert.

2. Akesu Oasis. The width of the Akesu Oasis is about 135 km. Hence, the evolution of precipitation in Akesu Oasis is at the first half of the second stage currently. The second stage means largest increase rate of precipitation with oasis width increase but the middle level precipitation amount. Hence, if the oasis expands towards the desert, the precipitation would increase substantially. However, the total increment of precipitation should be less than the increment precipitation in Kashi Oasis.

3. Kuerle Oasis. As shown in Figure 8 and Table 3, the current width of the oasis is at the first half of the second stage. Hence, it would exhibit similar characteristics to AKesu Oasis.

4. Hetian Oasis. Different from the other three oases, the Hetian Oasis is at the end of the first stage. With the oasis width increase, the precipitation would increase at a slope of approximately 0.65 mm/km. That means the total increment of precipitation is the least if the oasis expanded toward desert in Hetian Oasis because of the much smaller initial condition of precipitation.

5. Discussions

5.1. The Simulated Results for Desertification Prevention

In the Tarim Basin, affected by the substantial increase of local human activities, the destruction of oasis environments is increasing, resulting from the irrational reclamation of land and overuse of natural resources [62]. In fact, the irrigated oasis area has exceeded about 33.6% of the maximum carrying capacity of local precipitation [52]. The increase of the arable land is at the expense of the destruction of vegetation and shrinkage of the water area [53,54]. Even worse, the expansion of artificial oases has led to the degradation of natural oases and the oasis–desert ecotone, which may threaten the security and sustainable development of oases [48]. These problems require understanding the interaction between the oasis and local hydrological cycle, and to selecting the more appropriate areas for desertification prevention and layout of oasis expansion.

According to the simulated results shown in Figure 8 and the current stage of the oasis precipitation, it can be concluded that Kashi Oasis is at the beginning of the fourth stage, which implies that the increment of precipitation would be about 120 mm/year. The increment of precipitation in Akesu and Heian Oasis are the middle level because both oases are at the first half of the second stage. The least promotion would appear in the southwestern basin of Hetian Oasis because it is at the end of the first stage. Therefore, Kashi Oasis of the western basin is the most appropriate area for desertification prevention, while the southwestern basin, i.e., Hetian Oasis, presents a greater challenge.

5.2. The Model for Multisource Data Fusion

The considerable challenge to the data assimilation is the resolution [22]. In this study, the spatial resolution of the assimilated data is 0.1 degrees [23,24], and many researchers have used this dataset to analyze the regional hydrological processes, such as the impact of lake effects on the temporal and spatial distribution of precipitation in the Nam Co basin of the Tibetan Plateau [25], evaporation from the lake [26] and the impact of climatic factors on permafrost in the Tibetan Plateau [27]. The good spatial continuity of the dataset is reflected by gradual changes of the local hydrological processes.

However, the performance of the dataset is still not clear in the desert areas. In this study, the dataset agrees well with simulated results of the model based on information entropy, which is not only demonstrated by the good performance of the model, but also reflected by the good spatial continuity of annual precipitation evolution in the desert–oasis system. Furthermore, the poor performance of the model without the calibration is also an indicator of the different interaction between the oasis and water vapor transportation in the Hetian Oasis. In fact, opposite direction between the water vapor transportation and cold effect of the oasis leads to water vapor from local evapotranspiration accumulating near the desert boundary, where the higher temperature reduces the promotion of oasis on local precipitation substantially [49]. Hence, the model is a useful tool to test the spatial continuity of the dataset obtained from merging multisource data, especially in those areas with extreme natural conditions, for example, desert areas.

6. Conclusions

Based on the established model, parameter inversion, forward simulation, and the discussions on the potential applications of the method, the conclusions are as follows.

1. The model describing precipitation evolution from desert to oasis is based on information entropy theory: the evolution of precipitation is considered as a linear transformed cumulative distribution function, and the probability density function is obtained by solving the Lagrange conditioned extreme value with the objective function of maximum information entropy. The constraints are the constant arithmetic mean and geometric mean of precipitation evolution from desert to oasis when the oasis is wide enough. The general form of the model is a linear transformed cumulative distribution function of a gamma distribution.

2. The model performs well when the oasis is wide enough. The R^2 between the simulated result and the dataset is 0.99, 0.93 and 0.97 for the Kashi Oasis, Akesu Oasis and Kuerle Oasis, respectively. In the Hetian Oasis, however, the model performs poorly with the R^2 of 0.47. The reason is mainly the opposite direction between the oasis cold effect and the water vapor transportation that results in a buffer zone in the first several tens of kilometers. After removing the buffer zone, the model shows a good accuracy with the R^2 of 0.87. However, the model fails to describe the seasonal precipitation evolution because of the non-convergence problem solved by nonlinear fitting and the influence of upper boundary condition solved by the least square method.

3. The simulated results indicate that the evolution of annual precipitation from desert to oasis includes four main stages: slow increase first, then fast increase, followed by a slow increase again and finally a stable stage. The threshold value of the first stage is 30–50 km, implying the promotion of local precipitation is very small if the width of the oasis is less than this value. The second stage can be approximated linearly, with the threshold value of 155–185 km and the slope of 3.32, 2.86, 1.76 and 0.65 mm/10 km for the Kashi Oasis, Akesu Oasis, Kuerle Oasis and Hetian Oasis, respectively. The upper boundary of the third stage is about 220–300 km, after which the precipitation evolution is almost constant.

4. Currently, the four main oases in the Tarim basin are at different stages. The Kashi Oasis has reached the fourth stage, the Akesu and Kuerle Oases are at the first half of the second stage, and the Hetian Oasis is at the end of the first stage. These current stages reflect different local hydrological effect when prevent desertification in the four main oases: maximum promotion on precipitation in the Kashi Oasis, middle promotion in the Akesu Oasis and Kuerle Oasis, and the least promotion in the Hetian Oasis.

Therefore, the theory of the model and simulated results would provide a deeper insight from the perspective of informatics into understanding the precipitation evolution from desert to oasis, which is not only helpful in preventing desertification but also helpful in merging multisource data in the extreme drought desert areas.

Author Contributions: X.Z. as the first author was responsible for establishing the model, collecting the data to calibrate the model, plotting the figures and writing of the main body of the manuscript. Z.N., the second author and corresponding author, searched the background of the hydrological effect of oasis and proposed the idea of this study, and helped to revise this manuscript. W.L., the third author, contributed to the data analysis and manuscript approving. All authors have read and approved the final manuscript.

Funding: The research was funded by National Natural Science Foundation of China (grant number 41701558), Science and Technology Funding of Water Resources Department of Guizhou Province (grant number KT201707), Science and Technology Funding of Guizhou Province (grant number LH [2017]7290) and the open funding of Guizhou Provincial Key Laboratory of Public Big Data (grant number NO2017BDKFJJ021).

Acknowledgments: The forcing dataset used in this study was developed by Data Assimilation and Modeling Center for Tibetan Multi-spheres, Institute of Tibetan Plateau Research, Chinese Academy of Sciences.

Conflicts of Interest: The authors declare no conflict of interest.

References

1. Taha, H.; Akbari, H.; Rosenfeld, A. Heat island and oasis effects of vegetative canopies: Micro-meteorological field-measurements. *Theor. Appl. Climatol.* **1991**, *44*, 123–138. [CrossRef]
2. Kai, K.; Matsuda, M.; Sato, R. Oasis Effect Observed at Zhangye Oasis in the Hexi Corridor, China. *J. Meteorol. Soc. Jpn.* **1997**, *75*, 1171–1178. [CrossRef]
3. Chu, P.C.; Lu, S.; Chen, Y. A numerical modeling study on desert oasis self-supporting mechanisms. *J. Hydrol.* **2005**, *312*, 256–276. [CrossRef]
4. Feng, Q.; Si, J.; Zhang, Y.; Yao, J.; Liu, W.; Su, Y.H. Microclimatic characteristics of the Heihe oasis in the hyperarid zone of China. *J. Geogr. Sci.* **2006**, *16*, 34–44. [CrossRef]
5. Potchter, O.; Goldman, D.; Kadish, D.; Iluz, D. The oasis effect in an extremely hot and arid climate: The case of southern Israel. *J. Arid. Environ.* **2008**, *72*, 1721–1733. [CrossRef]
6. Li, X.M.; Yang, J.S.; Liu, M.X.; Liu, G.M.; Yu, M. Spatiotemporal changes of soil salinity in arid areas of south Xinjiang using electromagnetic induction. *J. Integr. Agr.* **2012**, *11*, 1365–1376. [CrossRef]
7. Hao, X.; Li, W. Oasis cold island effect and its influence on air temperature: A case study of Tarim Basin, Northwest China. *J. Arid. Land* **2016**, *8*, 172–183. [CrossRef]
8. Li, X.; Yang, K.; Zhou, Y. Progress in the study of oasis-desert interactions. *Agric. Forest. Meteorol.* **2016**, *230–231*, 1–7. [CrossRef]
9. Li, X.; Cheng, G.D.; Ge, Y.M.; Li, H.Y.; Han, F.; Hu, X.L.; Tian, W.; Tian, Y.; Pan, X.D.; Nian, Y.Y. Hydrological Cycle in the Heihe River Basin and Its Implication for Water Resource Management in Endorheic Basins. *J. Geophys. Res. Atmos.* **2018**, *123*, 890–914. [CrossRef]

10. Zhang, H.; Wu, J.W.; Zheng, Q.H.; Yu, Y.Y. A preliminary study of oasis evolution in the Tarim Basin, Xinjiang, China. *J. Arid. Environ.* **2003**, *55*, 545–553.

11. Qiu, G.Y.; Li, H.Y.; Zhang, Q.T.; Chen, W.; Liang, X.J.; Li, X.Z. Effects of evapotranspiration on mitigation of urban temperature by vegetation and urban agriculture. *J. Integr. Agric.* **2013**, *12*, 1307–1315. [CrossRef]

12. Jin, X.M.; Hu, G.C.; Li, W.M. Hysteresis effect of runoff of the Heihe River on vegetation cover in the Ejina Oasis in Northwestern China. *Earth Sci. Front.* **2008**, *15*, 198–203. [CrossRef]

13. Kong, Y.; Pang, Z.; Froehlich, K. Quantifying recycled moisture fraction in precipitation of an arid region using deuterium excess. *Tellus B* **2013**, *65*, 19251. [CrossRef]

14. Gat, J.R.; Klein, B.; Kushnir, Y.; Roether, W.; Wernli, H.; Yam, R.; Shemesh, A. Isotope composition of air moisture over the Mediterranean Sea: An index of the air-sea interaction pattern. *Tellus B* **2003**, *55*, 953–965. [CrossRef]

15. Worden, J.; Noone, D.; Bowman, K.; Beer, R.; Eldering, A.; Fisher, B.; Gunson, M.; Goldman, A.; Herman, R.; Kulawik, S.S. Importance of rain evaporation and continental convection in the tropical water cycle. *Nature* **2007**, *445*, 528–532. [CrossRef] [PubMed]

16. Ueta, A.; Sugimoto, A.; Iijima, Y.; Yabuki, H.; Maximov, T.C. Contribution of transpiration to the atmospheric moisture in eastern Siberia estimated with isotopic composition of water vapor. *Ecohydrology* **2014**, *7*, 197–208. [CrossRef]

17. Vallet-Coulomb, C.; Gasse, F.; Sonzogni, C. Seasonal evolution of the isotopic composition of atmospheric water vapour above a tropical lake: Deuterium excess and implication for water recycling. *Geochim. Cosmochim. Acta* **2008**, *72*, 4661–4674. [CrossRef]

18. Li, X.; Liu, S.M.; Xiao, Q.; Ma, M.G.; Jin, R.; Che, T.; Wang, W.Z.; Hu, X.L.; Xu, Z.W.; Wen, J.G. A multiscale dataset for understanding complex eco-hydrological processes in a heterogeneous oasis system. *Sci. Data* **2017**, *4*, 170083. [CrossRef] [PubMed]

19. Robinson, A.R.; Leslie, W.G. Estimation and prediction of oceanic eddy fields. *Prog. Oceanogr.* **1985**, *14*, 385–510. [CrossRef]

20. Moore, A.M.; Cooper, N.S.; Anderson, D.L.T. Initialization and data assimilation in models of the Indian Ocean. *J. Phys. Oceanogr.* **1987**, *17*, 1965–1977. [CrossRef]

21. Balsamo, G.; Mahfouf, J.F.; Belair, S. A land data assimilation system for soil moisture and temperature: An information content study. *J. Hydrometeorol.* **2007**, *8*, 1225–1242. [CrossRef]

22. Moore, A.M. *Some Challenges and Advances in Regional Ocean Data Assimilation. Seminar on Data Assimilation for Atmosphere and Ocean*; European Centre for Medium-Range Weather Forecasts: Reading, UK, 2011.

23. Yang, K.; He, J.; Tang, W.; Qin, J.; Cheng, C.C.K. On downward shortwave and long wave radiations over high altitude regions: Observation and modeling in the Tibetan Plateau. *Agric. For. Meteorol.* **2010**, *150*, 38–46. [CrossRef]

24. Chen, Y.; Yang, K.; He, J.; Qin, J.; Shi, J.C.; Du, J.Y.; He, Q. Improving land surface temperature modeling for dry land of China. *J. Geophys. Res. Atmos.* **2011**, *116*, 132–147. [CrossRef]

25. Dai, Y. *The Impact of Lake Effects on the Temporal and Spatial Distribution of Precipitation at Nam Co, on the Tibetan Plateau*; AGU: Washington, DC, USA, 2015.

26. Ma, N.; Szilagyi, J.; Niu, G.Y.; Zhang, Y.S.; Zhang, T.; Wang, B.B.; Wu, Y.H. Evaporation variability of Nam Co Lake in the Tibetan Plateau and its role in recent rapid lake expansion. *J. Hydrol.* **2016**, *537*, 27–35. [CrossRef]

27. Gao, S.; Wu, Q.; Zhang, Z.; Xu, X. Impact of climatic factors on permafrost of the Qinghai-Xizang Plateau in the time-frequency domain. *Quatern. Int.* **2015**, *374*, 110–117. [CrossRef]

28. Singh, V.P. The use of entropy in hydrology and water resources. *J. Hydrol. Process* **1997**, *11*, 587–626. [CrossRef]

29. Avseth, P.; Mukerji, T.; Mavko, G. Quantitative Seismic Interpretation. *Episodes* **2005**, *3*, 236–237.

30. Ebrahimi, N. Stochastic properties of a cumulative damage threshold crossing model. *J. Appl. Probab.* **1999**, *36*, 720–732. [CrossRef]

31. Crutchfield, J.; Feldman, D.P. Regularities unseen, randomness observed: Levels of entropy convergence. *Chaos* **2003**, *13*, 25–54. [CrossRef] [PubMed]

32. Chapman, T.G. Entropy as a measure of hydrologic data uncertainty and model performance. *J. Hydrol.* **1986**, *85*, 111–126. [CrossRef]

33. Maruyama, T.; Kawachi, T.; Singh, V.P. Entropy-based assessment and clustering of potential water resources availability. *J. Hydrol.* **2005**, *309*, 104–113. [CrossRef]

34. Brunsell, N.A. A multiscale information theory approach to assess spatial-temporal variability of daily precipitation. *J. Hydrol.* **2010**, *385*, 165–172. [CrossRef]

35. Hasan, M.M.; Dunn, P.K. Entropy consistency in rainfall distribution and potential water resource availability in Australia. *J. Hydrol. Process.* **2011**, *25*, 2613–2622. [CrossRef]

36. Yang, Y.; Burn, D.H. An entropy approach to data collection network design. *J. Hydrol.* **1994**, *157*, 307–324. [CrossRef]

37. Yoo, C.; Jung, K.; Lee, J. Evaluation of rain gauge network using entropy theory: Comparison of mixed and continuous distribution function applications. *J. Hydrol. Eng.* **2008**, *13*, 226–235. [CrossRef]

38. Singh, V.P. Entropy theory for derivation of infiltration equations. *Water Resour. Res.* **2010**, *46*, W03527. [CrossRef]

39. Singh, V.P. Entropy theory for movement of moisture in soils. *Water Resour. Res.* **2010**, *46*, W03516. [CrossRef]

40. Zhou, X.; Lei, W.; Ma, J. Entropy Base Estimation of Moisture Content of the Top 10-m Unsaturated Soil for the Badain Jaran Desert in Northwestern China. *Entropy* **2016**, *18*, 323. [CrossRef]

41. Zanette, D. Generalized Kolmogorov entropy in the dynamics of the multifractal generation. *Phys. A Stat. Mech. Appl.* **1996**, *223*, 87–98. [CrossRef]

42. Guariglia, E. Entropy and Fractal Antennas. *Entropy* **2016**, *18*, 84. [CrossRef]

43. Peng, Y.; Zhou, J.G.; Burrows, R. Modelling the free surface flow in rectangular shallow basins by lattice Boltzmann method. *J. Hydraul. Eng.* **2011**, *137*, 1680–1685. [CrossRef]

44. Peng, Y.; Zhou, J.G.; Burrows, R. Modelling solute transport in shallow water with the lattice Boltzmann method. *Comput. Fluids* **2011**, *50*, 181–188. [CrossRef]

45. Baumer, C. *Southern Silk Road: In the Footsteps of Sir Aurel Stein and Sven Hedin*; Orchid Press: Krung Thep Maha Nakhon, Thailand, 2001.

46. Liu, J.Q.; Qin, X.G. Evolution of the environmental framework and oasis in the Tarim Basin. *J. Quat. Sci.* **2005**, *25*, 533–539.

47. Zhu, Z.D.; Wu, Z.; Liu, S.; Di, X.M. *An Outline of Chinese Deserts*; Science Press: Beijing, China, 1980; Volume 107. (In Chinese)

48. Ren, X.; Mu, G.J.; Xu, L.S.; Lin, Y.C.; Zhao, X. Characteristics of artificial oasis expansion in south Tarim Basin from 2000 to 2013. *Arid Land Res.* **2015**, *38*, 1022–1030. (In Chinese)

49. Zhou, X.; Lei, W.J. Hydrological interaction between oases and water vapor transportation in the Tarim Basin, northwestern China. *Sci. Rep.* **2018**, *8*, 13431. [CrossRef] [PubMed]

50. Li, B.; Chen, Y.; Shi, X.; Chen, Z.; Li, W. Temperature and precipitation changes in different environments in the arid region of northwest China. *Theor. Appl. Climatol.* **2013**, *112*, 589–596. [CrossRef]

51. Halik, W.; Mamat, A.; Dang, J.H.; Deng, B.S.H.; Tiyip, T. Suitability analysis of human settlement environment within the Tarim Basin in Northwestern China. *Quatern. Int.* **2013**, *311*, 175–180. [CrossRef]

52. Zhao, X.F.; Xu, H.L.; Wang, M.; Zhang, P.; Ling, H. Overloading analysis of irrigation area in basins of Tarim River in different years. *Trans. CSAE* **2015**, *31*, 77–81. (In Chinese)

53. Yan, Z.L.; Huang, Q.; Chang, J.X.; Wang, Y. Analysis and dynamic monitoring of land use in the main stream of Tarim river valley using remote sensing. *Trans. CSAE* **2008**, *24*, 119–123. (In Chinese)

54. Zhang, J.S.; Yan, Z.L.; Wang, X.G.; Huang, Q.A.; Gao, F. Remote sensing analysis of spatial-temporal changes of desertification land in Lower Reaches of Tarim River. *Trans. CSAE* **2009**, *25*, 161–165. (In Chinese)

55. Yang, K. User's Guide for China Meteorological Forcing Dataset. 2010. Available online: http://dam.itpcas.ac.cn/data/User_Guide_for_China_Meteorological_Forcing_Dataset.htm (accessed on 14 September 2018).

56. Shannon, C.E. A mathematical theory of communication. *Bell Syst. Technol. J.* **1948**, *27*, 379–423. [CrossRef]

57. Ben-Naim, A. Entropy, Shannon's Measure of Information and Boltzmann's H-Theorem. *Entropy* **2017**, *19*, 48. [CrossRef]

58. Conrad, K. Probability distributions and maximum entropy. *Entropy* **2004**, *6*, 10.

59. Shi, Y.F. Discussion on the present climate change from warm-dry to warm-wet in northwest China. *J. Quat. Sci.* **2003**, *23*, 152–164. (In Chinese)

60. Zhang, J.C. *Monsoon and Precipitation in China*; China Meteorological Press: Beijing, China, 2010; Volume 16. (In Chinese)

61. Shen, L.L.; He, J.H.; Zhou, X.Y.; Chen, L.; Zhu, C.W. The regional variabilities of the summer rainfall in China and its relation with anomalous moisture transport during the recent 50 years. *Acta Meteorol. Sin.* **2010**, *68*, 918–931.

62. Lei, Z.D.; Hu, H.P.; Yang, S.X.; Tian, F.Q. Analysis on water consumption in oases of the Tarim Basin. *J. Hydraul. Eng.* **2006**, *37*, 1470–1475.

Article

Modeling the Application Depth and Water Distribution Uniformity of a Linearly Moved Irrigation System

Junping Liu, Xingye Zhu *, Shouqi Yuan and Alexander Fordjour

Research Center of Fluid Machinery Engineering and Technology, Jiangsu University, Zhenjiang 212013, China; liujp@ujs.edu.cn (J.L.); shouqiy@ujs.edu.cn (S.Y.); fordjouralexander27@yahoo.com (A.F.)

* Correspondence: zhuxy@ujs.edu.cn; Tel.: +86-511-887-80284

Received: 19 March 2019; Accepted: 17 April 2019; Published: 19 April 2019

Abstract: A model of a linearly moved irrigation system (LMIS) has been developed to calculate the water application depth and coefficient of uniformity (CU), and an experimental sample was used to verify the accuracy of the model. The performance testing of the LMIS equipped with 69-kPa and 138-kPa sprinkler heads was carried out in an indoor laboratory. The LMIS was towed by a winch with a 1.0 cycle/min pulsing frequency while operating at percent-timer settings of 30, 45, 60, 75, and 90%, corresponding to average moving speeds of 1.5, 2.3, 3.3, 4.0, and 4.7 m min^{-1}, respectively. The application depth and CU obtained under various speed conditions were compared between the measured and model-simulated data. The model calculation accuracy was high for both operating pressures of 69 and 138 kPa. The measured application depths were much larger than the triangular-shaped predictions of the simulated application depth and were between the parabolic-shaped predictions and the elliptical-shaped predictions of the simulated application depth. The results also indicate that the operating pressure and moving speed were not significant factors that affected the resulting CU values. For the parabolic- and elliptical-shaped predictions, the deviations between the measured and model-simulated values were within 5%, except for several cases at moving speeds of 2.3 and 4.0 m min^{-1}. The measured water distribution pattern of the individual sprinklers could be represented by both elliptical- and parabolic-shaped predictions, which are accurate and reliable for simulating the application performances of the LMIS. The most innovative aspect of the proposed model is that the water application depths and CU values of the irrigation system can be determined at any moving speed.

Keywords: linearly moved irrigation system; application depth; moving speed; uniformity coefficient

---

1. Introduction

With the development of science and technology and the need for practical production, the rapid development of the linearly moved irrigation system (LMIS) has been popular in various countries. In China, the LMIS was introduced in the 1980s, with increasing use in recent years due to technological innovations, management convenience, economics and water savings [1–3]. Linear systems, also known as mobile side, have structures similar to center pivot; however, circular movement is replaced by linear movement so that the water application rate is constant along the entire length of the lateral move type. The basic elements to determine LMIS operation include the moving speed, V (m/min), and the maximum range of the sprinkler head, R max (m). The moving speed of LMIS directly affects how water is sprayed in the unit area; when the moving speed is slow, the unit area of spraying is greater, and when the moving speed is fast, the unit area of water spraying is lower. Effective irrigation is not the application of water without control or planning, but is the application of the correct amount of water at the right time, especially with uniformity, during the irrigation process. Thus, to achieve water optimization in agriculture, it is important to frequently evaluate the performance of irrigation systems by using some parameters that express and quantify the operation quality. The application depth and water distribution uniformity coefficient are often used as indicators of problems concerning irrigation distribution.

Irrigation uniformity is defined as the variation in irrigation depths over an irrigated area and is an important performance characteristic of the sprinkler irrigation system [4–10]. The importance of sprinkler irrigation uniformity was recognized as early as 1942 [11]. Widespread research has been conducted on the factors affecting sprinkler irrigation uniformity. Such factors include nozzle size and pressure, the type of diffuser device, sprinkler spacing, riser height, field topography, discharge angle, number and configuration of the sprinklers and wind speed and direction, all of which can influence water application uniformity [12–23]. In recent years, several studies have been carried out on irrigation performance by center pivot to identify the main problems of irrigation efficiency [24–28]. However, there have been few studies concerning the application depth and uniformity of the LMIS at different operating speeds [29,30]. In the current studies, none of the existing methods can be conducted to calculate the application depth and uniformity of the LMIS. In some certain special cases, the surface runoff appears in the field as the application of the LMIS. It is difficult to determine a mismatch of the farmland soil infiltration capacity and the hydraulic performance of LMIS. Some targeted solutions for solving the problem are unavailable in the scientific research. Therefore, it is very important to study the calculation method of the application depth and the combination uniformity of the LMIS, which is of great theoretical value and practical significance. The objective of this study is to put forward a calculation method of application depth and uniformity of LMIS and to verify the accuracy of the calculation results through experimental tests.

2. Materials and Methods

2.1. LMIS and Spray Sprinkler

The sprinkler irrigation system used in this study was specifically manufactured as an experimental sample by the Research Center of Fluid Machinery Engineering and Technology (Jiangsu University, Zhenjiang City, Jiangsu Province, China). It was an intermittent linear moving system towed by a winch and installed with a fixed spraying plate package as shown in Figure 1. The fixed spray plate sprinkler studied in this research was the Nelson D 3000 sprayhead (Nelson Irrigation Co., Walla Walla WA, USA). A single-strut pressure regulator (Figure 2a) was part of the package for a low-pressure deflection sprinkler. The shapes of the nozzles were circular (Figure 2b). The nozzle diameter of the fixed spray plate sprinkler was 5.16 mm, corresponding to number #26 specified by the Nelson Company. The sprinkler deflects 36 water streams uniformly, centered from itself, to form a full-circle spray pattern. The pressure regulator and sprinkler were connected immediately with a dedicated screw-type connector (Figure 2c). The reason for using such sprinklers was that prior tests had shown that this package and pressure combination had a good coefficient of uniformity [31–33]. All sprinklers were on flexible drop hoses that set the D 3000 sprayhead approximately 1.0 m above the soil surface, and spaced 3.0 m apart. A pressure regulator of 10 or 20 psi (69 or 138 kPa) was installed just upstream of the spray sprinkler. The discharge of the LMIS was measured with an electrical flow gauge (Model E-mag/DN25, manufactured by Kaifeng Electronic Instrument Company, China) with an accuracy of 0.3%.

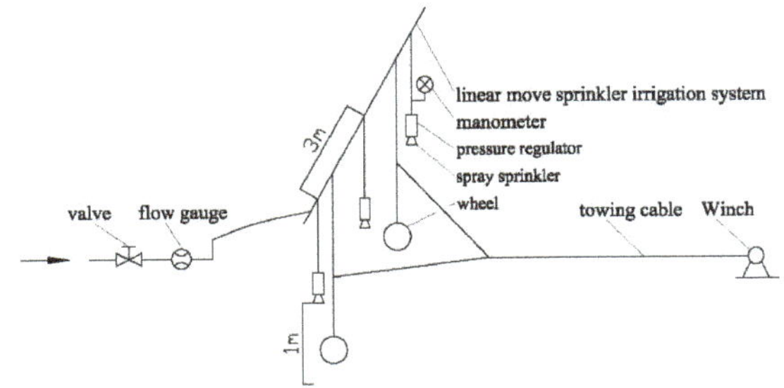

Figure 1. Description of the linearly moved irrigation system (LMIS).

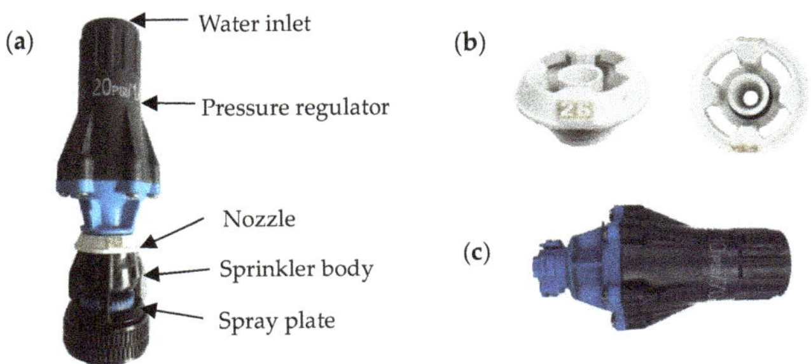

Figure 2. Pressure regulator and sprinkler used in the LMIS (**a**), nozzles used in the sprinkler (**b**), and dedicated screw-type connector of pressure regulator (**c**).

2.2. Modeling for Calculation of Application Depth

As the structure parameters of sprinkler type, installed height and spacing were determined, with the water distribution along the axis direction of the LMIS known. First, the LMIS was maintained at a constant working condition. Then, according to the maximum wetted radius, R max, and the water application rate of the sprinklers, the figure of water application depth could be obtained. The relationship between the water application and spraying distance may be described by a geometric pattern. An elliptical, parabolic or triangular pattern was chosen to represent it as shown in Figure 3, where the curve of abc represents the distribution line of sprinkler irrigation intensity, the vertical axis represents the water application rate p (mm/h), and the horizontal axis represents the spraying distance L (m). In theory, under the conditions of an indoor experiment without any wind effects, the value of L is two times the maximum wetted radius, R max.

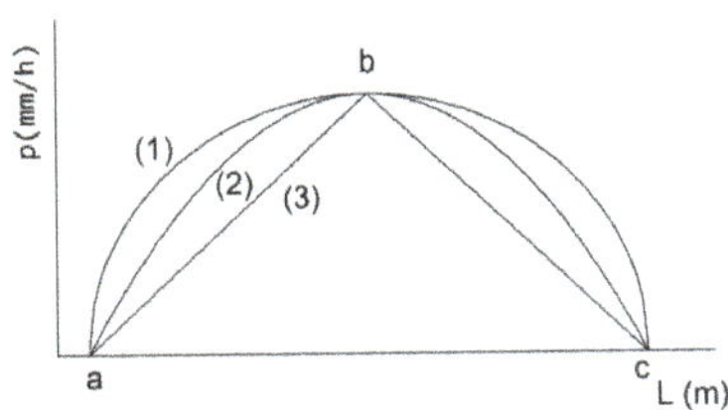

Figure 3. Water application depth: (1) elliptical, (2) parabolic and (3) triangular.

The catch device was supposed to be set at point O, away from the LMIS, with a distance larger than R max, to study the average application depth of the system. The working phenomenon was as following: first, the LMIS approached point O, and point O started to receive the water sprayed out by the LMIS; then, the LMIS gradually receded from point O while moving until the system passed over the catch devices entirely and point O did not receive any water. The LMIS-fulfilled total water application depth was the sum of passing section area of point O. Figure 4 represents the depiction of water application distribution of point O. Combined with Figure 3, when the LMIS was passing through point O at a speed of v, it means that the curve of abc was passing through point O at a speed of v and the passing time could be calculated as t = L_{ac}/v.

At this time, the curve of abc also represents the distribution line of sprinkler irrigation intensity, and the vertical axis also represents the water application rate p (mm/h); however, the horizontal axis represents the passing time t (s). When the LMIS was passing through any point of O, the receiving water of point O was the sum of the water application section area passing point O. Figure 4 represents the distribution section S, and S is the superimposed distribution map of all water sprayed to point O. When the travel speed is v, it can be seen that t = L_{ac}/v. The relationship between the water application rate ρ and the passing through time T was as follows:

Elliptical shape:

$$\rho = P_k \sqrt{1 - \frac{4(T - t/2)^2}{t^2}}, \ P_k = \frac{4WDP}{\pi t} \tag{1}$$

Parabolic shape:

$$\rho = -\frac{4P_k}{t^2}(T - t/2)^2 + P_k, \ P_k = \frac{1.5WDP}{t} \tag{2}$$

Triangular shape:

$$\left\{ \begin{array}{l} \rho = \frac{2P_k}{t}T \quad (0 \leq T \leq t/2) \\ \rho = \frac{2P_k}{t}(t - T) \quad (t/2 \leq T \leq t) \end{array} \right., \ P_k = \frac{2WDP}{t} \tag{3}$$

where

P_k = peak water application rate (mm h^{-1})
t = water application time (h).

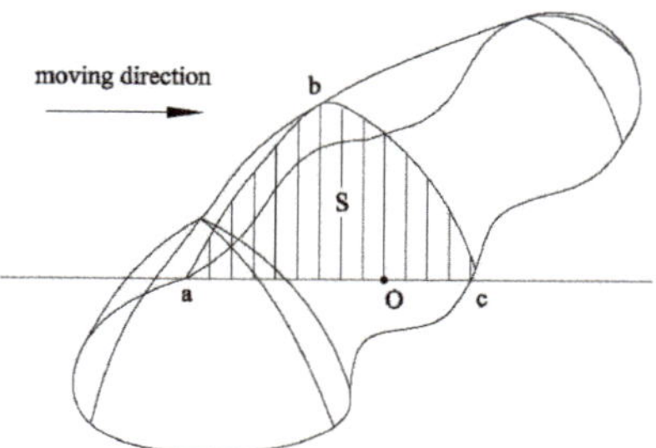

Figure 4. Depiction of water application distribution of point O.

Then, the wetted area S was calculated as follows:

$$S = \int_0^t \rho(T)dt \tag{4}$$

Supposing that the travel speed of LMIS was set at different values of v1, v2, and vn, respectively, t1 = Lac/v1, t2 = Lac/v2, and tn = Lac/vn. The parabolic pattern was selected and was assumed to be representative of LMIS irrigation water, whose pattern for the same water application depth and time has a peak rate between the elliptical and triangular patterns. Figure 5 shows an elliptical application shape. Notice the increase of water application depth (proportional to larger geometric areas) related to high speed, average speed, and low speed, as well as the constant peak water application rate.

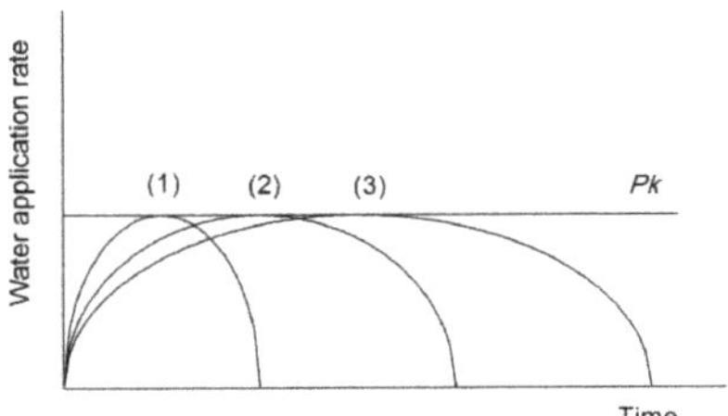

Figure 5. Water application patterns with decreasing speed at the same irrigation point: (1) high speed, (2) average speed, and (3) low speed.

2.3. Modeling for Calculation of Uniformity

The spraying irrigation state of LMIS can be regarded as a limit state of mobile spraying for fixed-type pipe with an infinite narrow width. Supposing that the LMIS traveled at a uniform speed, and the water distribution for every sprinkler was the same and does not change with time, it could be determined that the depth of irrigation water along the direction of the sprinkler is the same. Therefore, only the branch direction sprinkling uniformity degree needed to be considered, which can be representative of the entire area of the spraying uniformity. In this way, the concept of linear spraying uniformity is put forward. The uniformity of the LMIS is equal to the uniformity of the axial direction of the unit. The coefficient of uniformity CU (%), developed by Christiansen (1942), was calculated using the following equation:

$$CU = 1 - \frac{\Delta \overline{h}}{\overline{h}} \tag{5}$$

When the point was adopted by affirmative grid,

$$\Delta \overline{h} = \frac{\sum_{i=1}^{n} \left| h_i - \overline{h} \right|}{n}, \ \overline{h} = \frac{\sum_{i=1}^{n} h_i}{n} \tag{6}$$

where h_i = water depth of calculated point i, mm/h; $\overline{h}$ = mean water depth of all calculated points, mm/h; and n = total number of calculated points used in the evaluation.

Figure 6 represents the calculated sketch of spraying uniformity for LMIS. The K direction is the running direction and the j direction is the axial of the unit. The water depth of h_{jk} was as follows:

$$
\begin{aligned}
h_{11} &= h_{12} = h_{13} = \cdots = h_{1k} = h_1 \\
h_{21} &= h_{22} = h_{23} = \cdots = h_{2k} = h_2 \\
&\quad \cdots \cdots \\
h_{j1} &= h_{j2} = h_{j3} = \cdots = h_{jk} = h_j
\end{aligned} \tag{7}
$$

$$\text{Then, } \overline{h} = \frac{h_{11}+h_{12}+h_{13}+...h_{1k}+h_{21}+h_{22}+...+h_{2k}+h_{j1}+h_{j2}+...h_{jk}}{n_k \cdot n_j}$$

$$\text{Then, } \overline{h} = \frac{h_1+h_2+h_3+...+h_j}{n_j} = \sum_{i=1}^{j} h_i / n_j$$

Here, $\overline{h}$ has been simplified to the average value of the depth of the axial point irrigation water. In the same way:

$$\Delta \overline{h} = \frac{\left|\overline{h}-h_{11}\right|+\left|\overline{h}-h_{12}\right|+...+\left|\overline{h}-h_{1k}\right|+...+\left|\overline{h}-h_{j1}\right|+...+\left|\overline{h}-h_{jk}\right|}{n_k \cdot n_j}$$

$$= \frac{n_k\left|\overline{h}-h_1\right|+n_k\left|\overline{h}-h_2\right|+...+n_k\left|\overline{h}-h_j\right|}{n_k \cdot n_j} = \sum_{i=1}^{j} \frac{\left|\overline{h}-h_i\right|}{n_j} \tag{8}$$

This type has also been turned into an axial $\Delta\bar{h}$. The spray uniformity coefficient can be obtained after substitution of the axial $\bar{h}$ and $\Delta\bar{h}$ into Equation (5).

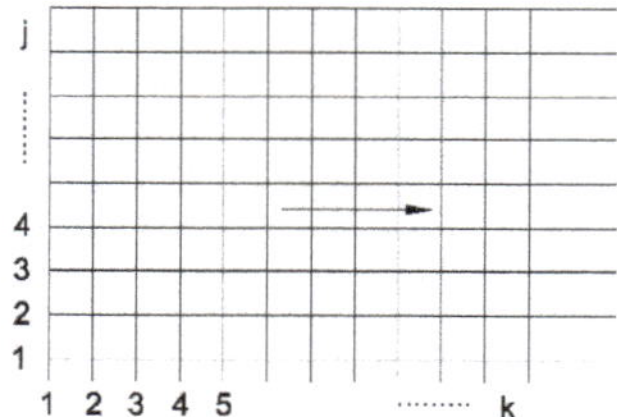

Figure 6. Calculated sketch of spraying uniformity for the LMIS.

2.4. Setup and Procedures of Indoor Experiment

The study site was located at the indoor facilities of the Research Center of Fluid Machinery Engineering and Technology, Jiangsu University (Jiangsu Province, China). Figure 7 shows the test plot for the LMIS mobile water distribution. The structure, built with several metal 40-mm-long and 20-mm-wide rectangular bars, was 12.0 m wide and the spray sprinkler could be adjusted from 0 to 2.5 m above the surface. Because only a short width (12.0 m) was tested, pressure variation through the measured width of the system was assumed to be negligible. The frame corners were equipped with four wheels and a winch was used to tow the system. The water source was a reservoir with a capacity of 60 m^3. A 3.0-kW electric centrifugal pump was connected to the water supplying pipe and a 36-mm external-diameter hose pipe was used to supply water to the spray sprinkler. Manometers and valves were installed as required to control water supply during the experiments (as shown in Figure 1). Catch cans were used to collect the applied water. They were constructed from transparent plastic, with an inverted conical shape. The catch can opening was 200 mm for the inside diameter and the catch can height was 250 mm. Nine rows of catch cans were distributed along the direction of the vertical moving unit, and the distance and spacing of catch cans were 1 m each. The LMIS and the catch cans were installed in a plot with cement flooring. After the spraying test, the weighing method was used to calculate the depth of irrigation water for every catch can. The measured irrigation depth of the point values was taken with an average irrigation depth of the 9 rows of catch cans.

Figure 7. Test plot for the LMIS mobile water distribution.

Experiments were carried out at a constant pressure of 69 and 138 kPa, respectively, maintained by a 10- and 20-psi pressure regulator. The pulsing frequency towed by the winch was 1.0 cycle/min while operating the winch at percent-timer settings of 30, 45, 60, 75%, and 90%, corresponding to the average moving speed of 1.5, 2.3, 3.3, 4.0, and 4.7 m min^{-1}, respectively. These selected five values cover the range of duty cycles that are expected to be used in the field. A value of 100% is the same as normal operating conditions (constant sprinkler discharge) and zero percent is the same as operating the system dry. All operating pressures and moving speeds were within the manufacturer's recommendations. The following standards [34–37] were adopted in the design of the experimental setup and in the experiment itself. The duration of each test was approximately 30 to 45 min. The applied water in the catch cans was read in a measuring cup with a volume of 500 mL and an accuracy of 5 mL. The read data were then converted to average irrigation depth by dividing the cross-sectional area of the catch can. A minimum of three replications were conducted for each pressure and moving speed combination, and data were averaged and used as the final experimental data. Flow rates from the two operating pressures used for the application depth tests were measured three times under the same operating conditions. A metal pipe was positioned over the sprinkler nozzles and the discharge water directed into a bucket for 2 min. Discharge volumes were then weighed with an electronic balance (Otimpa Corp, China) with 1 gram accuracy and converted into flow rate.

The average air temperature during testing was 20.8 °C and ranged from 17.8 °C to 22.4 °C. The average relative humidity was 41% and ranged from 36% to 47%. To verify the accuracy of the model, a moving superposed water quantity verification test was carried out. Taking into account the condition that the work at the beginning of the LMIS may not be stable, the moving procedure of the system was start-up and test data were collected after operating at the working pressure for more than 10 min.

3. Results and Discussion

The stable working state of the LMIS means the spraying speed is stable, and the working pressure of the spray head on the system is stable. The wind speed during the entire tests ranged from 0.0 to 0.12 m s^{-1} and was usually less than 0.1 m s^{-1}. These data are not presented or discussed as the values were less than the lower threshold of the measuring equipment. The potential source of error in this study is any differences in evaporative conditions across the time period of the various tests. Although many studies have been conducted on evaporation during sprinkler irrigation, Schneider [38] noted that no more than 2% losses resulted from evaporation during use of sprinkler irrigation systems. For example, even under a condition of an average temperature of 26 °C, a relative humidity of 64%, and a wind speed of 6.4 m/s, the measured evaporation was only 0.8% of total sprinkler discharge [39]. Therefore, due to the indoor experimental conditions without any wind effects in the study, this particular potential source of error was minimized. An indoor experiment was conducted to verify the accuracy of the model using the Nelson D 3000 spray head with a nozzle diameter of 5.16 mm at working pressures of 69 kPa and 138 kPa and at an elevation of 1.0 m above the soil surface.

3.1. Coefficient of Discharge

The results of measured flow rates of sprinkler irrigation nozzles used in this study are shown in Table 1. Analysis of the measured data were performed to find the influence of the geometrical parameters as well as the operating pressure on discharge of the sprinkler head, which can be expressed in terms of the discharge coefficient, C. The coefficient of discharge of the sprinkler is generally expressed as follows:

$$C = \frac{Q}{d^2 \frac{\pi}{4} \sqrt{2gH}} \tag{9}$$

where C is the coefficient of discharge, Q is the flow rate of the sprinkler (m^3 h^{-1}), d is the diameter of the nozzle (m), g is the gravitational acceleration (m s^{-2}), and H is the pressure head (m).

As shown in Table 1, when using the fixed spray plate sprinkler, the measured nozzle flow rates ranged from 0.77 to 0.84 m^3 h^{-1}, with a mean value of 0.808 m^3 h^{-1}, and 1.15 to 1.22 m^3 h^{-1} with a mean value of 1.183 m^3 h^{-1} for the pressure regulators of specification 69 and 138 kPa, respectively. After calculating using Equation (8), the nozzle coefficients of discharge ranged from 0.880 to 0.960 with a mean value of 0.923, and 0.929 to 0.986 with a mean value of 0.956, for 69 and 138 kPa, respectively. From the aforementioned analysis, it was found that the coefficients of discharge fluctuated within a small acceptable range under the same operating pressure, which could be attributed to the acceptable experimental error. Additionally, the coefficients of discharge obtained using the 138 kPa pressure regulator were higher than those obtained using the 69 kPa pressure regulator, which means that using a pressure regulator with a higher outlet pressure produced higher coefficients of discharge.

Table 1. Measured flow rates of sprinkler irrigation nozzles.

Flow Rate, m³ h⁻¹ Replications	69 kPa			138 kPa		
	Nozzle I	Nozzle II	Nozzle III	Nozzle I	Nozzle II	Nozzle III
1	0.8	0.79	0.8	1.16	1.21	1.22
2	0.84	0.83	0.81	1.18	1.17	1.15
3	0.77	0.8	0.83	1.21	1.16	1.19
Mean	0.80	0.81	0.81	1.18	1.18	1.19

3.2. Measured Water Distributions

Figure 8 represents the measured sprinkler intensity and fitting curve of radial points with a nozzle diameter of 5.16 mm and a height of 1.0 m at operating pressures of 69 kPa and 138 kPa, respectively.

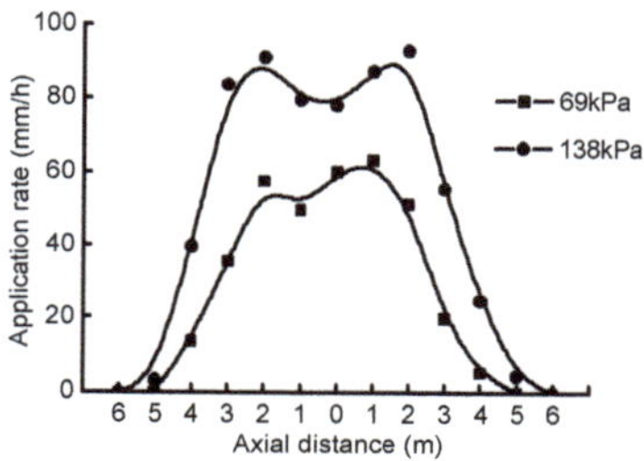

Figure 8. Measured radial application rate and fitting curve for operating pressures of 69 kPa and 138 kPa.

As shown in Figure 8, the sprinkler nozzle wetting radius for 69 kPa and 138 kPa was 4.4 m and 5.7 m, respectively. The sprinkler head produced quite similar water application profiles under different operating pressures. The average values of the sprinkler head application rates varied from 3.2 mm h⁻¹ to 92.8 mm h⁻¹. The water application rate increased to a maximum value first and then decreased approximately linearly as the distance from the sprinkler increased. The maximum application rate was determined for the two analyzed pressures: 63.2 mm h⁻¹ at distances of 1 m for 69 kPa, and 92.8 mm h⁻¹ at 2 m from the sprinkler for 138 kPa. Starting from this distance, the application rate decreased until it reached the minima. For 69 kPa at 4 m, and 138 kPa at 5 m, the minimum values were 5.3 and 4.3 mm h⁻¹, respectively. The application rates had large variability for all measured points. The applied water depths were extremely similar to the radial water depths, indicating that the water distribution pattern of individual sprinklers had a primary influence on the overall water application depth and distribution uniformity. The radial water distribution curve appeared as a saddle type, and had two sprinkler intensity peaks close to the sprinkler. The least squares method was used for curve fitting to simulate the water distribution characteristics, and the fitting degree was as high as 0.89. Table 2 summarizes the measured application depths and corresponding CU values obtained with an LMIS average moving speed of 1.5, 2.3, 3.3, 4.0 and 4.7 m/min under different operating pressures of 69 kPa and 138 kPa. For the various LMIS speeds, the application depth of 69 kPa varied from 1.3 to 4.2 mm with an average of 2.38 mm, the CU varied from 84.5% to 88.9% with an average of 86.58%, and the standard deviation had an average of 0.16; the application depth of 138 kPa varied from 2.6 to 7.7 mm with an average of 4.08 mm, the CU varied from 85.9% to 89.8% with an average of 87.28%, and the standard deviation had an average of 0.68. It appeared that the application depth decreased as the LMIS speed increased, which could be understood easily. The CU value did not linearly relate to the LMIS speed, which was interpreted as adjusting for unimportant variability on a small scale.

Table 2. Measured irrigation application depths and corresponding coefficient of uniformities under different operating pressures and LMIS speeds.

Operating Pressure (kPa)	LMIS Speed (m min^{-1})	Application Depth (mm)	CU (%)	Standard Deviation
69 kPa	1.5	4.2	86.0	0.15
	2.3	2.8	84.5	0.26
	3.3	2.0	88.7	0.12
	4.0	1.6	88.9	0.12
	4.7	1.3	84.8	0.15
	Mean	2.38	86.58	0.16
138 kPa	1.5	7.7	87.0	0.75
	2.3	4.2	89.8	0.82
	3.3	3.1	87.2	1.00
	4.0	2.8	85.9	0.47
	4.7	2.6	86.5	0.36
	Mean	4.08	87.28	0.68

3.3. Comparison of Application Depth and CU Values between Experimental Measured Data and Modeling Simulations

As shown in Sections 2.2 and 2.3, in the modeling for calculation of application depth and CU values of the sprinkler irrigation, the data from Figure 8 were used as the basic foundation, and the fitting curves were supposed to be regulated as elliptical, parabolic, and triangular shapes, respectively.

3.3.1. Application Depth

The simulated application depths were obtained with a constant spacing of 3.0 m for comparison. The data from Table 2 were used as the experimental measured data. As a further illustration of model validation of the application depth calculation, Figure 9 presents a comparison of application depth between measured data and model-simulated results at an operating pressure of 69 kPa and 138 kPa, respectively.

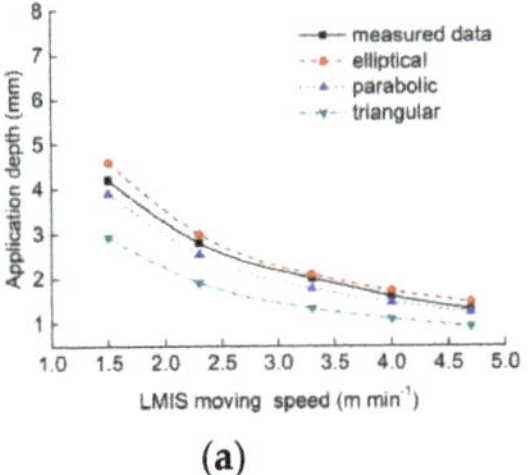

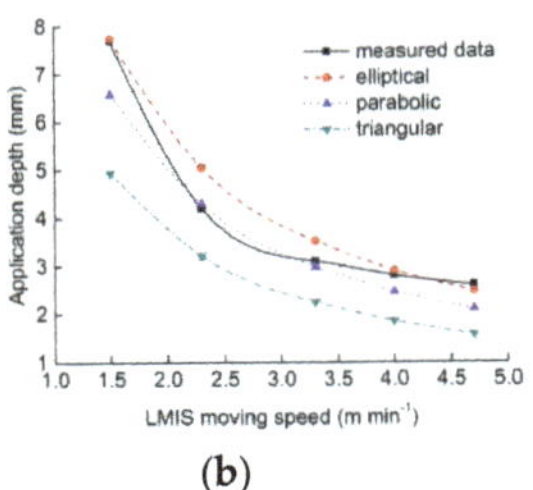

Figure 9. Comparison of application depth between experimental measured data and model-simulated results: (**a**) 69 kPa, (**b**) 138 kPa.

As seen from Figure 9, there appears agreement between the measured and simulated mean application depths. The measured data and model-simulated application depths for all the elliptical-, parabolic-, and triangular-shaped predictions decreased with an increasing average moving speed, which was in accordance with the analysis that slower moving speeds will result in more spraying water. The measured application depths were much larger than the triangular-shaped prediction simulated application depths and were just between the parabolic-shaped prediction and the elliptical-shaped prediction simulated application depths.

Compared to the experimental measured data at an operating pressure of 69 kPa, the simulated result was from 28.0% to 33.3% lower with an average of 30.9% for the triangular-shaped prediction, from 4.0% to 11.1% lower with an average of 7.8% for the parabolic-shaped prediction, and from 4.7% to 13.1% higher with an average of 8.5% for the elliptical-shaped prediction, respectively. Compared to the experimental measured data at an operating pressure of 138 kPa, the simulated result was from 23.3% to 39.3% lower with an average of 32.0% for the triangular-shaped prediction, from 2.3% to 19.1% lower with an average of 10.2% for the parabolic-shaped prediction, and from 3.9% to 20.4% higher with an average of 8.7% for the elliptical-shaped prediction, respectively. Normally, the measured water depth is generally a slight deviation from the calculated value, caused by the fitting error, the testing error, evaporation and drift during the spraying process [40–42]. Therefore, it was determined that both the parabolic- and elliptical-shaped predictions were the acceptable shapes of the water distribution pattern, as they were closer to the measured application depth.

The measured application depth and model-simulated results based on overlapping the measured water distribution data of individual sprinklers were curved lines. Special attention was given to the development of empirical equations for water application depth with regard to moving speed. The curvilinear Equation (10) was regressed and Table 3 presents the equations of those profiles. The coefficient of determination (R^2) for measured data ranged from 0.9698 to 0.9954 and for all the model simulations was 0.9946.

$$p = Av^2 - Bv + C \tag{10}$$

where p is the application depth (mm) and v is the LMIS speed (m min^{-1}).

Table 3. Equations of water application depth with regard to moving speed.

	69 kPa				138 kPa			
	A	**B**	**C**	**R²**	**A**	**B**	**C**	**R²**
Measured data	0.186	1.814	5.780	0.9954	0.529	4.331	11.26	0.9698
Elliptical prediction	0.231	2.140	6.461	0.9946	0.389	3.605	10.88	0.9946
Parabolic prediction	0.196	1.818	5.487	0.9946	0.330	3.061	9.240	0.9946
Triangular prediction	0.147	1.363	4.115	0.9946	0.248	2.300	6.930	0.9946

3.3.2. Coefficient of Uniformity

The simulated CU values were obtained using MATLAB calculations based on the basic foundation of application depth values from Figure 9. The combined CU values were calculated using the aforementioned method. The data from Table 2 were used as the experimental measured data. As a further illustration of model validation of CU calculations, Figure 10 presents the comparison of CU between experimental measured data and model-simulated results at an operating pressure of 69 kPa and 138 kPa, respectively.

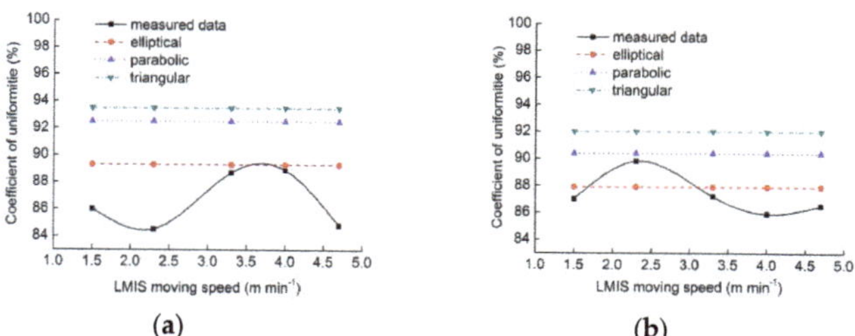

Figure 10. Comparison of CU between experimental measured data and model-simulated results. (**a**) 69 kPa, (**b**) 138 kPa.

As seen in Figure 10, there appears agreement between the measured and simulated CU values. The simulated CU for all the elliptical-, parabolic-, and triangular-shaped predictions maintains a constant value and does not vary with increasing speed. It was determined that a different moving speed has an influence on the depth of irrigation water along the direction of the sprinkler, which varies at a same percentage and does not affect CU values; only the branch direction data affect the sprinkling CU degrees, which can be representative of the entire area of the spraying uniformity. Therefore, when the branch water distribution shape was determined as an elliptical, parabolic, or triangular prediction, the simulated CU keeps a constant value at different moving speeds. However, the measured CUs were variable for all moving speeds. The range of combined CU values at different pressures were as follows: 84.5% at a moving speed of 2.3 m min^{-1} to 88.9% at a moving speed of 4.0 m min^{-1} (69 kPa), and 85.9% at a moving speed of 4.0 m min^{-1} to 89.8% at moving speed of 2.3 m min^{-1} (138 kPa). It was determined that the measured CU values with regard to moving speed fluctuated within a small range, and could be attributed to the acceptable fitting and experimental error.

Compared to the experimental measured data at an operating pressure of 69 kPa, the simulated CU value was 93.5% and of an absolute deviation rate from 5.2% to 10.7% with an average of 8.0% for the triangular-shaped prediction, 92.5% and of an absolute deviation rate from 4.0% to 9.5% with an average of 6.9% for the parabolic-shaped prediction, and 89.3% and of an absolute deviation rate from 0.4% to 5.7% with an average of 3.2% for the elliptical-shaped prediction, respectively. Compared to the experimental measured data at an operating pressure of 138 kPa, the simulated CU value was 92.0% and of an absolute deviation rate from 2.4% to 7.1% with an average of 5.4% for the triangular-shaped prediction, 90.4% and of an absolute deviation rate from 0.7% to 5.2% with an average of 3.6% for the parabolic-shaped prediction, and 87.9% and of an absolute deviation rate from 0.8% to 2.3% with an average of 1.6% for the elliptical-shaped prediction, respectively.

From the aforementioned analysis, it was determined that for the parabolic- and elliptical-shaped predictions, the deviations of the points are within 5% except for several cases at a moving speed of 2.3 and 4.0 m min^{-1}. This deviation may result from inaccurate measurement of the effective wetted widths. Generally, the simplified model has a high computational accuracy, which could be brought into Equation (5) to further calculate the LMIS spraying uniformity. The comparison of measured and model-simulated data indicated that the measured water distribution pattern of individual sprinklers could be represented both as an elliptical- or parabolic-shaped prediction, which was accurate and reliable for simulating the application performance of the LMIS and verified that the calculation of application depth and CU values shown in this work is applicable in practice.

In short, a new model of the LMIS has been developed and a sprinkler irrigation system was specifically manufactured as an experimental sample to verify the accuracy of the model. The differences in model accuracy owing to different operating pressures were not significant, and the moving speed of the LMIS did not appear to influence model accuracy either. Comparisons between experimental data and model simulations revealed that the model can accurately predict water application depth and CU values along the LMIS. Although the model was developed and validated for linear moving systems, it could be readily used for center pivots, which constitute a simplification of the calculation approach adopted.

4. Conclusions

This study presents a model of application depth and uniformity for the LMIS. Based on the obtained results and the conditions in which this trial was carried out, the following can be concluded:

At an operating pressure of 69 kPa and 138 kPa, the sprinkler nozzle wetting radius was 4.4 m and 5.7 m, respectively. The sprinkler head produced quite similar water application profiles, which appeared as a saddle type and had two sprinkler intensity peaks close to the sprinkler.

Compared to the experimental measured data at operating pressures of 69 kPa or 138 kPa, the simulated application depth was on average 30.9% or 32.0% lower for the triangular-shaped prediction, 7.8% or 10.2% lower for the parabolic-shaped prediction, and 8.5% or 8.7% for the elliptical-shaped prediction, respectively. The curvilinear equations for the measured application depth and model-simulated results were regressed and the coefficient of determination (R^2) was from 0.9698 to 0.9954. The simulated CU was an average deviation rate of 8.0% or 5.4% for the triangular-shaped prediction, 6.9% or 3.6% for the parabolic-shaped prediction, and 3.2% or 1.6% for the elliptical-shaped prediction, respectively.

This study indicates that a sprinkler irrigation system for the LMIS that is both elliptical and parabolic in shape could be considered as far as application depths and CU values are concerned. The model can therefore be further developed to provide a useful tool for LMIS design and management.

Author Contributions: J.L. was the supervision of this manuscript; X.Z. was writing the original draft and making all revisions; S.Y. was providing the funding acquisition; and A.F. was doing the literature resources.

Funding: This research was funded by the National Key Research and Development Program of China (No. 2016YFC0400202), the Key R&D Project of Jiangsu Province (Modern Agriculture) (No. BE2018313), and the Priority Academic Program Development of Jiangsu Higher Education Institutions (PAPD).

Acknowledgments: The authors are greatly indebted to the supports from students of the Research Centre of Fluid Machinery and Engineering, Jiangsu University for their assistances in conducting the experiment.

Conflicts of Interest: The authors declare no conflict of interest.

References

1. Yuan, S.Q.; Li, H.; Wang, X.K. Status, problems, trends and suggestions for water-saving irrigation equipment in China. *J. Drain. Irrig. Mach. Eng. (JDIME)* **2015**, *33*, 78–92.
2. Zhu, X.Y.; Chikangaise, P.; Shi, W.D.; Chen, W.H.; Yuan, S.Q. Review of intelligent sprinkler irrigation technologies for remote autonomous system. *Int. J. Agric. Biol. Eng.* **2018**, *11*, 23–30. [CrossRef]
3. Xu, D.; Li, Y.N.; Gong, S.H.; Zhang, B.Z. Experiment on sweet pepper nitrogen detection based on near infrared reflectivity spectral ridge regression. *J. Drain. Irrig. Mach. Eng. (JDIME)* **2019**, *37*, 63–72.
4. Hart, W.E. Sprinkler distribution analysis with a digital computer. *Trans ASAE* **1963**, *6*, 206–208.
5. Vories, E.D.; Von Bernuth, R.D. Single nozzle sprinkler performance in wind. *Trans. ASAE* **1986**, *29*, 1325–1330.
6. Li, J.; Kawano, H. Sprinkler rotation nonuniformity and water distribution. *Trans. ASAE* **1996**, *39*, 2027–2031. [CrossRef]
7. Zhu, X.Y.; Yuan, S.Q.; Jiang, J.Y.; Liu, J.P.; Liu, X.F. Comparison of fluidic and impact sprinklers based on hydraulic performance. *Irrig. Sci.* **2015**, *33*, 367–374. [CrossRef]
8. Liu, J.P.; Yuan, S.Q.; Li, H.; Zhu, X.Y. Experimental and combined calculation of variable fluidic sprinkler in agriculture irrigation. *AMA-Agr. Mech. Asia Afr. Lat. Am.* **2016**, *47*, 82–88.
9. Xiang, Q.J.; Xu, Z.D.; Chen, C. Experiments on air and water suction capability of 30PY impact sprinkler. *J. Drain. Irrig. Mach. Eng. (JDIME)* **2018**, *36*, 82–87.
10. Hu, G.; Zhu, X.Y.; Yuan, S.Q.; Zhang, L.G.; Li, Y.F. Comparison of ranges of fluidic sprinkler predicted with BP and RBF neural network models. *J. Drain. Irrig. Mach. Eng. (JDIME)* **2019**, *37*, 263–269.
11. Christiansen, J.E. *Irrigation by Sprinkling*; Bulletin 670; California Agricultural Experiment Station, University of California: Berkeley, CA, USA, 1942.
12. Solomon, K. Variability of sprinkler coefficient of uniformity test results. *Trans. ASAE* **1979**, *22*, 1078–1080. [CrossRef]
13. Fukui, Y.; Nakanishi, K.; Okamura, S. Computer evaluation of sprinkler irrigation uniformity. *Irrig. Sci.* **1980**, *2*, 23–32. [CrossRef]
14. Seginer, I.; Kantz, D.; Bernuth, R.D. Indoor measurement of single-radius sprinkler patterns. *Trans. ASAE* **1992**, *35*, 523–533. [CrossRef]
15. Tarjuelo, J.M.; Montero, J.; Valiente, M.; Honrubia, F.T.; Ortiz, J. Irrigation uniformity with medium size sprinklers. Part I: Characterization of water distribution in no-wind conditions. *Trans. ASAE* **1999**, *42*, 665–675. [CrossRef]
16. Michael, J.L.; John, S.S. Sprinkler head maintenance effects on water application uniformity. *J. Irrig. Drain. Eng. ASCE* **2000**, *126*, 142–148.

17. Playán, E.; Zapata, N.; Faci, J.M.; Tolosa, D.; Lacueva, J.L.; Pelegrin, J.; Salvador, R.; Sanchez, I.; Lafita, A. Assessing sprinkler irrigation uniformity using a ballistic simulation model. *Agric. Water Manag.* **2006**, *84*, 89–100. [CrossRef]

18. Zhu, X.Y.; Yuan, S.Q.; Liu, J.P. Effect of sprinkler head geometrical parameters on hydraulic performance of fluidic sprinkler. *J. Irrig. Drain. Eng. ASCE* **2012**, *138*, 1019–1026. [CrossRef]

19. Zhang, L.; Merkley, G.P.; Pinthong, K. Assessing whole-field sprinkler irrigation application uniformity. *Irrig. Sci.* **2013**, *31*, 87–105. [CrossRef]

20. Liu, J.P.; Yuan, S.Q.; Darko, R.O. Characteristics of water and droplet size distributions from fluidic sprinklers. *Irrig. Drain.* **2016**, *65*, 522–529. [CrossRef]

21. Yuan, S.Q.; Darko, R.O.; Zhu, X.Y.; Liu, J.P.; Tian, K. Optimization of movable irrigation system and performance assessment of distribution uniformity under varying conditions. *Int. J. Agric. Biol. Eng.* **2017**, *10*, 72–79.

22. Tang, L.D.; Yuan, S.Q.; Qiu, Z.P. Development and research status of water turbine for hose reel irrigator. *J. Drain. Irrig. Mach. Eng. (JDIME)* **2018**, *36*, 963–968.

23. Lu, M.Y.; Lu, K.J.; Hu, G.; Zhu, X.Y. Experiment on hydraulic performance of type SD-03 pop-up sprinkler. *J. Drain. Irrig. Mach. Eng. (JDIME)* **2018**, *36*, 1120–1124.

24. Buchleiter, G.W. Performance of LEPA equipment on center pivot machines. *Appl. Eng. Agric.* **1992**, *8*, 631–637. [CrossRef]

25. King, B.A.; Kincaid, D.C. Optimal performance from center pivot sprinkler systems. *Trans. ASABE* **1995**, *38*, 1737–1747.

26. Luz, P.B. A graphical solution to estimate potential runoff in center-pivot irrigation. *Trans. ASABE* **2011**, *54*, 81–92. [CrossRef]

27. Martin, D.L.; Kranz, W.L.; Thompson, A.L.; Liang, H. Selecting sprinkler packages for center pivots. *Trans. ASABE* **2012**, *55*, 513–523. [CrossRef]

28. Lu, J.; Cheng, J. Numerical simulation analysis of energy conversion in hydraulic turbine of hose reel irrigator JP75. *J. Drain. Irrig. Mach. Eng. (JDIME)* **2018**, *36*, 448–453.

29. Zhu, X.Y.; Peters, T.; Neibling, H. Hydraulic performance assessment of LESA at low pressure. *Irrig. Drain.* **2016**, *65*, 530–536. [CrossRef]

30. Tian, K.; Zhu, X.Y.; Wan, J.H.; Bao, Y. Development and performance test of lateral move irrigation system. *J. Drain. Irrig. Mach. Eng. (JDIME)* **2017**, *35*, 357–361.

31. Yan, H.J.; Jin, H.Z. Study on the discharge coefficient of nonrotatable sprays for center-pivot system. *J. Irrig. Drain. Eng. ASCE* **2004**, *23*, 55–58.

32. Dogana, E.; Kirnaka, H.; Dogan, Z. Effect of varying the distance of collectors below a sprinkler head and travel speed on measurements of mean water depth and uniformity for a linear move irrigation sprinkler system. *Biosyst. Eng.* **2008**, *99*, 190–195. [CrossRef]

33. Yan, H.J.; Jin, H.Z.; Qian, Y.C. Characterizing center pivot irrigation with fixed spray plate sprinklers. *Sci. Chin.* **2010**, *53*, 1398–1405. [CrossRef]

34. ASAE Standards. *Procedure for Sprinkler Distribution Testing for Research Purposes*, 32nd ed.; S330.1; ASAE: St. Joseph, MI, USA, 1985.

35. ASAE Standards. *Procedure for Sprinkler Testing and Performance Reporting*, 32nd ed.; S398.1; ASAE: St. Joseph, MI, USA, 1985.

36. ASAE. *Test Procedure for Determining the Uniformity of Water Distribution of Center Pivot and Lateral Move Irrigation Machines Equipped with Spray or Sprinkler Nozzles*; ANSI/ASAE S436.1 MAR01; ASAE Standards: St. Joseph, MI, USA, 2001.

37. MOD GB/T 19795.2. *Agricultural Irrigation Equipment—Rotating Sprinklers—Part 2: Uniformity of Distribution and Test Methods*; International Organization for Standardization: Geneva, Switzerland, 2005.

38. Schneider, A.D. Efficiency and uniformity of the LEPA and spray sprinkler methods: A review. *Trans. ASAE* **2000**, *43*, 937–944. [CrossRef]

39. Kohl, K.D.; Kohl, R.A.; DeBoer, D.W. Measurement of low pressure sprinkler evaporation loss. *Trans. ASAE* **1987**, *30*, 1071–1074. [CrossRef]

40. Dechmi, F.; Playán, E.; Cavero, J.; Faci, J.M.; Martínez-Cob, A. Wind effects on solid set sprinkler irrigation depth and yield of maize (*Zea mays*). *Irrig. Sci.* **2003**, *22*, 67–77. [CrossRef]

41. Playán, E.; Salvador, R.; Faci, J.M. Day and night wind drift and evaporation losses in sprinkler solid-sets and moving laterals. *Agric. Water Manag.* **2005**, *76*, 139–159. [CrossRef]
42. Zhang, M.; Zhao, W.X.; Li, J.S.; Li, Y.F. Fertigation uniformity and evaporation drift losses of center pivot irrigation system. *J. Drain. Irrig. Mach. Eng. (JDIME)* **2018**, *36*, 1125–1130.

MDPI

St. Alban-Anlage 66

4052 Basel

Switzerland

Tel. +41 61 683 77 34

Fax +41 61 302 89 18

www.mdpi.com

Water Editorial Office

E-mail: water@mdpi.com

www.mdpi.com/journal/water